TEACHER'S EDITION

McDougal Littell Science

Waves, Sound, and Light

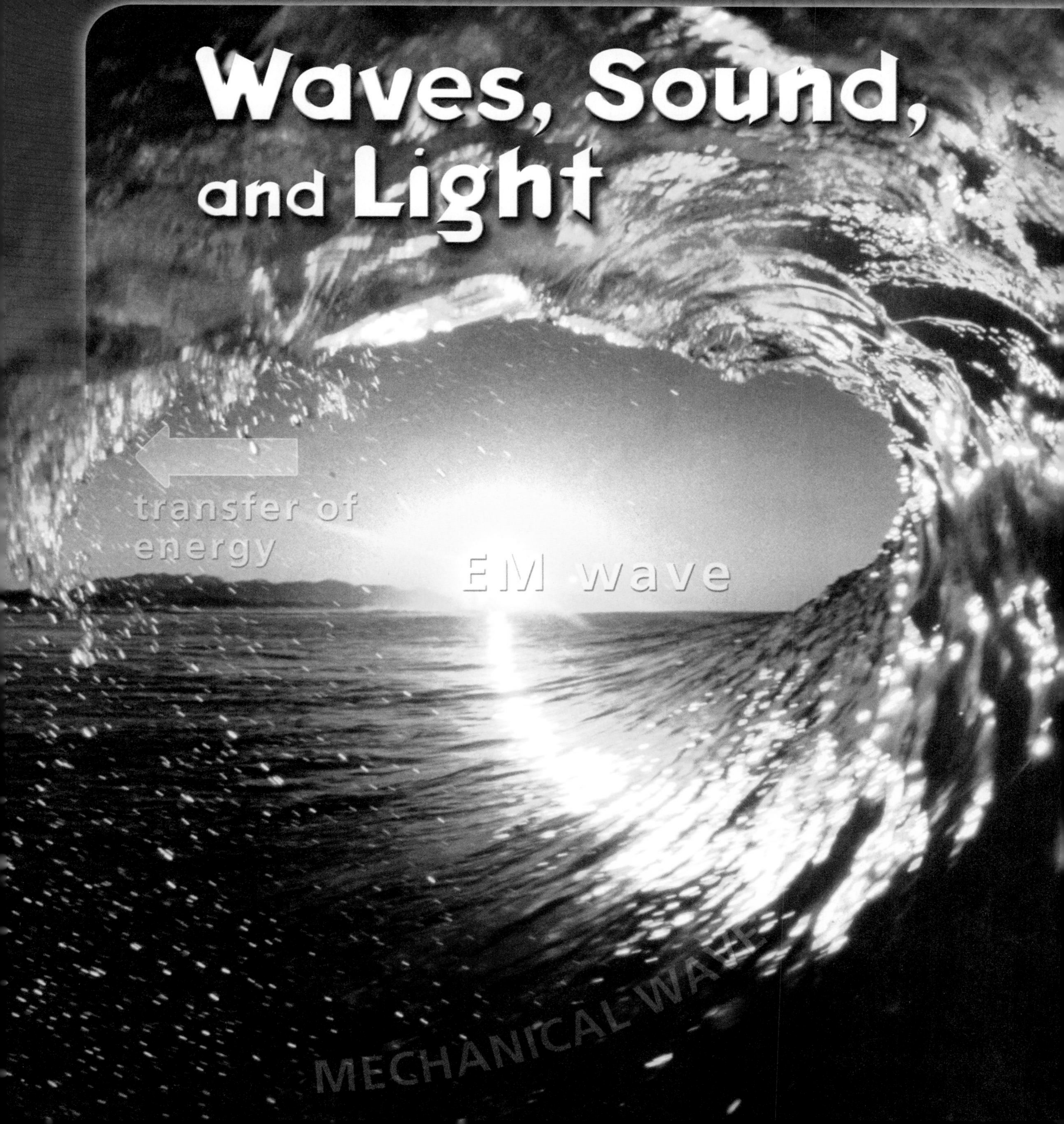

transfer of energy
EM wave
MECHANICAL WAVE

Credits
69C *left* Illustrations by Dan Stuckenschneider; *right* Illustration by Eric Chadwick; **109B** Photographs by Sharon Hoogstraten; **109C** © Kim Heacox/Getty Images; **109B** Illustration by Bart Vallecoccia.

Acknowledgements
Excerpts and adaptations from National Science Education Standards by the National Academy of Sciences. Copyright © 1996 by the National Academy of Sciences. Reprinted with permission from the National Academies Press, Washington, D.C.

ISBN: 0-618-33447-5   1 2 3 4 5 6 7 8 VJM 08 07 06 05 04

Internet Web Site: http://www.mcdougallittell.com

McDougal Littell Science

Effective Science Instruction Tailored for Middle School Learners

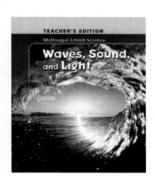

Waves, Sound, and Light
Teacher's Edition Contents

Consultants and Reviewers

Science Consultants

Chief Science Consultant

James Trefil, Ph.D. is the Clarence J. Robinson Professor of Physics at George Mason University. He is the author or co-author of more than 25 books, including *Science Matters* and *The Nature of Science*. Dr. Trefil is a member of the American Association for the Advancement of Science's Committee on the Public Understanding of Science and Technology. He is also a fellow of the World Economic Forum and a frequent contributor to *Smithsonian* magazine.

Rita Ann Calvo, Ph.D. is Senior Lecturer in Molecular Biology and Genetics at Cornell University, where for 12 years she also directed the Cornell Institute for Biology Teachers. Dr. Calvo is the 1999 recipient of the College and University Teaching Award from the National Association of Biology Teachers.

Kenneth Cutler, M.S. is the Education Coordinator for the Julius L. Chambers Biomedical Biotechnology Research Institute at North Carolina Central University. A former middle school and high school science teacher, he received a 1999 Presidential Award for Excellence in Science Teaching.

Instructional Design Consultants

Douglas Carnine, Ph.D. is Professor of Education and Director of the National Center for Improving the Tools of Educators at the University of Oregon. He is the author of seven books and over 100 other scholarly publications, primarily in the areas of instructional design and effective instructional strategies and tools for diverse learners. Dr. Carnine also serves as a member of the National Institute for Literacy Advisory Board.

Linda Carnine, Ph.D. consults with school districts on curriculum development and effective instruction for students struggling academically. A former teacher and school administrator, Dr. Carnine also co-authored a popular remedial reading program.

Donald Steely, Ph.D. serves as principal investigator at the Oregon Center for Applied Science (ORCAS) on federal grants for science and language arts programs. His background also includes teaching and authoring of print and multimedia programs in science, mathematics, history, and spelling.

Sam Miller, Ph.D. is a middle school science teacher and the Teacher Development Liaison for the Eugene, Oregon, Public Schools. He is the author of curricula for teaching science, mathematics, computer skills, and language arts.

Vicky Vachon, Ph.D. consults with school districts throughout the United States and Canada on improving overall academic achievement with a focus on literacy. She is also co-author of a widely used program for remedial readers.

Content Reviewers

John Beaver, Ph.D.
Ecology
Professor, Director of Science Education Center
College of Education and Human Services
Western Illinois University
Macomb, IL

Donald J. DeCoste, Ph.D.
Matter and Energy, Chemical Interactions
Chemistry Instructor
University of Illinois
Urbana-Champaign, IL

Dorothy Ann Fallows, Ph.D., MSc
Diversity of Living Things, Microbiology
Partners in Health
Boston, MA

Michael Foote, Ph.D.
The Changing Earth, Life Over Time
Associate Professor
Department of the Geophysical Sciences
The University of Chicago
Chicago, IL

Lucy Fortson, Ph.D.
Space Science
Director of Astronomy
Adler Planetarium and Astronomy Museum
Chicago, IL

Elizabeth Godrick, Ph.D.
Human Biology
Professor, CAS Biology
Boston University
Boston, MA

Isabelle Sacramento Grilo, M.S.
The Changing Earth
Lecturer, Department of the Geological Sciences
Montana State University
Bozeman, MT

David Harbster, MSc
Diversity of Living Things
Professor of Biology
Paradise Valley Community College
Phoenix, AZ

Richard D. Norris, Ph.D.
Earth's Waters
Professor of Paleobiology
Scripps Institution of Oceanography
University of California, San Diego
La Jolla, CA

Donald B. Peck, M.S.
*Motion and Forces; Waves, Sound, and Light;
  Electricity and Magnetism*
Director of the Center for Science Education (retired)
Fairleigh Dickinson University
Madison, NJ

Javier Penalosa, Ph.D.
Diversity of Living Things, Plants
Associate Professor, Biology Department
Buffalo State College
Buffalo, NY

Raymond T. Pierrehumbert, Ph.D.
Earth's Atmosphere
Professor in Geophysical Sciences (Atmospheric Science)
The University of Chicago
Chicago, IL

Brian J. Skinner, Ph.D.
Earth's Surface
Eugene Higgins Professor of Geology and Geophysics
Yale University
New Haven, CT

Nancy E. Spaulding, M.S.
Earth's Surface, The Changing Earth, Earth's Waters
Earth Science Teacher (retired)
Elmira Free Academy
Elmira, NY

Steven S. Zumdahl, Ph.D.
Matter and Energy, Chemical Interactions
Professor Emeritus of Chemistry
University of Illinois
Urbana-Champaign, IL

Susan L. Zumdahl, M.S.
Matter and Energy, Chemical Interactions
Chemistry Education Specialist
University of Illinois
Urbana-Champaign, IL

Safety Consultant

Juliana Texley, Ph.D.
Former K–12 Science Teacher and School Superintendent
Boca Raton, FL

English Language Advisor

Judy Lewis, M.A.
Director, State and Federal Programs for reading proficiency
and high risk populations
Rancho Cordova, CA

Research-Based Solutions for Your Classroom

The distinguished program consultant team and a thorough, research-based planning and development process assure that *McDougal Littell Science* supports all students in learning science concepts, acquiring inquiry skills, and thinking scientifically.

Standards-Based Instruction

Concepts and skills were selected based on careful analysis of national and state standards.

• National Science Education Standards

• Project 2061 Benchmarks for Science Literacy

• Comprehensive database of state science standards

CHAPTER

1 Waves

the BIG idea

Waves transfer energy and interact in predictable ways.

Key Concepts

SECTION

1.1 Waves transfer energy.
Learn about forces and energy in wave motion.

SECTION

1.2 Waves have measurable properties.
Learn how the amplitude, wavelength, and frequency of a wave are measured.

SECTION

1.3 Waves behave in predictable ways.
Learn about reflection, refraction, diffraction, and interference.

Internet Preview

CLASSZONE.COM
Chapter 1 online resources: Content Review, Simulation, Visualization, two Resource Centers, Math Tutorial, Test Practice

What is moving these surfers?

Standards and Benchmarks

Each chapter in **Waves, Sound, and Light** covers some of the learning goals that are described in the *National Science Education Standards* (NSES) and the Project 2061 *Benchmarks for Science Literacy*. Selected content and skill standards are shown below in shortened form. The following National Science Education Standards are covered on pages xii–xxvii, in Frontiers in Science, and in Timelines in Science, as well as in chapter features and laboratory investigations: Understandings About Scientific Inquiry (A.9), Understandings About Science and Technology (E.6), Science and Technology in Society (F.5), Science as a Human Endeavor (G.1), Nature of Science (G.2), and History of Science (G.3).

Content Standards

1 Waves

	National Science Education Standards
B.3.a	Energy is transferred in many ways. Energy is often associated with • sound • light • mechanical motion
	Project 2061 Benchmarks
4.F.4	Vibrations in materials set up wavelike disturbances that spread away from the source. Sound and earthquake waves are examples. These and other waves move at different speeds through different materials.

2 Sound

	Project 2061 Benchmarks
4.F.4	Vibrations start waves. Waves move at different speeds in different materials.

3 Electromagnetic Waves

	National Science Education Standards
B.3.c	Light interacts with matter by • transmission • absorption • scattering
B.3.f	Energy from the Sun is transferred to Earth in the form of • visible light • infrared light • ultraviolet light
	Project 2061 Benchmarks
4.F.1	Light from the sun is made up of a mixture of many different colors of light, even though to the eye the light looks almost white. Other things that give off or reflect light have a different mix of colors.
4.F.4	Waves move at different speeds through different materials.
4.F.5	Human eyes detect only a small range of waves in the electromagnetic spectrum. Different wavelengths of visible light are perceived as different colors.

x Unit: Waves, Sound, and Light

Internet Activity: Waves

Go to **ClassZone.com** to simulate the effect that different degrees of force have on a wave.

Observe and Think
What do you think would happen to the wave if you increased the number of times the flapper moved? What other ways could you affect the wave in the pool?

NSTA
scilinks.org SCLINKS

Seismic Waves Code: MDL027

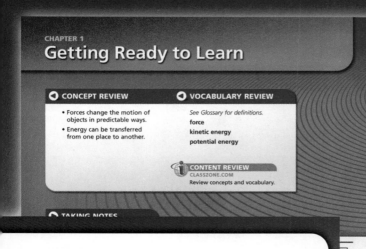

CHAPTER 1

Getting Ready to Learn

CONCEPT REVIEW
- Forces change the motion of objects in predictable ways.
- Energy can be transferred from one place to another.

VOCABULARY REVIEW
See Glossary for definitions.
force
kinetic energy
potential energy

CONTENT REVIEW
CLASSZONE.COM
Review concepts and vocabulary.

TAKING NOTES

Effective Instructional Strategies

McDougal Littell Science incorporates strategies that research shows are effective in improving student achievement. These strategies include

- Notetaking and nonlinguistic representations (Marzano, Pickering, and Pollock)
- A focus on big ideas (Kameenui and Carnine)
- Background knowledge and active involvement (Project CRISS)

Robert J. Marzano, Debra J. Pickering, and Jane E. Pollock, *Classroom Instruction that Works; Research-Based Strategies for Increasing Student Achievement* (ASCD, 2001)

Edward J. Kameenui and Douglas Carnine, *Effective Teaching Strategies that Accommodate Diverse Learners* (Pearson, 2002)

Project CRISS (Creating Independence through Student Owned Strategies)

Light and Optics

National Science Education Standards

B.3.c	To see an object, light from that object must enter the eye.

Project 2061 Benchmarks

4.F.2	Something can be "seen" when light waves emitted or reflected by it enter the eye—just as something can be "heard" when sound waves from it enter the ear.
4.F.5	Human eyes respond to only a narrow range of wavelengths of electromagnetic radiation—visible light. Differences of wavelength within that range are perceived as differences in color.

Process and Skill Standards

	National Science Education Standards		Project 2061 Benchmarks
A.1	Identify questions that can be answered through investigation.	1.A.3	Some knowledge in science is very old and yet is still used today.
A.2	Design and conduct a scientific investigation.	1.C.1	Contributions to science and technology have been made by different people, in different cultures, at different times.
A.3	Use appropriate tools and techniques to gather and interpret data.	6.A.5	People use technology to match or go beyond the abilities of other species.
A.4	Use evidence to describe, predict, explain, and model.	9.B.3	Graphs can show the relationship between two variables.
A.5	Use critical thinking to find relationships between results and interpretations.	9.C.4	Graphs show patterns and can be used to make predictions.
A.6	Consider alternative explanations and predictions.	9.D.3	The mean, median, and mode are different ways of finding the middle of a set of data.
A.7	Communicate procedures, results, and conclusions.	11.C.4	Use equations to summarize observed changes.
A.8	Use mathematics in scientific investigations.	12.B.4	Find the mean and median of a set of data.
E.1	Identify a problem to be solved.	12.C.3	Use appropriate units, use and read instruments that measure quantities such as length, volume, weight, time, rate, and temperature.
E.2	Design a solution or product.	12.D.1	Use tables and graphs to organize information and identify relationships.
E.3	Implement the proposed solution.	12.D.2	Read, interpret, and describe tables and graphs.
E.4	Evaluate the solution or design.	12.D.4	Understand information that includes different types of charts and graphs, including circle charts, bar graphs, line graphs, data tables, diagrams, and symbols.
F.4.c	Use systematic thinking to estimate risks.		
F.4.d	Decisions are made based on estimated risks and benefits.	12.E.4	There may be more than one good way to interpret scientific findings.

Standards and Benchmarks **xi**

VOCA
wave p
mediu
mechan
transve
longitu

A wave is a disturbance.

You experience the effects of waves every day. Every sound you hear depends on sound waves. Every sight you see depends on light waves. A tiny wave can travel across the water in a glass, and a huge wave can travel across the ocean. Sound waves, light waves, and water waves seem very different from one another. So what, exactly, is a wave?

READING TIP
To *disturb* means to agitate or unsettle.

A **wave** is a disturbance that transfers energy from one place to another. Waves can transfer energy over distance without moving matter the entire distance. For example, an ocean wave can travel many kilometers without the water itself moving many kilometers. The water moves up and down—a motion known as a disturbance. It is the disturbance that travels in a wave, transferring energy.

CHECK YOUR READING How does an ocean wave transfer energy across the ocean?

Chapter 1: Waves **9**

Comprehensive Research, Review, and Field Testing

An ongoing program of research and review guided the development of *McDougal Littell Science.*

- Program plans based on extensive data from classroom visits, research surveys, teacher panels, and focus groups
- All pupil edition activities and labs classroom-tested by middle school teachers and students
- All chapters reviewed for clarity and scientific accuracy by the Content Reviewers listed on page T5
- Selected chapters field-tested in the classroom to assess student learning, ease of use, and student interest

Content Organized Around Big Ideas

Each chapter develops a big idea of science, helping students to place key concepts in context.

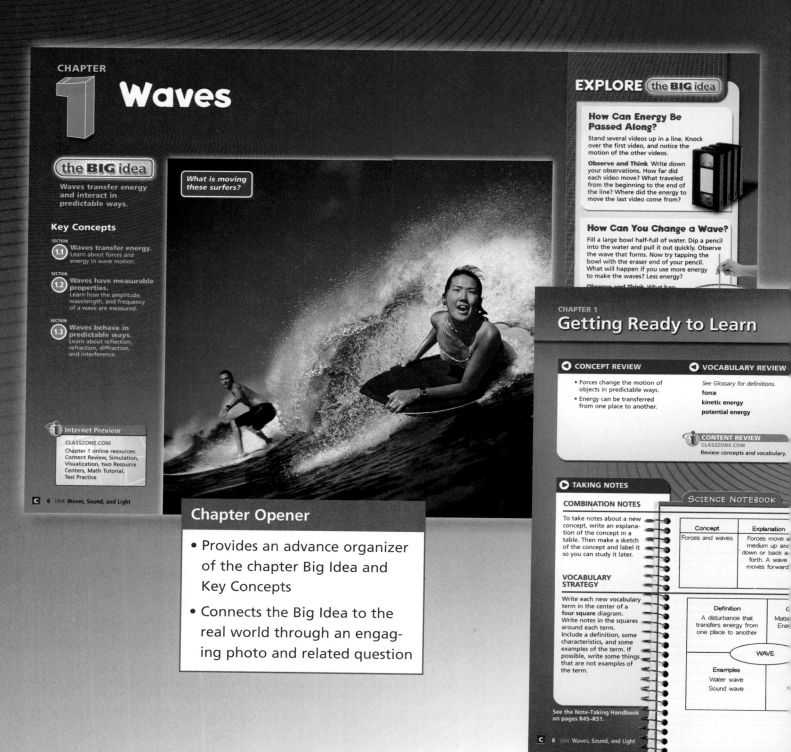

CHAPTER

1 Waves

the BIG idea

Waves transfer energy and interact in predictable ways.

Key Concepts

SECTION
1.1 Waves transfer energy.
Learn about forces and energy in wave motion.

SECTION
1.2 Waves have measurable properties.
Learn how the amplitude, wavelength, and frequency of a wave are measured.

SECTION
1.3 Waves behave in predictable ways.
Learn about reflection, refraction, diffraction, and interference.

Internet Preview

CLASSZONE.COM
Chapter 1 online resources:
Content Review, Simulation, Visualization, two Resource Centers, Math Tutorial, Test Practice

C 6 Unit: Waves, Sound, and Light

What is moving these surfers?

EXPLORE the BIG idea

How Can Energy Be Passed Along?
Stand several videos up in a line. Knock over the first video, and notice the motion of the other videos.

Observe and Think Write down your observations. How far did each video move? What traveled from the beginning to the end of the line? Where did the energy to move the last video come from?

How Can You Change a Wave?
Fill a large bowl half-full of water. Dip a pencil into the water and pull it out quickly. Observe the wave that forms. Now try tapping the bowl with the eraser end of your pencil. What will happen if you use more energy to make the waves? Less energy?

Observe and Think What hap...

CHAPTER 1
Getting Ready to Learn

CONCEPT REVIEW
• Forces change the motion of objects in predictable ways.
• Energy can be transferred from one place to another.

VOCABULARY REVIEW
See Glossary for definitions.
force
kinetic energy
potential energy

CONTENT REVIEW
CLASSZONE.COM
Review concepts and vocabulary.

TAKING NOTES

COMBINATION NOTES
To take notes about a new concept, write an explanation of the concept in a table. Then make a sketch of the concept and label it so you can study it later.

VOCABULARY STRATEGY
Write each new vocabulary term in the center of a four square diagram. Write notes in the squares around each term. Include a definition, some characteristics, and some examples of the term. If possible, write some things that are not examples of the term.

See the Note-Taking Handbook on pages R45–R51.

SCIENCE NOTEBOOK

Concept	Explanation
Forces and waves	Forces move a medium up and down or back and forth. A wave moves forward

Definition	
A disturbance that transfers energy from one place to another	Matter Energy

WAVE

Examples
Water wave
Sound wave

C 8 Unit: Waves, Sound, and Light

Chapter Opener

• Provides an advance organizer of the chapter Big Idea and Key Concepts

• Connects the Big Idea to the real world through an engaging photo and related question

Visual Summary

- Summarizes Key Concepts using both text and visuals
- Reinforces the connection of Key Concepts to the Big Idea

---

the BIG idea

Waves transfer energy and interact in predictable ways.

CONTENT REVIEW
CLASSZONE.COM

KEY CONCEPTS SUMMARY

1.1 Waves transfer energy.

Transverse Wave

direction of disturbance

direction of wave

transfer of energy

Longitudinal Wave

direction of disturbance

direction of wave

transfer of energy

VOCABULARY
wave p. 9
medium p. 11
mechanical wave p. 11
transverse wave p. 13
longitudinal wave p. 14

1.2 Waves have measurable properties.

water level at rest

wavelength

crest

amplitude

trough

fixed point

VOCABULARY
crest p. 17
trough p. 17
amplitude p. 17
wavelength p. 17
frequency p. 17

Frequency is the number of wavelengths passing a fixed point in a certain amount of time.

1.3 Waves behave in predictable ways.

Reflection

Refraction

VOCABULARY

Reviewing Vocabulary

Draw a triangle for each of the terms below. On the wide bottom of the triangle, write the term and your own definition of it. Above that, write a sentence in which you use the term correctly. At the top of the triangle, draw a small picture to show what the term looks like. The first triangle is completed for you.

The amplitude of the wave was 30 cm.

Amplitude is the distance between a line through the middle of a wave and a crest or trough.

1. amplitude	6. interference
2. diffraction	7. reflection
3. frequency	8. trough
4. medium	9. refraction
5. crest	10. wavelength

Reviewing Key Concepts

Multiple Choice *Choose the letter of the best answer.*

Section Opener

- Highlights the Key Concept
- Connects new learning to prior knowledge
- Previews important vocabulary

Thinking Critically

Use the diagram below to answer the next two questions.

20. What two letters in the diagram measure the same thing? What do they both measure?

21. In the diagram above, what does the letter *c* measure?

Use the diagram below to answer the next three questions. The diagram shows waves passing a fixed point.

fixed point   fixed point

0 seconds   1 second

22. At 0 seconds, no waves have passed. How many waves have passed after 1 second?

23. What is the measurement shown in the diagram?

24. How would you write the measurement taken in the diagram?

25. EVALUATE Do you think the following is an accurate definition of medium? Explain your answer.
A **medium** is any solid through which waves travel.

26. APPLY Picture a pendulum. The pendulum is swinging back and forth at a steady rate. How could you make it swing higher? How is swinging a pendulum like making a wave?

27. PREDICT What might happen to an ocean wave that encounters a gap or hole in a cliff along the shore?

28. EVALUATE Do you think *interference* is an appropriate name for the types of wave interaction you read about in Se... your answer.

Using Math in Science

29. At what speed is the wave below traveling if it has a frequency of 2 wavelengths/s?

1.2 m

wave

30. An ocean wave has a wavelength of 9 m and a frequency of 0.42 wavelengths/s. What is the wave's speed?

31. Suppose a sound wave has a frequency of 10,000 wavelengths/s. The wave's speed is 340 m/s. Calculate the wavelength of this sound wave.

32. A water wave is traveling at a speed of 2.5 m/s. The wave has a wavelength of 4 m. Calculate the frequency of this water wave.

the BIG idea

33. INTERPRET Look back at the photograph at the start of the chapter on pages 6–7. How does this photograph illustrate a transfer of energy?

34. SYNTHESIZE Describe three situations in which you can predict the behavior of waves.

35. SUMMARIZE Write a paragraph summarizing this chapter. Use the big idea from page 6 as the topic sentence. Then write an example from each of the key concepts listed under the big idea.

UNIT PROJECTS

If you are doing a unit project, make a folder for your project. Include in your folder a list of the resources you will need, the date on which the project is due, and a schedule to track your progress. Begin gathering data.

KEY CONCEPT

1.1 Waves transfer energy.

▶ **BEFORE, you learned**
- Forces can change an object's motion
- Energy can be kinetic or potential

▶ **NOW, you will learn**
- How forces cause waves
- How waves transfer energy
- How waves are classified

VOCABULARY
wave p. 9
medium p. 11
mechanical wave p. 11
transverse wave p. 13
longitudinal wave p. 14

EXPLORE Waves

How will the rope move?

PROCEDURE

① Tie a ribbon in the middle of a rope. Then tie one end of the rope to a chair.

② Holding the loose end of the rope in your hand, stand far enough away from the chair that the rope is fairly straight.

③ Flick the rope by moving your hand up and down quickly. Observe what happens.

WHAT DO YOU THINK?
- How did the rope move? How did the ribbon move?
- What do you think starts a wave, and what keeps it going?

MATERIALS
- ribbon
- rope
- chair

A wave is a disturbance.

You experience the effects of waves every day. Every sound you hear depends on sound waves. Every sight you see depends on light waves. A tiny wave can travel across the water in a glass, and a huge wave can travel across the ocean. Sound waves, light waves, and water waves seem very different from one another. So what, exactly, is a wave?

A **wave** is a disturbance that transfers energy from one place to another. Waves can transfer energy over distance without moving matter the entire distance. For example, an ocean wave can travel many kilometers without the water itself moving many kilometers. The water moves up and down—a motion known as a disturbance. It is the disturbance that travels in a wave, transferring energy.

READING TiP
To *disturb* means to agitate or unsettle.

CHECK YOUR READING How does an ocean wave transfer energy across the ocean?

Chapter 1: **Waves** 9 C

The Big Idea Questions

- Help students connect their new learning back to the Big Idea
- Prompt students to synthesize and apply the Big Idea and Key Concepts

Many Ways to Learn

Because students learn in so many ways, *McDougal Littell Science* gives them a variety of experiences with important concepts and skills. Text, visuals, activities, and technology all focus on Big Ideas and Key Concepts.

Considerate Text

- Clear structure of meaningful headings
- Information clearly connected to main ideas
- Student-friendly writing style

Integrated Technology

- Interaction with Key Concepts through Simulations and Visualizations
- Easy access to relevant Web resources through Resource Centers and SciLinks
- Opportunities for review through Content Review and Math Tutorials

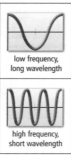

low frequency,
long wavelength

high frequency,
short wavelength

REMINDER
Frequency is the number of wavelengths that pass a given point in a certain amount of time.

VISUALIZATION
CLASSZONE.COM
Watch the graph of a wave form.

How Frequency and Wavelength Are Related

The frequency and wavelength of a wave are related. When frequency increases more wave crests pass a fixed point each second. That means the wavelength shortens. So, as frequency increases, wavelength decreases. The opposite is also true—as frequency decreases, wavelength increases.

Suppose you are making waves in a rope. If you make one wave crest every second, the frequency is one wavelength per second. Now suppose you want to increase the frequency to more than one wavelength per second. You flick the rope up and down faster. The wave crests are now closer together. In other words, their wavelengths have decreased.

Graphing Wave Properties

The graph of a transverse wave looks much like a wave itself. The illustration on page 19 shows the graph of an ocean wave. The measurements for the graph come from a float, or buoy (BOO-ee), that keeps track of how high or low the water goes. The graph shows the positioning of the buoy at three different points in time. These points are numbered. Since the graph shows what happens over time, you can see the frequency of the waves.

Unlike transverse waves, longitudinal waves look different from their graphs. The graph of a longitudinal wave in a spring is drawn below. The coils of the spring get closer and then farther apart as the wave moves through them.

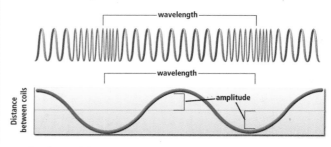

wavelength

wavelength

Distance between coils

amplitude

The shape of the graph resembles the shape of a transverse wave. The wavelength on a longitudinal wave is the distance from one compression to the next. The amplitude of a longitudinal wave measures how compressed the medium gets. Just as in a transverse wave, frequency in a longitudinal wave is the number of wavelengths passing a fixed point in a certain amount of time.

CHECK YOUR READING How are longitudinal waves measured?

INVESTIGATE Frequency

How can you change frequency?

PROCEDURE

① Tie 3 washers to a string. Tape the string to the side of your desk so that it can swing freely. The swinging washers can model wave action.

② Pull the washers slightly to the side and let go. Find the frequency by counting the number of complete swings that occur in 1 minute.

③ Make a table in your notebook to record both the length of the string and the frequency.

④ Shorten the string by moving and retaping it. Repeat for 5 different lengths. Keep the distance you pull the washers the same each time.

WHAT DO YOU THINK?

length of the string affect the frequency?

represent a wave? How does it differ from a wave?

uld you vary the amplitude of this model?
e amplitude would affect the frequency.

SKILL FOCUS
Collecting data

MATERIALS
• 3 metal washers
• piece of string
• tape
• stopwatch

TIME
30 minutes

Wave speed can be measured.

In addition to amplitude, wavelength, and frequency, a wave's speed can be measured. One way to find the speed of a wave is to time how long ... ther way ... ave can b... know...

D... ... ple, light ... ... und wave ... ve speed ... mes with ... the light ... ... es reach...

H... ...el at di... ge to fi...

Light

Graphing a Wave

The graph of a transverse wave looks like a wave itself. The graph shows what happens over time.

The buoy moves up and down as the waves pass.

① **Time: 0 s** The buoy is below the rest position.

② **Time: 1 s** The buoy is equal with the rest position.

③ **Time: 2 s** The buoy is above the rest position.

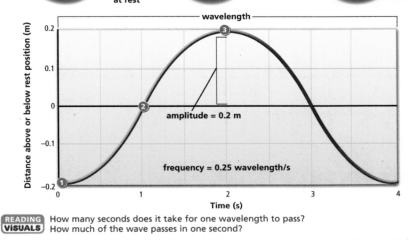

water level at rest

wavelength

amplitude = 0.2 m

frequency = 0.25 wavelength/s

Distance above or below rest position (m) / Time (s)

READING VISUALS How many seconds does it take for one wavelength to pass? How much of the wave passes in one second?

Chapter 1: Waves 19 **C**

Hands-on Learning

• Activities that reinforce Key Concepts

• Skill Focus for important inquiry and process skills

• Multiple activities in every chapter, from quick Explores to full-period Chapter Investigations

Visuals that Teach

• Information-rich visuals directly connected to the text

• Thoughtful pairing of diagrams and real-world photos

• Reading Visuals questions to support student learning

Differentiated Instruction

A full spectrum of resources for differentiating instruction supports you in reaching the wide range of learners in your classroom.

1.1 INSTRUCT

Teach from Visuals

To help students interpret the rope wave visual, ask:

- What represents the forces? *the up and down arrows*
- What produces the forces? *the person's hand and arm moving up and down*
- What is the result of the force? *a wave traveling down the rope*

Ask similar questions for the other visuals on the page.

Real World Example

A telephone cord wave can occur when you stretch out the cord and shake one end of the cord. The wave moves along the cord and reaches the handset.

EXPLORE (the BIG idea)

Revisit "Internet Activity: Waves" on p. 7. Have students explain the reasons for their results.

Ongoing Assessment

Explain how forces cause waves.

When a diver hits the smooth surface of water in a swimming pool, series of waves spread in all directions. Ask: How do forces make this happen? *The force of the diver striking the surface pushes water briefly out of the way; then the water rushes back in. These movements in the water set off the series of ripples, or waves, across the pool.*

Forces and Waves

You know that a force is required to change the motion of an object. Forces can also start a disturbance, sending a wave through a material. The following examples describe how forces cause waves.

READING TIP
As you read each example, think of how it is similar to and different from the other examples.

Example 1 Rope Wave Think of a rope that is tied to a doorknob. You apply one force to the rope by flicking it upward and an opposite force when you snap it back down. This sends a wave through the rope. Both forces—the one that moves the rope up and the one that moves the rope down—are required to start a wave.

wave / force

Example 2 Water Wave Forces are also required to start a wave in water. Think of a calm pool of water. What happens if you apply a force to the water by dipping your finger into it? The water rushes back after you remove your finger. The force of your finger and the force of the water rushing back send waves across the pool.

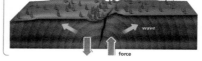
wave / force

Example 3 Earthquake Wave An earthquake is a sudden release of energy that has built up in rock as a result of the surrounding rock pushing and pulling on it. When these two forces cause the rock to suddenly break, the energy is transferred as a wave through the ground.

wave / force

Materials and Waves

A rope tied to a doorknob, water, and the ground all have something in common. They are all materials through which waves move. A **medium** is any substance that a wave moves through. Water is the medium for an ocean wave; the ground is the medium for an earthquake wave; the rope is the medium for the rope wave. In the next chapter, you will learn that sound waves can move through many mediums, including air.

Waves that transfer energy through matter are known as **mechanical waves**. All of the waves you have read about so far, even sound waves, are mechanical waves. Water, the ground, a rope, and the air are all made up of matter. Later, you will learn about waves that can transfer energy through empty space. Light is an example of a wave that transfers energy through empty space.

VOCABULARY
Add a four square for *medium* to your notebook.

CHECK YOUR READING How are all mechanical waves similar?

Energy and Waves

The waves caused by an earthquake are good examples of energy transfer. The disturbed ground shakes from side to side and up and down as the waves move through it. Such waves can travel kilometers away from their source. The ground does not travel kilometers away from where it began; it is the energy that travels in a wave. In the case of an earthquake, it is kinetic energy, or the energy of motion that is transferred.

This photograph was taken after a 1995 earthquake in Japan. A seismic wave transferred enough energy through the ground to bend the railroad tracks, leaving them in the shape of a wave.

DIFFERENTIATE INSTRUCTION

? More Reading Support

A What is required to make a disturbance that will set off a wave? *forces*

English Learners English learners may not always understand certain abbreviated instructions. For example, when a student is asked to "study a sketch," the word *study* is vague to an English learner. Use precise instructions like *examine*, *analyze*, or *compare*, and make sure students know the exact meaning of these instructions.

DIFFERENTIATE INSTRUCTION

? More Reading Support

B What is the medium for an ocean wave? *water*

C Does a mechanical wave transfer material or energy? *energy*

Below Level Students may find the term *medium* confusing in the context of wave science. Students are likely to be familiar with *medium* meaning "halfway" or "in between." Write the following sentences on the board:

- I like my steak medium rare.
- Water is the medium for an ocean wave . . . (p. 11).

Ask: Which of these sentences involves the scientific definition of *medium*? *the second sentence*

Teacher's Edition

- More Reading Support for below-level readers
- Strategies for below-level and advanced learners, English learners, and inclusion students

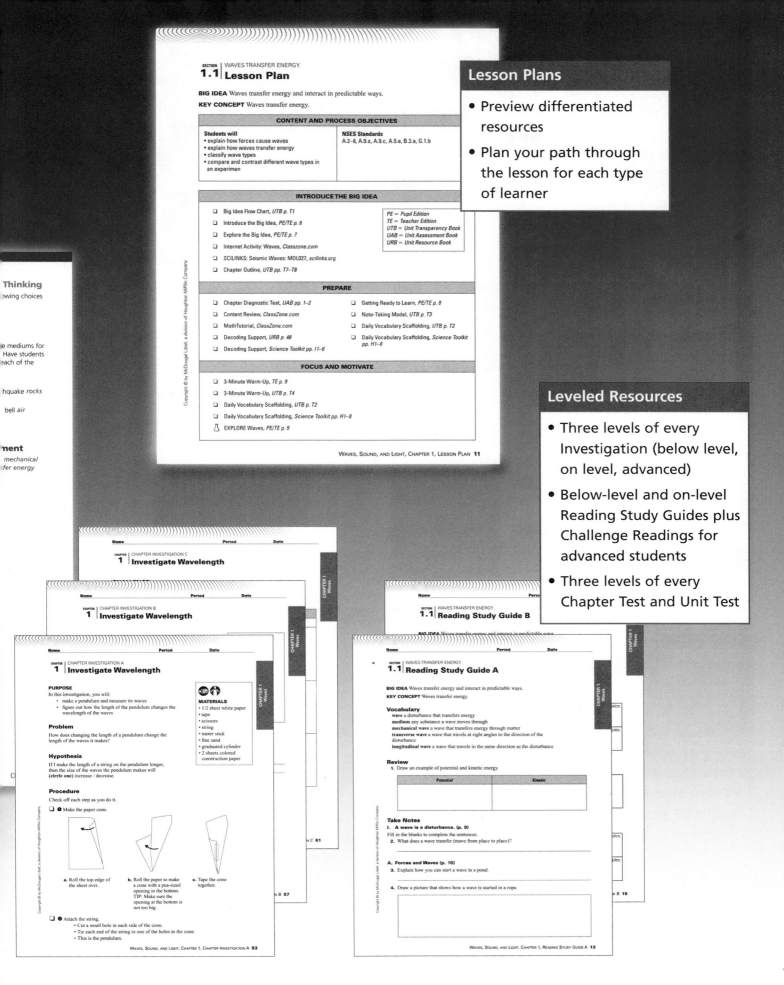

Lesson Plans

- Preview differentiated resources
- Plan your path through the lesson for each type of learner

Leveled Resources

- Three levels of every Investigation (below level, on level, advanced)
- Below-level and on-level Reading Study Guides plus Challenge Readings for advanced students
- Three levels of every Chapter Test and Unit Test

Effective Assessment

McDougal Littell Science incorporates a comprehensive set of resources for assessing student knowledge and performance before, during, and after instruction.

Diagnostic Tests

- Assessment of students' prior knowledge
- Readiness check for concepts and skills in the upcoming chapter

---

Ongoing Assessment

Classify wave types.

Ask: How do a rope wave and the waves represented on this page compare and contrast? *Both start with a disturbance caused by a force; both transfer energy. A rope wave is a transverse wave; waves on this page are longitudinal waves.*

Teach from Visuals

Remind students that, in a transverse wave such as an ocean wave, a medium that alternates between high and low points forms the wave. Ask: What makes up the wave pattern in the spring toy? *alternating regions of bunched-up and spread-out spring coils*

Reinforce (the BIG idea)

Have students relate the section to the Big Idea.

Reinforcing Key Concepts, p. 21

1.1 ASSESS & RETEACH

Assess

Section 1.1 Quiz, p. 3

Reteach

Have students picture a tub with a toy boat floating on the water. Tell them to imagine plunging a fist into the water.

- Have students predict what would happen to the water's surface and the boat. *waves would move out from the disturbance; boat would bob up and down*
- Have students explain how waves start, move, and transfer energy.

Technology Resources

Have students visit ClassZone.com for reteaching of Key Concepts.

- CONTENT REVIEW
- CONTENT REVIEW CD-ROM

C 14 Unit: Waves, Sound, and Light

Longitudinal Waves

READING TIP
The word *long* can help you remember longitudinal waves. The disturbance moves along the length of the spring.

Another type of wave is a longitudinal wave. In a **longitudinal wave** (LAHN-jih-TOOD-n-uhl), the wave travels in the same direction as the disturbance. A longitudinal wave can be started in a spring by moving it forward and backward. The coils of the spring move forward and bunch up and then move backward and spread out. This forward and backward motion is the disturbance. Longitudinal waves are sometimes called compressional waves because the bunched-up area is known as a compression. How is a longitudinal wave similar to a transverse wave? How is it different?

Longitudinal Wave

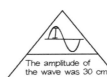

direction of disturbance
Time 1

compression
Time 2

direction of wave
Time 3

RESOURCE CENTER
CLASSZONE.COM
Learn more about waves.

Sound waves are examples of longitudinal waves. Imagine a bell ringing. The clapper inside the bell strikes the side and makes it vibrate, or move back and forth rapidly. The vibrating bell pushes and pulls on nearby air molecules, causing them to move forward and backward. These air molecules, in turn, set more air molecules into motion. A sound wave pushes forward. In sound waves, the vibrations of the air molecules are in the same direction as the movement of the wave.

1.1 Review

KEY CONCEPTS
1. Describe how forces start waves.
2. Explain how a wave can travel through a medium and yet the medium stays in place. Use the term *energy* in your answer.
3. Describe two ways in which waves travel, and give an example of each.

CRITICAL THINKING
4. **Analyze** Does water moving through a hose qualify as a wave? Explain why or why not.
5. **Classify** Suppose you drop a cookie crumb in your milk. At once, you see ripples spreading across the surface of the milk. What type of wave are these? What is the disturbance?

CHALLENGE
6. **Predict** Suppose you had a rope long enough to extend several blocks down the street. If you were to start a wave in the rope, do you think it would continue all the way to the other end of the street? Explain why or why not.

14 Unit: Waves, Sound, and Light

ANSWERS

1. Forces start a disturbance that travels through a medium.
2. Energy travels through the medium, leaving the matter in place.
3. Transverse waves travel at right angles to the disturbance. Longitudinal waves

travel in the same direction as the disturbance. Examples are ocean waves and sound waves.
4. No; the water travels from one end to the other end of the hose. In a wave, the disturbance moves, but the medium does not.

5. Transverse waves; the falling crumb
6. A rope wave would probably not continue all the way down the street. Gravity would be pulling down on the rope along the way, eventually stopping it.

Ongoing Assessment

- Check Your Reading questions for student self-check of comprehension
- Consistent Teacher Edition prompts for assessing understanding of Key Concepts

---

Reviewing Vocabulary

Draw a triangle for each of the terms below. On the wide bottom of the triangle, write the term and your own definition of it. Above that, write a sentence in which you use the term correctly. At the top of the triangle, draw a small picture to show what the term looks like. The first triangle is completed for you.

The amplitude of the wave was 30 cm.

Amplitude is the distance between a line through the middle of a wave and a crest or trough.

1. amplitude
2. diffraction
3. frequency
4. medium
5. crest
6. interference
7. reflection
8. trough
9. refraction
10. wavelength

Reviewing Key Concepts

Multiple Choice *Choose the letter of the best answer.*

11. The direction in which a transverse wave travels is
    a. in the same direction as the disturbance
    b. toward the disturbance
    c. from the disturbance downward
    d. at right angles to the disturbance

12. An example of a longitudinal wave is a
    a. water wave
    b. stadium wave
    c. sound wave
    d. rope wave

13. Which statement best defines a wave medium?
    a. the material through which a wave travels
    b. a point half-way between the crest and trough of a wave
    c. the distance from one wave crest to the next
    d. the speed at which waves travel in water

14. As you increase the amplitude of a wave, you also increase the
    a. frequency   c. speed
    b. wavelength   d. energy

15. To identify the amplitude in a longitudinal wave, you would look for areas of
    a. reflection   c. crests
    b. compression   d. refraction

16. Which statement describes the relationship between frequency and wavelength?
    a. When frequency increases, wavelength increases.
    b. When frequency increases, wavelength decreases.
    c. When frequency increases, wavelength remains constant.
    d. When frequency increases, wavelength varies unpredictably.

17. For wave refraction to take place, a wave must
    a. increase in velocity
    b. enter a new medium
    c. increase in frequency
    d. merge with another wave

18. Which setup in a wave tank would enable you to demonstrate diffraction?
    a. water only
    b. water and sand
    c. water and food coloring
    d. water and a barrier with a small gap

19. Two waves come together and interact to form a new, smaller wave. This process is called
    a. destructive interference
    b. constructive interference
    c. reflective interference
    d. positive interference

Section and Chapter Reviews

- Focus on Key Concepts and critical thinking skills
- A full range of question types and levels of thinking

Leveled Chapter and Unit Tests

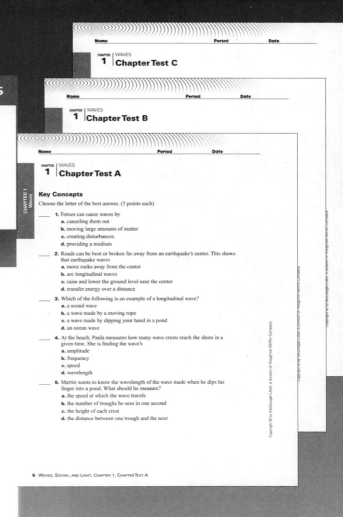

- Three levels of test for every chapter and unit
- Same Big Ideas, Key Concepts, and essential skills assessed on all levels

CHAPTER 1 WAVES
1 Chapter Test C

CHAPTER 1 WAVES
1 Chapter Test B

CHAPTER 1 WAVES
1 Chapter Test A

Key Concepts

Choose the letter of the best answer. (5 points each)

_____ **1.** Forces can cause waves by
  - **a.** canceling them out
  - **b.** moving large amounts of matter
  - **c.** creating disturbances
  - **d.** providing a medium

_____ **2.** Roads can be bent or broken far away from an earthquake's center. This shows that earthquake waves
  - **a.** move rocks away from the center
  - **b.** are longitudinal waves
  - **c.** raise and lower the ground level near the center
  - **d.** transfer energy over a distance

_____ **3.** Which of the following is an example of a longitudinal wave?
  - **a.** a sound wave
  - **b.** a wave made by a moving rope
  - **c.** a wave made by dipping your hand in a pond
  - **d.** an ocean wave

_____ **4.** At the beach, Paula measures how many wave crests reach the shore in a given time. She is finding the wave's
  - **a.** amplitude
  - **b.** frequency
  - **c.** speed
  - **d.** wavelength

_____ **5.** Martin wants to know the wavelength of the wave made when he dips his finger into a pond. What should he measure?
  - **a.** the speed at which the wave travels
  - **b.** the number of troughs he sees in one second
  - **c.** the height of each crest
  - **d.** the distance between one trough and the next

6 WAVES, SOUND, AND LIGHT, CHAPTER 1, CHAPTER TEST A

Thinking Critically

Use the diagram below to answer the next two questions.

20. What two letters in the diagram measure the same thing? What do they both measure?

21. In the diagram above, what does the letter c measure?

Use the diagram below to answer the next three questions. The diagram shows waves passing a fixed point.

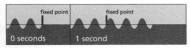

22. At 0 seconds, no waves have passed. How many waves have passed after 1 second?

23. What is the measurement shown in the diagram?

24. How would you write the measurement taken in the diagram?

25. EVALUATE Do you think the following is an accurate definition of medium? Explain your answer.

A **medium** is any solid through which waves travel.

26. APPLY Picture a pendulum. The pendulum is swinging back and forth at a steady rate. How could you make it swing higher? How is swinging a pendulum like making a wave?

27. PREDICT What might happen to an ocean wave that encounters a gap or hole in a cliff along the shore?

28. EVALUATE Do you think *interference* is an appropriate name for the types of wave interaction you read about in Section 1.3? Explain your answer.

32 Unit: Waves, Sound, and Light

Using Math in Science

29. At what speed is the wave below traveling if it has a frequency of 2 wavelengths/s?

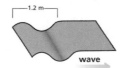

wave

30. An ocean wave has a wavelength of 9 m and a frequency of 0.42 wavelengths/s. What is the wave's speed?

31. Suppose a sound wave has a frequency of 10,000 wavelengths/s. The wave's speed is 340 m/s. Calculate the wavelength of this sound wave.

32. A water wave is traveling at a speed of 2.5 m/s. The wave has a wavelength of 4 m. Calculate the frequency of this water wave.

the BIG idea

33. INTERPRET Look back at the photograph at the start of the chapter on pages 6–7. How does this photograph illustrate a transfer of energy?

34. SYNTHESIZE Describe three situations in which you can predict the behavior of waves.

35. SUMMARIZE Write a paragraph summarizing this chapter. Use the big idea from page 6 as the topic sentence. Then write an example from each of the key concepts listed under the big idea.

UNIT PROJECTS

If you are doing a unit project, make a folder for

Rubrics

- Rubrics in Teacher Edition for all extended response questions
- Rubrics for all Unit Projects
- Alternative Assessment with rubric for each chapter
- A wide range of additional rubrics in the Science Toolkit

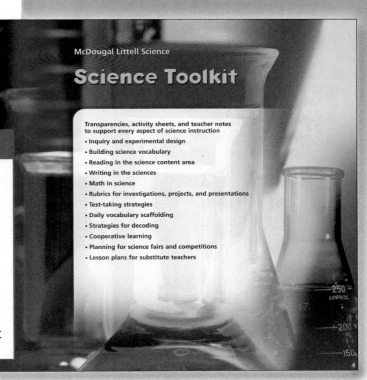

McDougal Littell Science

Science Toolkit

Transparencies, activity sheets, and teacher notes to support every aspect of science instruction
- Inquiry and experimental design
- Building science vocabulary
- Reading in the science content area
- Writing in the sciences
- Math in science
- Rubrics for investigations, projects, and presentations
- Test-taking strategies
- Daily vocabulary scaffolding
- Strategies for decoding
- Cooperative learning
- Planning for science fairs and competitions
- Lesson plans for substitute teachers

McDougal Littell Science Modular Series

McDougal Littell Science lets you choose the titles that match your curriculum. Each module in this flexible 15-book series takes an in-depth look at a specific area of life, earth, or physical science.

- Flexibility to match your curriculum

- Convenience of smaller books

- Complete Student Resource Handbooks in every module

Life Science Titles

A ▶ Cells and Heredity
1. The Cell
2. How Cells Function
3. Cell Division
4. Patterns of Heredity
5. DNA and Modern Genetics

B ▶ Life Over Time
1. The History of Life on Earth
2. Classification of Living Things
3. Population Dynamics

C ▶ Diversity of Living Things
1. Single-Celled Organisms and Viruses
2. Introduction to Multicellular Organisms
3. Plants
4. Invertebrate Animals
5. Vertebrate Animals

D ▶ Ecology
1. Ecosystems and Biomes
2. Interactions Within Ecosystems
3. Human Impact on Ecosystems

E ▶ Human Biology
1. Systems, Support, and Movement
2. Absorption, Digestion, and Exchange
3. Transport and Protection
4. Control and Reproduction
5. Growth, Development, and Health

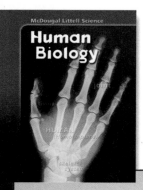

Earth Science Titles

A ▶ **Earth's Surface**
1. Views of Earth Today
2. Minerals
3. Rocks
4. Weathering and Soil Formation
5. Erosion and Deposition

B ▶ **The Changing Earth**
1. Plate Tectonics
2. Earthquakes
3. Mountains and Volcanoes
4. Views of Earth's Past
5. Natural Resources

C ▶ **Earth's Waters**
1. The Water Planet
2. Freshwater Resources
3. Ocean Systems
4. Ocean Environments

D ▶ **Earth's Atmosphere**
1. Earth's Changing Atmosphere
2. Weather Patterns
3. Weather Fronts and Storms
4. Climate and Climate Change

E ▶ **Space Science**
1. Exploring Space
2. Earth, Moon, and Sun
3. Our Solar System
4. Stars, Galaxies, and the Universe

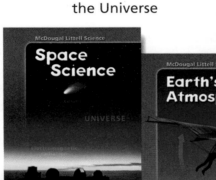

Physical Science Titles

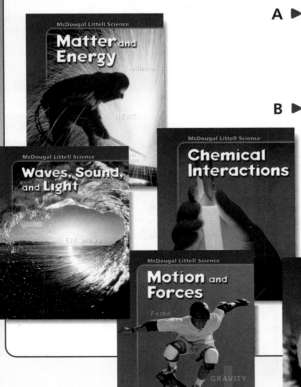

A ▶ **Matter and Energy**
1. Introduction to Matter
2. Properties of Matter
3. Energy
4. Temperature and Heat

B ▶ **Chemical Interactions**
1. Atomic Structure and the Periodic Table
2. Chemical Bonds and Compounds
3. Chemical Reactions
4. Solutions
5. Carbon in Life and Materials

C ▶ **Motion and Forces**
1. Motion
2. Forces
3. Gravity, Friction, and Pressure
4. Work and Energy
5. Machines

D ▶ **Waves, Sound, and Light**
1. Waves
2. Sound
3. Electromagnetic Waves
4. Light and Optics

E ▶ **Electricity and Magnetism**
1. Electricity
2. Circuits and Electronics
3. Magnetism

Teaching Resources

A wealth of print and technology resources help you adapt the program to your teaching style and to the specific needs of your students.

Book-Specific Print Resources

Unit Resource Book provides all of the teaching resources for the unit organized by chapter and section.

- Family Letters
- *Scientific American Frontiers* Video Guide
- Unit Projects
- Lesson Plans
- Reading Study Guides (Levels A and B)
- Spanish Reading Study Guides
- Challenge Readings
- Challenge and Extension Activities
- Reinforcing Key Concepts
- Vocabulary Practice
- Math Support and Practice
- Investigation Datasheets
- Chapter Investigations (Levels A, B, and C)
- Additional Investigations (Levels A, B, and C)
- Summarizing the Chapter

Unit Assessment Book contains complete resources for assessing student knowledge and performance.

- Chapter Diagnostic Tests
- Section Quizzes
- Chapter Tests (Levels A, B, and C)
- Alternative Assessments
- Unit Tests (Levels A, B, and C)

Unit Transparency Book includes instructional visuals for each chapter.

- Three-Minute Warm-Ups
- Note-Taking Models
- Daily Vocabulary Scaffolding
- Chapter Outlines
- Big Idea Flow Charts
- Chapter Teaching Visuals

Unit Lab Manual

Unit Note-Taking/Reading Study Guide

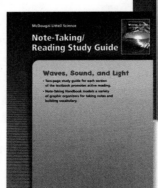

McDougal Littell Science

Unit Resource Book

Waves, Sound, and Light

- Family Letters (English and Spanish)
- *Scientific American Frontiers* Video Guides
- Unit Projects (with Rubrics)
- Lesson Plans
- Reading Study Guides (Levels A and B and Spanish)
- Challenge Activities and Readings
- Reinforcing Key Concepts
- Vocabulary Practice and Decoding Support
- Math Support and Practice
- Investigation Datasheets
- Chapter Investigations (Levels A, B, and C)
- Additional Investigations (Levels A, B, and C)

Program-Wide Print Resources

Process and Lab Skills

Problem Solving and Critical Thinking

Standardized Test Practice

Science Toolkit

City Science

Visual Glossary

Multi-Language Glossary

English Learners Package

Scientific American Frontiers Video Guide

How Stuff Works Express
This quarterly magazine offers opportunities to explore current science topics.

Technology Resources

Scientific American Frontiers **Video Program**
Each specially-tailored segment from this award-winning PBS series correlates to a unit; available on VHS and DVD

Audio CDs Complete chapter texts read in both English and Spanish

Lab Generator CD-ROM
A searchable database of all activities from the program plus additional labs for each unit; edit and print your own version of labs

Test Generator CD-ROM

eEdition CD-ROM

EasyPlanner CD-ROM

Content Review CD-ROM

Power Presentations CD-ROM

Online Resources

ClassZone.com

Content Review Online

eEdition Plus Online

EasyPlanner Plus Online

eTest Plus Online

Correlation to National Science Education Standards

This chart provides an overview of how the five Physical Science modules of *McDougal Littell Science* address the National Science Education Standards.

A Matter and Energy
B Chemical Interactions
C Motion and Forces
D Waves, Sound, and Light
E Electricity and Magnetism

A. Science as Inquiry	Book, Chapter, and Section
A.1– A.8 Abilities necessary to do scientific inquiry Identify questions for investigation; design and conduct investigations; use evidence; think critically and logically; analyze alternative explanations; communicate; use mathematics.	All books (pp. R2–R44), All Chapter Investigations, All Think Science features
A.9 Understandings about scientific inquiry Different kinds of investigations for different questions; investigations guided by current scientific knowledge; importance of mathematics and technology for data gathering and analysis; importance of evidence, logical argument, principles, models, and theories; role of legitimate skepticism; scientific investigations lead to new investigations.	All books (pp. xxii–xxv) A3.1, B2.2, C4.2, D3.2, E3.1

B. Physical Science	Book, Chapter, and Section
B.1 Properties and changes of properties in matter Physical properties; substances, elements, and compounds; chemical reactions.	A1.1, A1.2, A1.3, A1.4, A2.1, A2.2, B1, B3.2, B4.1, B4.2, B4.3, B5.1, B5.3, C3.4
B.2 Motions and forces Position, speed, direction of motion; balanced and unbalanced forces.	C1.1, C1.2, C1.3, C2.1, C2.3, C3.1, C3.2, C3.3, C3.4, C4.1, E5.1
B.3 Transfer of energy Energy transfer; forms of energy; heat and light; electrical circuits; sun as source of Earth's energy.	A3.1, A3.2, A3.3, A4.1, A4.2, A4.3, B3.3, C4.2, D1.1, D3.3, D3.4, D4.1, D4.2, D4.3, E1.1, E1.2, E1.3, E2.1, E2.2

C. Life Science	Book, Chapter, and Section
C.1 Structure and function in living systems Systems; structure and function; levels of organization; cells and cell activities; specialization; human body systems; disease.	B1.1 (Connecting Sciences), B5.2, C5.2 (Connecting Sciences)

D. Earth and Space Science	Book, Chapter, and Section
D.1 Earth's changing atmosphere	B4.2 (Connecting Sciences)
D.3 Earth in the solar system Sun, planets, asteroids, comets; regular and predictable motion and day, year, phases of the moon, and eclipses; gravity and orbits; sun as source of energy for earth; cause of seasons.	A3.2, A3.3, C3.1, D3.3

E. Science and Technology	Book, Chapter, and Section
E.1– E.5 Abilities of technological design Identify problems; design a solution or product; implement a proposed design; evaluate completed designs or products; communicate the process of technological design.	A2.3, A3.1, A4.3, B (p. 5), B4.2, C1.2, C2.1, C3.2, C5.3, D2.4, D3.1, D3.3, D4.4, E2.3
E.6 Understandings about science and technology Similarities and differences between scientific inquiry and technological design; contributions of people in different cultures; reciprocal nature of science and technology; nonexistence of perfectly designed solutions; constraints, benefits, and unintended consequences of technological designs.	All books (pp. xxvi–xxvii) All books (Frontiers in Science, Timelines in Science) A.1.2, A3.3, B3.4, B3.1, B3.3, C5.3, D4.4, E1.2, E1.3

F.	Science in Personal and Social Perspectives	Book, Chapter, and Section
F.1	**Personal health** Exercise; fitness; hazards and safety; tobacco, alcohol, and other drugs; nutrition; STDs; environmental health	B4.2, B5.2
F.2	**Populations, resources, and environments** Overpopulation and resource depletion; environmental degradation.	A3.1
F.3	**Natural hazards** Earthquakes, landslides, wildfires, volcanic eruptions, floods, storms; hazards from human activity; personal and societal challenges.	B3.2, D3.2, E1.2
F.4	**Risks and benefits** Risk analysis; natural, chemical, biological, social, and personal hazards; decisions based on risks and benefits.	B4.3
F.5	**Science and technology in society** Science's influence on knowledge and world view; societal challenges and scientific research; technological influences on society; contributions from people of different cultures and times; work of scientists and engineers; ethical codes; limitations of science and technology.	All books (Timelines in Science) A1.2, A3.2, A3.3, B3.4, B4.4, C5.3, D2, D3.2, D3.3, D4.4, E1.2, E1.3

G.	History and Nature of Science	Book, Chapter, and Section
G.1	**Science as a human endeavor** Diversity of people w.orking in science, technology, and related fields; abilities required by science	All books (pp. xxii–xxv; Frontiers in Science)
G.2	**Nature of science** Observations, experiments, and models; tentative nature of scientific ideas; differences in interpretation of evidence; evaluation of results of investigations, experiments, observations, theoretical models, and explanations; importance of questioning, response to criticism, and communication.	B1.2, B2.1, B2.3, E3.2
G.3	**History of science** Historical examples of inquiry and relationships between science and society; scientists and engineers as valued contributors to culture; challenges of breaking through accepted ideas.	All books (Frontiers in Science; Timelines in Science) B1.2, B3.2, C2.1, D2.4

Correlations to Benchmarks

This chart provides an overview of how the five Physical Science modules of *McDougal Littell Science* address the National Science Education Standards.

A Matter and Energy
B Chemical Interactions
C Motion and Forces
D Waves, Sound, and Light
E Electricity and Magnetism

1. The Nature of Science	Book, Chapter, and Section
	The Nature of Science (pp. xxii–xxv); E2.3; Think Science Features: A3.1, B2.2, C2.1, C4.2, D3.2, E3.1; Scientific Thinking Handbook (pp. R2–R9); Lab Handbook (pp. R10–R35)

3. The Nature of Technology	Book, Chapter, and Section
	The Nature of Technology (pp. xxvi–xxvii); A3.3, B4.4, D4.4, E1, E2.3, E3.2, E3.3, E3.4; Timelines in Science Features

4. The Physical Setting	Book, Chapter, and Section
4.B THE EARTH	A3.1, A4.3, C3.1
4.D STRUCTURE OF MATTER	
4.D.1 All matter is made of atoms; atoms of any element are alike but different from atoms of other elements; different arrangements of atoms into groups compose all substances.	A1.2, A1.3, B1.1, B2.1, B2.2
4.D.2 Equal volumes of different substances usually have different weights.	A2.1, A2.3
4.D.3 Atoms and molecules are perpetually in motion; increased temperature means greater average energy of motion; states of matter: solids, liquids, gases.	A1.2, A4.1
4.D.4 Temperature and acidity of a solution influence reaction rates. Many substances dissolve in water, which may facilitate reactions between them.	B3.1, B4.2, B4.3
4.D.5 Greek philosopheres' scientific ideas about elements; most elements tend to combine with others, so few elements are found in their pure form.	B1.1
4.D.6 Groups of elements have similar properties; oxidation; some elements, like carbon and hydrogen, don't fit into any category and are essential elements of living matter.	B1.2, B1.3, B3.1, B5.1, B5.2
4.D.7 Conservation of matter: the total weight of a closed system remains the same because the total number of atoms stays the same regardless of how they interact with one another.	B3.2
4.E ENERGY TRANSFORMATIONS	
4.E.1 Energy cannot be created or destroyed, but only changed from one form into another.	A3.2, C4.2
4.E.2 Most of what goes on in the universe involves energy transformations.	A3, A4.2, A4.3
4.E.3 Heat can be transferred through materials by the collisions of atoms or across space by radiation; convection currents transfer heat in fluid materials.	A4.2, A4.3
4.E.4 Energy appears in many different forms, including heat energy, chemical energy, mechanical energy, and gravitational energy.	A3.1, A4.2, A4.3, C4.2
4.F MOTION	
4.F.1 Light from the Sun is made up of many different colors of light; objects that give off or reflect light have a different mix of colors.	D3.3, D3.4
4.F.2 Something can be "seen" when light waves emitted or reflected by it enter the eye.	D4.1, D4.3

4.F.3 An unbalanced force acting on an object changes its speed or direction of motion, or both. If the force acts toward a single center, the object's path may curve into an orbit around the center.	C2.1, C2.2, C3.1
4.F.4 Vibrations in materials set up wavelike disturbances (such as sound) that spread away from the source; waves move at different speeds in different materials.	D1, D2.1, D2.2, D3.1, D3.4
4.F.5 Human eyes respond to only a narrow range of wavelengths of electromagnetic radiation—visible light. Differences of wavelengths within that range are perceived as differences in color.	D3.2, D3.4, D4.3
4.G FORCES OF NATURE	
4.G.1 Objects exerts gravitational forces on one another, but these forces depend on the mass and distance of objects, and may be too small to detect.	C3.1
4.G.2 The Sun's gravitational pull holds Earth and other planets in their orbits; planets' gravitational pull keeps their moons in orbit around them.	C3.1
4.G.3 Electric currents and magnets can exert a force on each other.	E3.1, E3.2, E3.3

5. The Living Environment	**Book, Chapter, and Section**
5.E Flow of Matter and Energy	B5.2

8. The Designed World	A2.1, A2.3, A3.3, B3.4, B4.4, B5.3, C5.3, E2.3, E3.2

9. The Mathematical World	All Math in Science Features, E2.3

10. Historical Perspectives	B1, B2, B3.2, C1.1, C2, D4.4

12. Habits of Mind	**Book, Chapter, and Section**
12.A VALUES AND ATTITUDES	Think Science Features: A3.1, B2.2, C2.1, C4.2, D3.2, E3.1
12.B Computation and Estimation	All Math in Science Features, Lab Handbook (pp. R10–R35)
12.C Manipulation and Observation	All Investigates and Chapter Investigations
12.D Communication Skills	All Chapter Investigations, Lab Handbook (pp. R10–R35)
12.E Critical-Response Skills	Think Science Features: A3.1, B2.2, C2.1, C4.2, D3.2, E3.1; Scientific Thinking Handbook (pp. R2–R9)

Planning the Unit

The Pacing Guide provides suggested pacing for all chapters in the unit as well as the two unit features shown below.

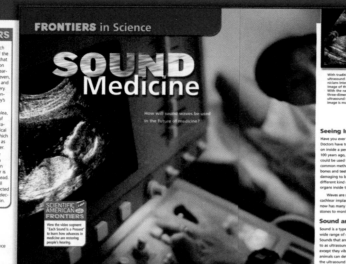

Frontiers in Science

- Features cutting-edge research as an engaging point of entry into the unit
- Connects to an accompanying *Scientific American Frontiers* video and viewing guide
- Introduces three options for unit projects.

Timelines in Science

- Traces the history of key scientific discoveries
- Highlights interactions between science and technology.

Waves, Sound, and Light Pacing Guide

The following pacing guide shows how the chapters in *Waves, Sound, and Light* can be adapted to fit your specific course needs.

	TRADITIONAL SCHEDULE (DAYS)	BLOCK SCHEDULE (DAYS)
Frontiers in Science: Sound Medicine	1	0.5
Chapter 1 Waves		
1.1 Waves transfer energy.	2	1
1.2 Waves have measurable properties.	2	1
1.3 Waves behave in predictable ways.	3	1.5
Chapter Investigation	1	0.5
Chapter 2 Sound		
2.1 Sound is a wave.	2	1
2.2 Frequency determines pitch.	2	1
2.3 Intensity determines loudness.	2	1
2.4 Sound has many uses.	3	1.5
Chapter Investigation	1	0.5
Chapter 3 Electromagnetic Waves		
3.1 Electromagnetic waves have unique traits.	2	1
3.2 Electromagnetic waves have many uses.	2	1
3.3 The Sun is the source of most visible light.	2	1
3.4 Light waves interact with materials.	3	1.5
Chapter Investigation	1	0.5
Timelines in Science: The Story of Light	1	0.5
Chapter 4 Light and Optics		
4.1 Mirrors form images by reflecting light.	2	1
4.2 Lenses form images by refracting light.	2	1
4.3 The eye is a natural optical tool.	2	1
4.4 Optical technology makes use of light waves.	3	1.5
Chapter Investigation	1	0.5
Total Days for Module	**40**	**20**

Planning the Chapter

Complete planning support precedes each chapter.

CHAPTER

1 Waves

Physical Science
UNIFYING PRINCIPLES

PRINCIPLE 1	PRINCIPLE 2	PRINCIPLE 3	PRINCIPLE 4
Matter is made of particles too small to see.	Matter changes form and moves from place to place.	Energy change one form to a it cannot be c destroyed.	

Unit: Waves, Sound, and Light
BIG IDEAS

CHAPTER 1 Waves	CHAPTER 2 Sound	CHAPTER 3 Electromagnetic Waves	C L O o
Waves transfer energy and interact in predictable ways.	Sound waves transfer energy through vibrations.	Electromagnetic waves transfer energy through radiation.	

CHAPTER 1 KEY CONCEPTS

SECTION 1.1	SECTION 1.2
Waves transfer energy.	**Waves have measurable properties**
1. A wave is a disturbance.	1. Waves have amplitude, wavelength, and frequency.
2. Waves can be classified by how they move.	2. Wave speed can be measured.

The Big Idea Flow Chart is available on p. T1 in the **UNIT TRANSPARENCY BOOK**

Previewing Content

SECTION
1.1 Waves transfer energy. pp. 9–15

1. A wave is a disturbance.
A **wave** is a disturbance that transfers energy from one place to another. **Mechanical waves** travel through a material, called a **medium,** transferring energy.
When a mechanical wave travels through a medium, such as water, ground, or air, the medium moves as the wave passes through it, but is not permanently moved. After the wave has

SECTION
1.2

1. **Wa**
A w
wav
(at
Wa
wav
cres

Previewing Content

SECTION
1.3 Waves behave in predictable ways. pp. 24–29

1. Waves interact with materials.
Waves behave predictably when they interact with barriers or other obstacles. Waves can undergo reflection, refraction, or diffraction.
- In **reflection,** waves meet a solid barrier and bounce back. For example, an echo occurs when a sound wave meets a wall and bounces back to the source of the sound.
- In **refraction,** waves move from one medium to another and bend, or refract. An example is a glass of water with a straw. Where the straw passes into the water, it looks broken. But the break is an illusion caused by refraction of light waves as they pass from air to water.
- In **diffraction,** waves interact with a partial barrier and a portion of the waves pass through and spread out. An example is the way sound waves spread around corners.

2. Waves interact with other waves.
In **constructive interference,** two waves combine in phase, so that a crest meets a crest or a trough meets a trough. In **destructive interference,** two waves of the same frequency meet up such that the trough of one wave joins with the crests of the other. If the amplitudes of the two original waves are equal, the two waves cancel each other out.

The following diagrams show how wave amplitudes can be added and subtracted as the waves interfere.

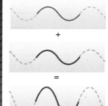

Constructive Interference

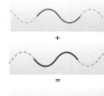

Destructive Interference

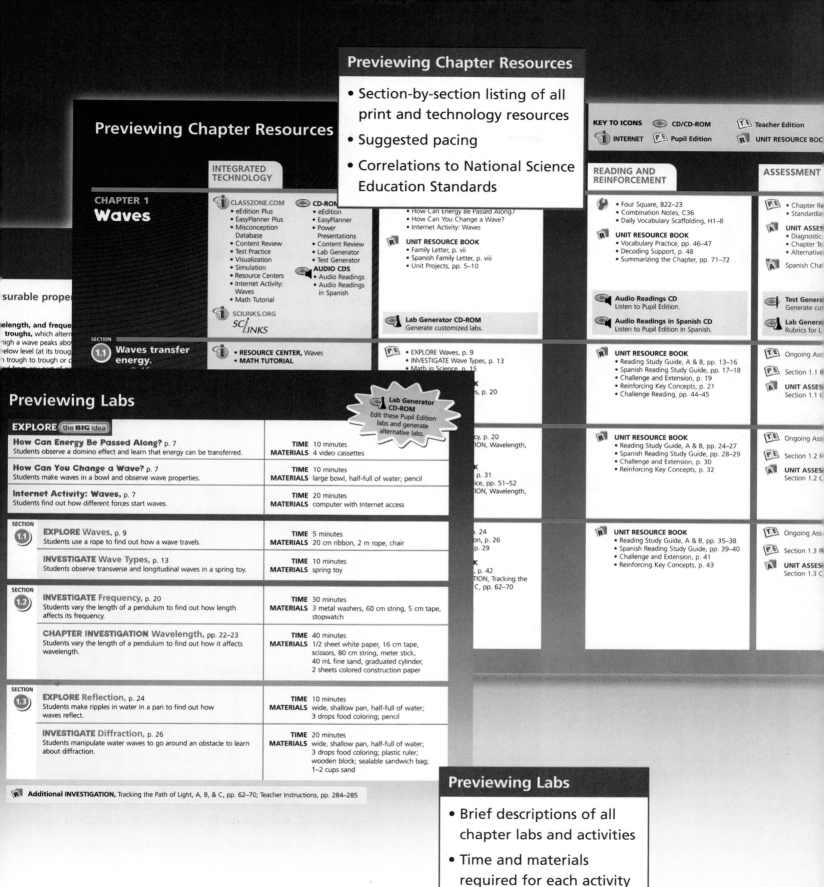

Previewing Chapter Resources

- Section-by-section listing of all print and technology resources
- Suggested pacing
- Correlations to National Science Education Standards

KEY TO ICONS 💿 CD/CD-ROM [TE] Teacher Edition
🕸 INTERNET [PE] Pupil Edition [R] UNIT RESOURCE BOO[K]

Previewing Chapter Resources

	INTEGRATED TECHNOLOGY			READING AND REINFORCEMENT	ASSESSMENT

CHAPTER 1
Waves

🕸 **CLASSZONE.COM**
- eEdition Plus
- EasyPlanner Plus
- Misconception Database
- Content Review
- Test Practice
- Visualization
- Simulation
- Resource Centers
- Internet Activity: Waves
- Math Tutorial

🕸 **SCILINKS.ORG**
SCI/LINKS

💿 **CD-ROM**
- eEdition
- EasyPlanner
- Power Presentations
- Content Review
- Lab Generator
- Test Generator

💿 **AUDIO CDS**
- Audio Readings
- Audio Readings in Spanish

- How Can Energy Be Passed Along?
- How Can You Change a Wave?
- Internet Activity: Waves

[R] **UNIT RESOURCE BOOK**
- Family Letter, p. vii
- Spanish Family Letter, p. viii
- Unit Projects, pp. 5–10

💿 **Lab Generator CD-ROM**
Generate customized labs.

- Four Square, B22–23
- Combination Notes, C36
- Daily Vocabulary Scaffolding, H1–8

[R] **UNIT RESOURCE BOOK**
- Vocabulary Practice, pp. 46–47
- Decoding Support, p. 48
- Summarizing the Chapter, pp. 71–72

💿 **Audio Readings CD**
Listen to Pupil Edition.

💿 **Audio Readings in Spanish CD**
Listen to Pupil Edition in Spanish.

[PE] • Chapter R[e]
 • Standardiz[e]

[A] **UNIT ASSES[S]**
 • Diagnostic
 • Chapter Te[st]
 • Alternative

[SP][A] Spanish Cha[

💿 **Test Genera[**
Generate cu[

💿 **Lab Genera[**
Rubrics for L[

SECTION
(1.1) **Waves transfer energy.**

🕸 • **RESOURCE CENTER**, Waves
 • **MATH TUTORIAL**

[PE] • EXPLORE Waves, p. 9
 • INVESTIGATE Wave Types, p. 13
 • Math in Science, p. 15

[R] **UNIT RESOURCE BOOK**
- Reading Study Guide, A & B, pp. 13–16
- Spanish Reading Study Guide, pp. 17–18
- Challenge and Extension, p. 19
- Reinforcing Key Concepts, p. 21
- Challenge Reading, pp. 44–45

[TE] Ongoing Ass[

[PE] Section 1.1 [

[A] **UNIT ASSES[S**
Section 1.1 C[

Previewing Labs

EXPLORE the BIG idea

How Can Energy Be Passed Along? p. 7
Students observe a domino effect and learn that energy can be transferred.
TIME 10 minutes
MATERIALS 4 video cassettes

How Can You Change a Wave? p. 7
Students make waves in a bowl and observe wave properties.
TIME 10 minutes
MATERIALS large bowl, half-full of water; pencil

Internet Activity: Waves, p. 7
Students find out how different forces start waves.
TIME 20 minutes
MATERIALS computer with Internet access

💿 **Lab Generator CD-ROM**
Edit these Pupil Edition labs and generate alternative labs.

SECTION (1.1)

EXPLORE Waves, p. 9
Students use a rope to find out how a wave travels.
TIME 5 minutes
MATERIALS 20 cm ribbon, 2 m rope, chair

INVESTIGATE Wave Types, p. 13
Students observe transverse and longitudinal waves in a spring toy.
TIME 10 minutes
MATERIALS spring toy

SECTION (1.2)

INVESTIGATE Frequency, p. 20
Students vary the length of a pendulum to find out how length affects its frequency.
TIME 30 minutes
MATERIALS 3 metal washers, 60 cm string, 5 cm tape, stopwatch

CHAPTER INVESTIGATION Wavelength, pp. 22–23
Students vary the length of a pendulum to find out how it affects wavelength.
TIME 40 minutes
MATERIALS 1/2 sheet white paper, 16 cm tape, scissors, 80 cm string, meter stick, 40 mL fine sand, graduated cylinder, 2 sheets colored construction paper

SECTION (1.3)

EXPLORE Reflection, p. 24
Students make ripples in water in a pan to find out how waves reflect.
TIME 10 minutes
MATERIALS wide, shallow pan, half-full of water; 3 drops food coloring; pencil

INVESTIGATE Diffraction, p. 26
Students manipulate water waves to go around an obstacle to learn about diffraction.
TIME 20 minutes
MATERIALS wide, shallow pan, half-full of water; 3 drops food coloring; plastic ruler; wooden block; sealable sandwich bag; 1–2 cups sand

[R] **Additional INVESTIGATION,** Tracking the Path of Light, A, B, & C, pp. 62–70; Teacher Instructions, pp. 284–285

(partial, center-overlapped section)

[PE] • EXPLORE Waves, p. 9
• INVESTIGATE Wave Types, p. 13
• Math in Science, p. 15

...s, p. 20

...cy, p. 20
...TION, Wavelength,

...p. 31
...ice, pp. 51–52
...TION, Wavelength,

[R] **UNIT RESOURCE BOOK**
- Reading Study Guide, A & B, pp. 24–27
- Spanish Reading Study Guide, pp. 28–29
- Challenge and Extension, p. 30
- Reinforcing Key Concepts, p. 32

[TE] Ongoing Ass[

[PE] Section 1.2 [R

[A] **UNIT ASSES[**
Section 1.2 C[

...p. 24
...on, p. 26
...p. 29

...p. 42
...TION, Tracking the
...C, pp. 62–70

[R] **UNIT RESOURCE BOOK**
- Reading Study Guide, A & B, pp. 35–38
- Spanish Reading Study Guide, pp. 39–40
- Challenge and Extension, p. 41
- Reinforcing Key Concepts, p. 43

[TE] Ongoing Ass[

[PE] Section 1.3 [R

[A] **UNIT ASSES[**
Section 1.3 C[

Previewing Labs

- Brief descriptions of all chapter labs and activities
- Time and materials required for each activity

Planning the Lesson

Point-of-use support for each lesson provides a wealth of teaching options.

1. Prepare

- Concept and vocabulary review
- Note-taking and vocabulary strategies

2. Focus

- Set Learning Goals
- 3-Minute Warm-up

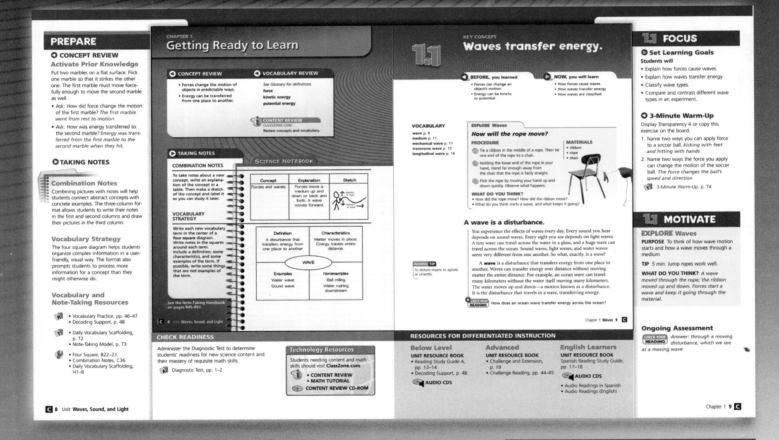

3. Motivate

- Engaging entry into the section
- Explore activity or Think About question

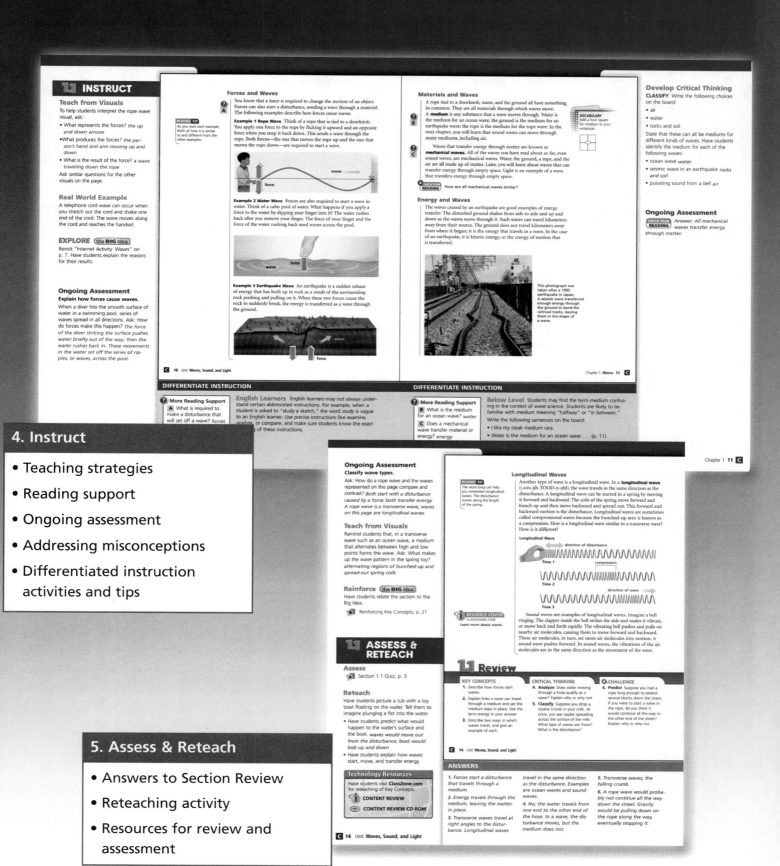

4. Instruct

- Teaching strategies
- Reading support
- Ongoing assessment
- Addressing misconceptions
- Differentiated instruction activities and tips

5. Assess & Reteach

- Answers to Section Review
- Reteaching activity
- Resources for review and assessment

Lab Materials List

The following charts list the consumables, nonconsumables, and equipment needed for all activities. Quantities are per group of four students. Lab aprons, goggles, water, books, paper, pens, pencils, and calculators are assumed to be available for all activities.

Materials kits are available. For more information, please call McDougal Littell at 1-800-323-5435.

Consumable

Description	Quantity per Group	Explore page	Investigate page	Chapter Investigation page
acetate, blue, 4" x 4"	1			100
acetate, green, 4" x 4"	1			100
acetate, red, 4" x 4"	1			100
aluminum foil, 12" x 12"	1	113		
bag, zip-top sandwich	1	26		
balloon	1		41	
cardboard tube, 12"	4	58	134	
cardboard, 10 cm x 25 cm	1		53	
carton, milk, half-gallon	1		115	
clay, modeling	3 sticks	131	128	124
coffee filter, basket	3		98	
cup, clear plastic	4	119	98	
food coloring	1 bottle	24	26	
index card	5	131	128	124
marker, colored	1			124
marker, permanent black	1		84	
marker, water soluble black, various brands	3		98	
milk	2 tsp	93		
mineral oil	3–4 oz	119		
paper, construction, 8.5" x 11"	2			22
paper, white, 8.5" x 11"	2		84	22
plate, white paper	1		128	
posterboard, white, 20 cm x 20 cm	1			124
rubber band, large	10		41, 48, 53	64
salt, table	pinch		41	
sand	2 cups	26		22
shoe box, cardboard	1		48	
shoebox, cardboard with lid	2			64, 100
string	250 cm	37	20	22

Description	Quantity per Group	Explore *page*	Investigate *page*	Chapter Investigation *page*
tape, duct	1 roll		134	
tape, electrical	1 roll	79		
tape, masking	1 roll	58	20, 48, 115	22, 100, 124
wire, copper, uninsulated	50 cm	79		

Nonconsumables

Description	Quantity per Group	Explore *page*	Investigate *page*	Chapter Investigation *page*
battery, D cell	1	79		
chair	1	9		
flashlight with batteries	1	93		124
fork, steel	1	79		
graduated cylinder, 100 mL	1			22
jar, 2 liter clear plastic with lid	1	93		
jar, baby food	1		41	
lamp, gooseneck desk	1		90, 128	100
lens, convex	2	131	128, 134	124
light bulb, compact fluorescent, 15 watt	1		90	
light bulb, incandescent, 100 watt	1		90	
light bulb, incandescent, 60 watt	1		90	
light bulb, incandescent, 75 watt	1		128	100
measuring spoon, teaspoon	1	93		
meter stick	1			22, 124
mirror on a stand	1	73		
mirror, 3 1/2" x 3 1/2"	2		115	
pan, glass, 10" x 14"	1	24	26	
prism	1		84	
protractor	1		115	
radio, portable with batteries	1	79		
radiometer	1		76	

Description	Quantity per Group	Explore page	Investigate page	Chapter Investigation page
ribbon	20 cm	9		
rope	2–3 m	9		
ruler, metric	1	26, 45	53	64, 100
scissors	1		41, 53, 98, 115	22, 64, 100
small object, black	1			100
small object, red	1			100
small object, white	1			100
small object, yellow	1			100
spoon, large metal	1	37		
spring toy	1		13	
stopwatch	1		20	
thermometer	3		84	
TV with infrared remote control	1	73		
washer, metal 1"	3		20	
wood block, 3"	1	26		

McDougal Littell Science

Waves, Sound, and Light

transfer of energy
EM wave
MECHANICAL WAVE

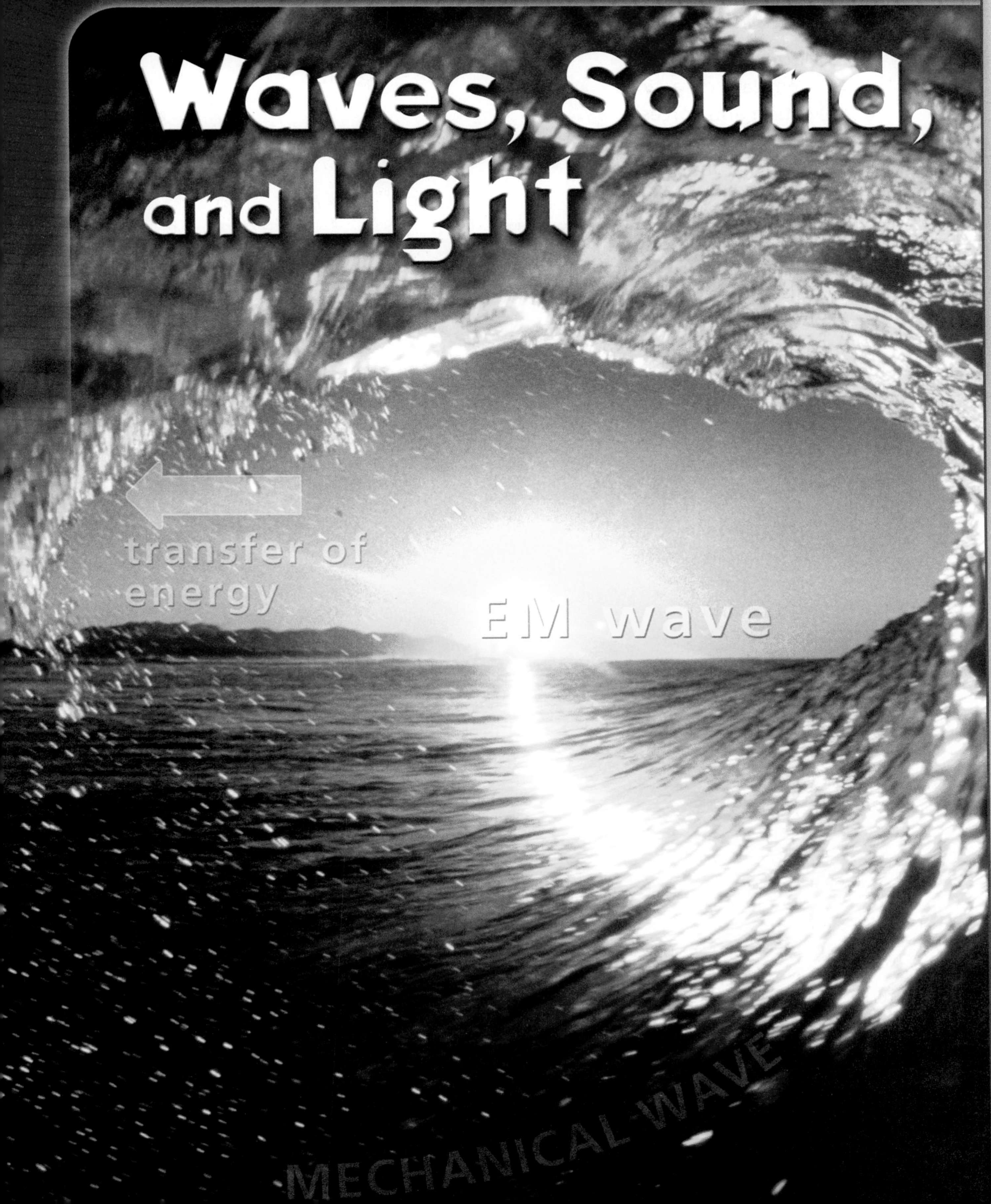

PHYSICAL SCIENCE

A ▶ Matter and Energy
B ▶ Chemical Interactions
C ▶ Motion and Forces
D ▶ Waves, Sound, and Light
E ▶ Electricity and Magnetism

LIFE SCIENCE

A ▶ Cells and Heredity
B ▶ Life Over Time
C ▶ Diversity of Living Things
D ▶ Ecology
E ▶ Human Biology

EARTH SCIENCE

A ▶ Earth's Surface
B ▶ The Changing Earth
C ▶ Earth's Waters
D ▶ Earth's Atmosphere
E ▶ Space Science

ISBN: 0-618-33446-7   1 2 3 4 5 6 7 8 VJM 08 07 06 05 04

Internet Web Site: http://www.mcdougallittell.com

Science Consultants

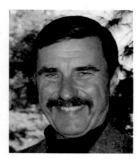

Chief Science Consultant

James Trefil, Ph.D. is the Clarence J. Robinson Professor of Physics at George Mason University. He is the author or co-author of more than 25 books, including *Science Matters* and *The Nature of Science*. Dr. Trefil is a member of the American Association for the Advancement of Science's Committee on the Public Understanding of Science and Technology. He is also a fellow of the World Economic Forum and a frequent contributor to *Smithsonian* magazine.

Rita Ann Calvo, Ph.D. is Senior Lecturer in Molecular Biology and Genetics at Cornell University, where for 12 years she also directed the Cornell Institute for Biology Teachers. Dr. Calvo is the 1999 recipient of the College and University Teaching Award from the National Association of Biology Teachers.

Kenneth Cutler, M.S. is the Education Coordinator for the Julius L. Chambers Biomedical Biotechnology Research Institute at North Carolina Central University. A former middle school and high school science teacher, he received a 1999 Presidential Award for Excellence in Science Teaching.

Instructional Design Consultants

Douglas Carnine, Ph.D. is Professor of Education and Director of the National Center for Improving the Tools of Educators at the University of Oregon. He is the author of seven books and over 100 other scholarly publications, primarily in the areas of instructional design and effective instructional strategies and tools for diverse learners. Dr. Carnine also serves as a member of the National Institute for Literacy Advisory Board.

Linda Carnine, Ph.D. consults with school districts on curriculum development and effective instruction for students struggling academically. A former teacher and school administrator, Dr. Carnine also co-authored a popular remedial reading program.

Donald Steely, Ph.D. serves as principal investigator at the Oregon Center for Applied Science (ORCAS) on federal grants for science and language arts programs. His background also includes teaching and authoring of print and multimedia programs in science, mathematics, history, and spelling.

Sam Miller, Ph.D. is a middle school science teacher and the Teacher Development Liaison for the Eugene, Oregon, Public Schools. He is the author of curricula for teaching science, mathematics, computer skills, and language arts.

Vicky Vachon, Ph.D. consults with school districts throughout the United States and Canada on improving overall academic achievement with a focus on literacy. She is also co-author of a widely used program for remedial readers.

Content Reviewers

John Beaver, Ph.D.
Ecology
Professor, Director of Science Education Center
College of Education and Human Services
Western Illinois University
Macomb, IL

Donald J. DeCoste, Ph.D.
Matter and Energy, Chemical Interactions
Chemistry Instructor
University of Illinois
Urbana-Champaign, IL

Dorothy Ann Fallows, Ph.D., MSc
Diversity of Living Things, Microbiology
Partners in Health
Boston, MA

Michael Foote, Ph.D.
The Changing Earth, Life Over Time
Associate Professor
Department of the Geophysical Sciences
The University of Chicago
Chicago, IL

Lucy Fortson, Ph.D.
Space Science
Director of Astronomy
Adler Planetarium and Astronomy Museum
Chicago, IL

Elizabeth Godrick, Ph.D.
Human Biology
Professor, CAS Biology
Boston University
Boston, MA

Isabelle Sacramento Grilo, M.S.
The Changing Earth
Lecturer, Department of the Geological Sciences
Montana State University
Bozeman, MT

David Harbster, MSc
Diversity of Living Things
Professor of Biology
Paradise Valley Community College
Phoenix, AZ

Richard D. Norris, Ph.D.
Earth's Waters
Professor of Paleobiology
Scripps Institution of Oceanography
University of California, San Diego
La Jolla, CA

Donald B. Peck, M.S.
*Motion and Forces; Waves, Sound, and Light;
  Electricity and Magnetism*
Director of the Center for Science Education (retired)
Fairleigh Dickinson University
Madison, NJ

Javier Penalosa, Ph.D.
Diversity of Living Things, Plants
Associate Professor, Biology Department
Buffalo State College
Buffalo, NY

Raymond T. Pierrehumbert, Ph.D.
Earth's Atmosphere
Professor in Geophysical Sciences (Atmospheric Science)
The University of Chicago
Chicago, IL

Brian J. Skinner, Ph.D.
Earth's Surface
Eugene Higgins Professor of Geology and Geophysics
Yale University
New Haven, CT

Nancy E. Spaulding, M.S.
Earth's Surface, The Changing Earth, Earth's Waters
Earth Science Teacher (retired)
Elmira Free Academy
Elmira, NY

Steven S. Zumdahl, Ph.D.
Matter and Energy, Chemical Interactions
Professor Emeritus of Chemistry
University of Illinois
Urbana-Champaign, IL

Susan L. Zumdahl, M.S.
Matter and Energy, Chemical Interactions
Chemistry Education Specialist
University of Illinois
Urbana-Champaign, IL

Safety Consultant

Juliana Texley, Ph.D.
Former K–12 Science Teacher and School Superintendent
Boca Raton, FL

English Language Advisor

Judy Lewis, M.A.
Director, State and Federal Programs for reading proficiency
and high risk populations
Rancho Cordova, CA

Teacher Panel Members

Carol Arbour
Tallmadge Middle School,
Tallmadge, OH

Patty Belcher
Goodrich Middle School,
Akron, OH

Gwen Broestl
Luis Munoz Marin Middle School,
Cleveland, OH

Al Brofman
Tehipite Middle School,
Fresno, CA

John Cockrell
Clinton Middle School,
Columbus, OH

Jenifer Cox
Sylvan Middle School,
Citrus Heights, CA

Linda Culpepper
Martin Middle School,
Charlotte, NC

Kathleen Ann DeMatteo
Margate Middle School,
Margate, FL

Melvin Figueroa
New River Middle School,
Ft. Lauderdale, FL

Doretha Grier
Kannapolis Middle School,
Kannapolis, NC

Robert Hood
Alexander Hamilton Middle School,
Cleveland, OH

Scott Hudson
Coverdale Elementary School,
Cincinnati, OH

Loretta Langdon
Princeton Middle School,
Princeton, NC

Carlyn Little
Glades Middle School,
Miami, FL

Ann Marie Lynn
Amelia Earhart Middle School,
Riverside, CA

James Minogue
Lowe's Grove Middle School,
Durham, NC

Joann Myers
Buchanan Middle School,
Tampa, FL

Barbara Newell
Charles Evans Hughes Middle School,
Long Beach, CA

Anita Parker
Kannapolis Middle School,
Kannapolis, NC

Greg Pirolo
Golden Valley Middle School,
San Bernardino, CA

Laura Pottmyer
Apex Middle School,
Apex, NC

Lynn Prichard
Booker T. Washington Middle Magnet
School, Tampa, FL

Jacque Quick
Walter Williams High School,
Burlington, NC

Robert Glenn Reynolds
Hillman Middle School,
Youngstown, OH

Theresa Short
Abbott Middle School,
Fayetteville, NC

Rita Slivka
Alexander Hamilton Middle School,
Cleveland, OH

Marie Sofsak
B F Stanton Middle School,
Alliance, OH

Nancy Stubbs
Sweetwater Union Unified School District,
Chula Vista, CA

Sharon Stull
Quail Hollow Middle School,
Charlotte, NC

Donna Taylor
Okeeheelee Middle School,
West Palm Beach, FL

Sandi Thompson
Harding Middle School,
Lakewood, OH

Lori Walker
Audubon Middle School & Magnet Center,
Los Angeles, CA

Teacher Lab Evaluators

Jill Brimm-Byrne
Albany Park Academy,
Chicago, IL

Gwen Broestl
Luis Munoz Marin Middle School,
Cleveland, OH

Al Brofman
Tehipite Middle School,
Fresno, CA

Michael A. Burstein
The Rashi School,
Newton, MA

Trudi Coutts
Madison Middle School,
Naperville, IL

Jenifer Cox
Sylvan Middle School,
Citrus Heights, CA

Larry Cwik
Madison Middle School,
Naperville, IL

Jennifer Donatelli
Kennedy Junior High School,
Lisle, IL

Paige Fullhart
Highland Middle School,
Libertyville, IL

Sue Hood
Glen Crest Middle School,
Glen Ellyn, IL

Ann Min
Beardsley Middle School,
Crystal Lake, IL

Aileen Mueller
Kennedy Junior High School,
Lisle, IL

Nancy Nega
Churchville Middle School,
Elmhurst, IL

Oscar Newman
Sumner Math and Science Academy,
Chicago, IL

Marina Penalver
Moore Middle School,
Portland, ME

Lynn Prichard
Booker T. Washington Middle Magnet
School, Tampa, FL

Jacque Quick
Walter Williams High School,
Burlington, NC

Seth Robey
Gwendolyn Brooks Middle School,
Oak Park, IL

Kevin Steele
Grissom Middle School,
Tinley Park, IL

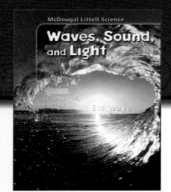

Waves, Sound, and Light

Unit Features

SCIENTIFIC AMERICAN

1 Waves   6

the **BIG** idea

Waves transfer energy and interact in predictable ways.

2 Sound   34

the **BIG** idea

Sound waves transfer energy through vibrations.

How is this guitar player producing sound? page 34

How does this phone stay connected? page 70

Features

Visual Highlights

Internet Resources @ ClassZone.com

INVESTIGATIONS AND ACTIVITIES

Standards and Benchmarks

Each chapter in **Waves, Sound, and Light** covers some of the learning goals that are described in the *National Science Education Standards* (NSES) and the Project 2061 *Benchmarks for Science Literacy*. Selected content and skill standards are shown below in shortened form. The following National Science Education Standards are covered on pages xii–xxvii, in Frontiers in Science, and in Timelines in Science, as well as in chapter features and laboratory investigations: Understandings About Scientific Inquiry (A.9), Understandings About Science and Technology (E.6), Science and Technology in Society (F.5), Science as a Human Endeavor (G.1), Nature of Science (G.2), and History of Science (G.3).

Content Standards

1 Waves

National Science Education Standards

B.3.a | Energy is transferred in many ways. Energy is often associated with
- sound
- light
- mechanical motion

Project 2061 Benchmarks

4.F.4 | Vibrations in materials set up wavelike disturbances that spread away from the source. Sound and earthquake waves are examples. These and other waves move at different speeds through different materials.

2 Sound

Project 2061 Benchmarks

4.F.4 | Vibrations start waves. Waves move at different speeds in different materials.

3 Electromagnetic Waves

National Science Education Standards

B.3.c | Light interacts with matter by
- transmission
- absorption
- scattering

B.3.f | Energy from the Sun is transferred to Earth in the form of
- visible light
- infrared light
- ultraviolet light

Project 2061 Benchmarks

4.F.1 | Light from the sun is made up of a mixture of many different colors of light, even though to the eye the light looks almost white. Other things that give off or reflect light have a different mix of colors.

4.F.4 | Waves move at different speeds through different materials.

4.F.5 | Human eyes detect only a small range of waves in the electromagnetic spectrum. Different wavelengths of visible light are perceived as different colors.

4 Light and Optics

National Science Education Standards

B.3.c | To see an object, light from that object must enter the eye.

Project 2061 Benchmarks

4.F.2 | Something can be "seen" when light waves emitted or reflected by it enter the eye—just as something can be "heard" when sound waves from it enter the ear.

4.F.5 | Human eyes respond to only a narrow range of wavelengths of electromagnetic radiation—visible light. Differences of wavelength within that range are perceived as differences in color.

Process and Skill Standards

National Science Education Standards

A.1 | Identify questions that can be answered through investigation.

A.2 | Design and conduct a scientific investigation.

A.3 | Use appropriate tools and techniques to gather and interpret data.

A.4 | Use evidence to describe, predict, explain, and model.

A.5 | Use critical thinking to find relationships between results and interpretations.

A.6 | Consider alternative explanations and predictions.

A.7 | Communicate procedures, results, and conclusions.

A.8 | Use mathematics in scientific investigations.

E.1 | Identify a problem to be solved.

E.2 | Design a solution or product.

E.3 | Implement the proposed solution.

E.4 | Evaluate the solution or design.

F.4.c | Use systematic thinking to estimate risks.

F.4.d | Decisions are made based on estimated risks and benefits.

Project 2061 Benchmarks

1.A.3 | Some knowledge in science is very old and yet is still used today.

1.C.1 | Contributions to science and technology have been made by different people, in different cultures, at different times.

6.A.5 | People use technology to match or go beyond the abilities of other species.

9.B.3 | Graphs can show the relationship between two variables.

9.C.4 | Graphs show patterns and can be used to make predictions.

9.D.3 | The mean, median, and mode are different ways of finding the middle of a set of data.

11.C.4 | Use equations to summarize observed changes.

12.B.4 | Find the mean and median of a set of data.

12.C.3 | Use appropriate units, use and read instruments that measure quantities such as length, volume, weight, time, rate, and temperature.

12.D.1 | Use tables and graphs to organize information and identify relationships.

12.D.2 | Read, interpret, and describe tables and graphs.

12.D.4 | Understand information that includes different types of charts and graphs, including circle charts, bar graphs, line graphs, data tables, diagrams, and symbols.

12.E.4 | There may be more than one good way to interpret scientific findings.

Introducing Physical Science

Scientists are curious. Since ancient times, they have been asking and answering questions about the world around them. Scientists are also very suspicious of the answers they get. They carefully collect evidence and test their answers many times before accepting an idea as correct.

In this book you will see how scientific knowledge keeps growing and changing as scientists ask new questions and rethink what was known before. The following sections will help get you started.

What Is Physical Science?

In the simplest terms, physical science is the study of what things are made of and how they change. It combines the studies of both physics and chemistry. Physics is the science of matter, energy, and forces. It includes the study of topics such as motion, light, and electricity and magnetism. Chemistry is the study of the structure and properties of matter, and it especially focuses on how substances change into different substances.

The text and pictures in this book will help you learn key concepts and important facts about physical science. A variety of activities will help you investigate these concepts. As you learn, it helps to have a big picture of physical science as a framework for this new information. The four unifying principles listed below will give you this big picture. Read the next few pages to get an overview of each of these principles and a sense of why they are so important.

- **Matter is made of particles too small to see.**

- **Matter changes form and moves from place to place.**

- **Energy changes from one form to another, but it cannot be created or destroyed.**

- **Physical forces affect the movement of all matter on Earth and throughout the universe.**

the **BIG** idea

Each chapter begins with a big idea. Keep in mind that each big idea relates to one or more of the unifying principles.

UNIFYING PRINCIPLE

Matter is made of particles too small to see.

This simple statement is the basis for explaining an amazing variety of things about the world. For example, it explains why substances can exist as solids, liquids, and gases, and why wood burns but iron does not. Like the tiles that make up this mosaic picture, the particles that make up all substances combine to make patterns and structures that can be seen. Unlike these tiles, the individual particles themselves are far too small to see.

What It Means

To understand this principle better, let's take a closer look at the two key words: *matter* and *particles.*

Matter

Objects you can see and touch are all around you. The materials that these objects are made of are called **matter.** All living things—even you—are also matter. Even though you can't see it, the air around you is matter too. Scientists often say that matter is anything that has mass and takes up space. **Mass** is a measure of the amount of matter in an object. We use the word **volume** to refer to the amount of space an object or a substance takes up.

Particles

The tiny particles that make up all matter are called **atoms.** Just how tiny are atoms? They are far too small to see, even through a powerful microscope. In fact, an atom is more than a million times smaller than the period at the end of this sentence.

There are more than 100 basic kinds of matter called **elements.** For example, iron, gold, and oxygen are three common elements. Each element has its own unique kind of atom. The atoms of any element are all alike but different from the atoms of any other element.

Many familiar materials are made of particles called molecules. In a **molecule,** two or more atoms stick together to form a larger particle. For example, a water molecule is made of two atoms of hydrogen and one atom of oxygen.

Why It's Important

Understanding atoms and molecules makes it possible to explain and predict the behavior of matter. Among other things, this knowledge allows scientists to

- explain why different materials have different characteristics
- predict how a material will change when heated or cooled
- figure out how to combine atoms and molecules to make new and useful materials

Matter changes form and moves from place to place.

You see matter change form every day. You see the ice in your glass of juice disappear without a trace. You see a black metal gate slowly develop a flaky, orange coating. Matter is constantly changing and moving.

What It Means

Remember that matter is made of tiny particles called atoms. Atoms are constantly moving and combining with one another. All changes in matter are the result of atoms moving and combining in different ways.

Matter Changes and Moves

You can look at water to see how matter changes and moves. A block of ice is hard like a rock. Leave the ice out in sunlight, however, and it changes into a puddle of water. That puddle of water can eventually change into water vapor and disappear into the air. The water vapor in the air can become raindrops, which may fall on rocks, causing them to weather and wear away. The water that flows in rivers and streams picks up tiny bits of rock and carries them from one shore to another. Understanding how the world works requires an understanding of how matter changes and moves.

Matter Is Conserved

No matter was lost in any of the changes described above. The ice turned to water because its molecules began to move more quickly as they got warmer. The bits of rock carried away by the flowing river were not gone forever. They simply ended up farther down the river. The puddles of rainwater didn't really disappear; their molecules slowly mixed with molecules in the air.

Under ordinary conditions, when matter changes form, no matter is created or destroyed. The water created by melting ice has the same mass as the ice did. If you could measure the water vapor that mixes with the air, you would find it had the same mass as the water in the puddle did.

Why It's Important

Understanding how mass is conserved when matter changes form has helped scientists to

- describe changes they see in the world
- predict what will happen when two substances are mixed
- explain where matter goes when it seems to disappear

Energy changes from one form to another, but it cannot be created or destroyed.

When you use energy to warm your food or to turn on a flashlight, you may think that you "use up" the energy. Even though the camp-stove fuel is gone and the flashlight battery no longer functions, the energy they provided has not disappeared. It has been changed into a form you can no longer use. Understanding how energy changes forms is the basis for understanding how heat, light, and motion are produced.

What It Means

Changes that you see around you depend on energy. **Energy,** in fact, means the ability to cause change. The electrical energy from an outlet changes into light and heat in a light bulb. Plants change the light energy from the Sun into chemical energy, which animals use to power their muscles.

Energy Changes Forms

Using energy means changing energy. You probably have seen electric energy changing into light, heat, sound, and mechanical energy in household appliances. Fuels like wood, coal, and oil contain chemical energy that produces heat when burned. Electric power plants make electrical energy from a variety of energy sources, including falling water, nuclear energy, and fossil fuels.

Energy Is Conserved

Energy can be converted into forms that can be used for specific purposes. During the conversion, some of the original energy is converted into unwanted forms. For instance, when a power plant converts the energy of falling water into electrical energy, some of the energy is lost to friction and sound.

Similarly, when electrical energy is used to run an appliance, some of the energy is converted into forms that are not useful. Only a small percentage of the energy used in a light bulb, for instance, produces light; most of the energy becomes heat. Nonetheless, the total amount of energy remains the same through all these conversions.

The fact that energy does not disappear is a law of physical science. The **law of conservation of energy** states that energy cannot be created or destroyed. It can only change form.

Why It's Important

Understanding that energy changes form but does not disappear has helped scientists to

- predict how energy will change form
- manage energy conversions in useful ways
- build and improve machines

Physical forces affect the movement of all matter on Earth and throughout the universe.

What makes the world go around? The answer is simple: forces. Forces allow you to walk across the room, and forces keep the stars together in galaxies. Consider the forces acting on the rafts below. The rushing water is pushing the rafts forward. The force from the people paddling helps to steer the rafts.

What It Means

A **force** is a push or a pull. Every time you push or pull an object, you're applying a force to that object, whether or not the object moves. There are several forces—several pushes and pulls—acting on you right now. All these forces are necessary for you to do the things you do, even sitting and reading.

- You are already familiar with the force of gravity. **Gravity** is the force of attraction between two objects. Right now gravity is at work pulling you to Earth and Earth to you. The Moon stays in orbit around Earth because gravity holds it close.

- A contact force occurs when one object pushes or pulls another object by touching it. If you kick a soccer ball, for instance, you apply a contact force to the ball. You apply a contact force to a shopping cart that you push down a grocery aisle or a sled that you pull up a hill.

- **Friction** is the force that resists motion between two surfaces pressed together. If you've ever tried to walk on an icy sidewalk, you know how important friction can be. If you lightly rub your finger across a smooth page in a book and then across a piece of sandpaper, you can feel how the different surfaces produce different frictional forces. Which is easier to do?

- There are other forces at work in the world too. For example, a compass needle responds to the magnetic force exerted by Earth's magnetic field, and objects made of certain metals are attracted by magnets. In addition to magnetic forces, there are electrical forces operating between particles and between objects. For example, you can demonstrate electrical forces by rubbing an inflated balloon on your hair. The balloon will then stick to your head or to a wall without additional means of support.

Why It's Important

Although some of these forces are more obvious than others, physical forces at work in the world are necessary for you to do the things you do. Understanding forces allows scientists to

- predict how objects will move
- design machines that perform complex tasks
- predict where planets and stars will be in the sky from one night to the next

The Nature of Science

You may think of science as a body of knowledge or a collection of facts. More important, however, science is an active process that involves certain ways of looking at the world.

Scientific Habits of Mind

Scientists are curious. They are always asking questions. Scientists have asked questions such as, "What is the smallest form of matter?" and "How do the smallest particles behave?" These and other important questions are being investigated by scientists around the world.

Scientists are observant. They are always looking closely at the world around them. Scientists once thought the smallest parts of atoms were protons, neutrons, and electrons. Later, protons and neutrons were found to be made of even smaller particles called quarks.

Scientists are creative. They draw on what they know to form possible explanations for a pattern, an event, or an interesting phenomenon that they have observed. Then scientists create a plan for testing their ideas.

Scientists are skeptical. Scientists don't accept an explanation or answer unless it is based on evidence and logical reasoning. They continually question their own conclusions and the conclusions suggested by other scientists. Scientists trust only evidence that is confirmed by other people or methods.

Scientists cannot always make observations with their own eyes. They have developed technology, such as this particle detector, to help them gather information about the smallest particles of matter.

Scientists ask questions about the physical world and seek answers through carefully controlled procedures. Here a researcher works with supercooled magnets.

Science Processes at Work

You can think of science as a continuous cycle of asking and seeking answers to questions about the world. Although there are many processes that scientists use, scientists typically do each of the following:

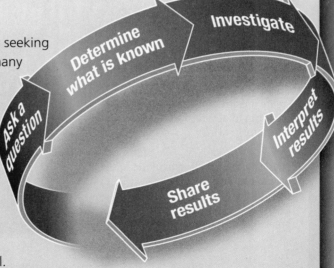

- Ask a question
- Determine what is known
- Investigate
- Interpret results
- Share results

Ask a Question

It may surprise you that asking questions is an important skill. A scientific process may start when a scientist asks a question. Perhaps scientists observe an event or a process that they don't understand, or perhaps answering one question leads to another.

Determine What Is Known

When beginning an inquiry, scientists find out what is already known about a question. They study results from other scientific investigations, read journals, and talk with other scientists. A scientist working on subatomic particles is most likely a member of a large team using sophisticated equipment. Before beginning original research, the team analyzes results from previous studies.

Investigate

Investigating is the process of collecting evidence. Two important ways of investigating are observing and experimenting.

Observing is the act of noting and recording an event, a characteristic, or anything else detected with an instrument or with the senses. A researcher may study the properties of a substance by handling it, finding its mass, warming or cooling it, stretching it, and so on. For information about the behavior of subatomic particles, however, a researcher may rely on technology such as scanning tunneling microscopes, which produce images of structures that cannot be seen with the eye.

An **experiment** is an organized procedure to study something under controlled conditions. In order to study the effect of wing shape on the motion of a glider, for instance, a researcher would need to conduct controlled studies in which gliders made of the same materials and with the same masses differed only in the shape of their wings.

Scanning tunneling microscopes create images that allow scientists to observe molecular structure.

Physical chemists have found a way to observe chemical reactions at the atomic level. Using lasers, they can watch bonds breaking and new bonds forming.

Forming hypotheses and making predictions are two of the skills involved in scientific investigations. A **hypothesis** is a tentative explanation for an observation, a phenomenon, or a scientific problem that can be tested by further investigation. For example, in the mid-1800s astronomers noticed that the planet Uranus departed slightly from its expected orbit. One astronomer hypothesized that the irregularities in the planet's orbit were due to the gravitational effect of another planet—one that had not yet been detected. A **prediction** is an expectation of what will be observed or what will happen. A prediction can be used to test a hypothesis. The astronomers predicted that they would discover a new planet in the position calculated, and their prediction was confirmed with the discovery of the planet Neptune.

Interpret Results

As scientists investigate, they analyze their evidence, or data, and begin to draw conclusions. **Analyzing data** involves looking at the evidence gathered through observations or experiments and trying to identify any patterns that might exist in the data. Scientists often need to make additional observations or perform more experiments before they are sure of their conclusions. Many times scientists make new predictions or revise their hypotheses.

Often scientists use computers to help them analyze data. Computers reveal patterns that might otherwise be missed.

Scientists use computers to create models of objects or processes they are studying. This model shows carbon atoms forming a sphere.

Share Results

An important part of scientific investigation is sharing results of experiments. Scientists read and publish in journals and attend conferences to communicate with other scientists around the world. Sharing data and procedures gives them a way to test one another's results. They also share results with the public through newspapers, television, and other media.

The Nature of Technology

When you think of technology, you may think of cars, computers, and cell phones, as well as refrigerators, radios, and bicycles. Technology is not only the machines and devices that make modern lives easier, however. It is also a process in which new methods and devices are created. Technology makes use of scientific knowledge to design solutions to real-world problems.

Science and Technology

Science and technology go hand in hand. Each depends upon the other. Even designing a device as simple as a toaster requires knowledge of how heat flows and which materials are the best conductors of heat. Just as technology based on scientific knowledge makes our lives easier, some technology is used to advance scientific inquiry itself. For example, researchers use a number of specialized instruments to help them collect data. Microscopes, telescopes, spectrographs, and computers are just a few of the tools that help scientists learn more about the world. The more information these tools provide, the more devices can be developed to aid scientific research and to improve modern lives.

The Process of Technological Design

The process of technology involves many choices. For example, how does an automobile engineer design a better car? Is a better car faster? safer? cheaper? Before designing any new machine, the engineer must decide exactly what he or she wants the machine to do as well as what may be given up for the machine to do it. A faster car may get people to their destinations more quickly, but it may cost more and be less safe. As you study the techno-logical process, think about all the choices that were made to build the technologies you use.

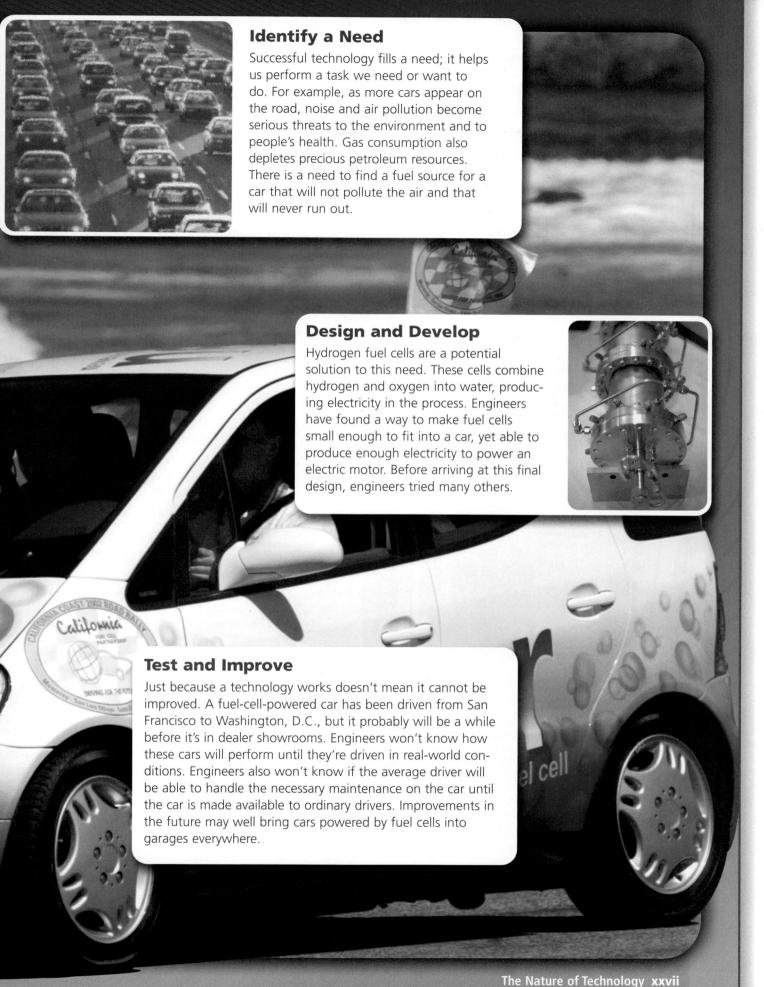

Identify a Need

Successful technology fills a need; it helps us perform a task we need or want to do. For example, as more cars appear on the road, noise and air pollution become serious threats to the environment and to people's health. Gas consumption also depletes precious petroleum resources. There is a need to find a fuel source for a car that will not pollute the air and that will never run out.

Design and Develop

Hydrogen fuel cells are a potential solution to this need. These cells combine hydrogen and oxygen into water, producing electricity in the process. Engineers have found a way to make fuel cells small enough to fit into a car, yet able to produce enough electricity to power an electric motor. Before arriving at this final design, engineers tried many others.

Test and Improve

Just because a technology works doesn't mean it cannot be improved. A fuel-cell-powered car has been driven from San Francisco to Washington, D.C., but it probably will be a while before it's in dealer showrooms. Engineers won't know how these cars will perform until they're driven in real-world conditions. Engineers also won't know if the average driver will be able to handle the necessary maintenance on the car until the car is made available to ordinary drivers. Improvements in the future may well bring cars powered by fuel cells into garages everywhere.

Using McDougal Littell Science

Reading Text and Visuals

This book is organized to help you learn. Use these boxed pointers as a path to help you learn and remember the **Big Ideas** and **Key Concepts**.

Take notes.

Use the strategies on the **Getting Ready to Learn** page.

Read the Big Idea.

As you read **Key Concepts** for the chapter, relate them to **the Big Idea**.

CHAPTER 2

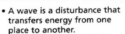

CHAPTER

2 Sou

the BIG idea

Sound waves transfer energy through vibrations.

Key Concepts

SECTION
2.1 Sound is a wave.
Learn how sound waves are produced and detected.

SECTION
2.2 Frequency determines pitch.
Learn about the relationship between the frequency of a sound wave and its pitch.

SECTION
2.3 Intensity determines loudness.
Learn how the energy of a sound wave relates to its loudness.

SECTION
2.4 Sound has many uses.
Learn how sound waves are used to detect objects and to make music.

Internet Preview

CLASSZONE.COM
Chapter 2 online resources: Content Review, two Visualizations, three Resource Centers, Math Tutorial, Test Practice

---

Getting Ready to Learn

CONCEPT REVIEW

- A wave is a disturbance that transfers energy from one place to another.
- Mechanical waves are waves that travel through matter.

VOCABULARY REVIEW

medium p. 11
longitudinal wave p. 14
amplitude p. 17
wavelength p. 17
frequency p. 17

CONTENT REVIEW
CLASSZONE.COM
Review concepts and vocabulary.

TAKING NOTES

OUTLINE

As you read, copy the headings on your paper in the form of an outline. Then add notes in your own words that summarize what you have read.

VOCABULARY STRATEGY

Place each vocabulary term at the center of a **description wheel** diagram. Write some words on the spokes describing it.

See the Note-Taking Handbook on pages R45–R51.

SCIENCE NOTEBOOK

I. Sound is a type of mechanical wave
  A. How sound waves are produced
    1.
    2.
    3.
  B. How sound waves are detected
    1.
    2.
    3.

rapid back-and-forth motion

can produce a sound

VIBRATION

usually too small to see

can make with vocal cords

KEY CONCEPT

2.1 Sound is a wave.

◀ **BEFORE, you learned**

- Waves transfer energy
- Waves have wavelength, amplitude, and frequency

▶ **NOW, you will learn**

- How sound waves are produced and detected
- How sound waves transfer energy
- What affects the speed of sound waves

VOCABULARY

sound p. 37
vibration p. 37
vacuum p. 41

EXPLORE Sound

What is sound?

PROCEDURE

1. Tie the middle of the string to the spoon handle.

2. Wrap the string ends around your left and right index fingers. Put the tips of these fingers gently in your ears and hold them there.

3. Stand over your desk so that the spoon dangles without touching your body or the desk. Then move a little to make the spoon tap the desk lightly. Listen to the sound.

WHAT DO YOU THINK?

- What did you hear when the spoon tapped the desk?
- How did sound travel from the spoon to your ears?

MATERIALS

- piece of string
- large metal spoon

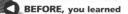

Sound is a type of mechanical wave.

In the last chapter, you read that a mechanical wave travels through a material medium. Such mediums include air, water, and solid materials. Sound is an example of a mechanical wave. **Sound** is a wave that is produced by a vibrating object and travels through matter.

The disturbances that travel in a sound wave are vibrations. A **vibration** is a rapid, back-and-forth motion. Because the medium vibrates back and forth in the same direction as the wave travels, sound is a longitudinal wave. Like all mechanical waves, sound waves transfer energy through a medium.

CHECK YOUR READING What do sound waves have in common with other mechanical waves? Your answer should include the word *energy*.

Chapter 2: **Sound** 37

Reading Text and Visuals

A Wave Model

When these fans do "the wave" in a stadium, they are modeling the way a disturbance travels through a medium.

Each person only moves up and down.

The wave can move all the way around the stadium.

Study the visuals.

- Read the title.
- Read all labels and captions.
- Figure out what the picture is showing. Notice colors, arrows, and lines.

READING VISUALS In which direction do people move when doing the stadium wave? In which direction does the wave move?

Look at the illustration of people modeling a wave in a stadium. In this model, the crowd of people represents a wave medium. The people moving up and down represent the disturbance. The transfer of the disturbance around the stadium represents a wave. Each person only moves up and down, while the disturbance can move all the way around the stadium.

Ocean waves are another good example of energy transfer. Ocean waves travel to the shore, one after another. Instead of piling up all the ocean water on the shore, however, the waves transfer energy. A big ocean wave transfers enough kinetic energy to knock someone down.

Read one paragraph at a time.

Look for a topic sentence that explains the main idea of the paragraph. Figure out how the details relate to that idea. One paragraph might have several important ideas; you may have to reread to understand.

CHECK YOUR READING How does the stadium wave differ from a real ocean wave?

Waves can be classified by how they move.

As you have seen, one way to classify waves is according to the medium through which they travel. Another way to classify waves is by how they move. You have read that some waves transfer an up-and-down or a side-to-side motion. Other waves transfer a forward-and-backward motion.

Answer the questions.

Check Your Reading questions will help you remember what you read.

2 Unit: **Waves, Sound, and Light**

Doing Labs

To understand science, you have to see it in action. Doing labs helps you understand how things really work.

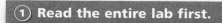

① Read the entire lab first.

② Form a hypothesis.

③ Follow the procedure.

④ Record the data.

CHAPTER INVESTIGATION

Wavelength

OVERVIEW AND PURPOSE The pendulum on a grandfather clock keeps time as it swings back and forth at a steady rate. The swings of a pendulum can be recorded as a wave with measurable properties. How do the properties of the pendulum affect the properties of the waves it produces? In this investigation you will use your understanding of wave properties to
- construct a pendulum and measure the waves it produces, and
- determine how the length of the pendulum affects the wavelength of the waves.

▶ **Problem** [Write It Up]

How does changing the length of a pendulum affect the wavelength?

▶ **Hypothesize** [Write It Up]

Write a hypothesis in "If . . . , then . . . , because . . ." form to answer the problem question.

▶ **Procedure**

1. Make a data table like the one shown on the sample notebook page.

2. Make a cone with the half-sheet of paper by rolling it and taping it as shown. The hole in the bottom of the cone should be no larger than a pea.

3. Cut a hole in each side of the cone and tie the ends of the string to the cone to make a pendulum.

4. Hold the string on the pendulum so that the distance from your fingers holding the string to the bottom of the cone is 20 cm.

5. Cover the bottom of the cone with your fingertip. While you hold the cone, have your partner pour about 40 ml of sand into the cone.

MATERIALS
- 1/2 sheet white paper
- tape
- scissors
- string
- meter stick
- fine sand
- graduated cylinder
- 2 sheets colored construction paper

6. Hold the pendulum about 5 cm above the construction paper as shown. Pull the pendulum from the bottom to one side of the construction paper. Be careful not to move the pendulum at the top, or to pull the pendulum over the edge of the paper.

7. Let the pendulum go while your partner gently pulls the paper forward so that the sand makes waves on the paper. Be sure to pull the paper at a steady rate. Let the remaining sand pile up on the end of the paper.

8. Measure the wavelength from crest to crest or trough to trough. Record the wavelength in your table.

9. Run two more trials, repeating steps 5–8. Be sure to pull the paper at the same speed for each trial. Calculate the average wavelength over all three trials, and record it in your table.

10. Repeat steps 4–8, changing the length of the pendulum to 30 cm and then to 40 cm.

▶ **Observe and Analyze** [Write It Up]

1. **RECORD OBSERVATIONS** Draw the setup of your procedure. Be sure your data table is complete.

2. **IDENTIFY VARIABLES AND CONSTANTS** Identify the variables and constants that affected the wave produced by the moving pendulum. List them in your notebook.

3. **ANALYZE** What patterns can you find in your data? For example, do the numbers increase or decrease as you read down each column?

▶ **Conclude** [Write It Up]

1. **INFER** Answer your problem question.

2. **INTERPRET** Compare your results with your hypothesis. Do your data support your hypothesis?

3. **IDENTIFY LIMITS** What possible limitations or sources of error could have affected your results?

4. **APPLY** Suppose you were examining the tracing made by a seismograph, a machine that records an earthquake wave. What would happen if you increased the speed at which the paper ran through the machine? What do you think the amplitude of the tracing represents?

▶ **INVESTIGATE Further**

CHALLENGE Revise your experiment to change one variable other than the length of the pendulum. Run a new trial, changing the variable you choose but keeping everything else constant. How did changing the variable affect the wave produced?

Wavelength

Problem How does changing the length of a pendulum affect the wavelength?

Hypothesize

Observe and Analyze

Table 1. Wavelengths Produced by Pendulums

Pendulum Length (cm)	Trial 1	Trial 2	Trial 3	Average Wavelength (cm)
20				
30				
40				

Conclude

⑤ Analyze your results.

⑥ Write your lab report.

Using Technology

The Internet is a great source of up-to-date science. The ClassZone Website and SciLinks have exciting sites for you to explore. Video clips and simulations can make science come alive.

Look for red banners.

Go to **classzone.com** to see simulations, visualizations, and content review.

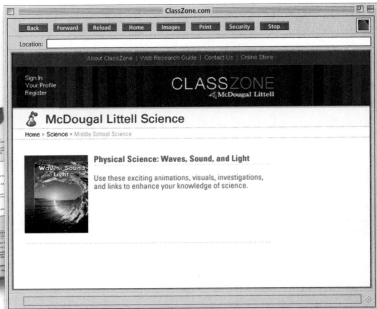

Watch the video.

See science at work in the **Scientific American Frontiers** video.

Look up SciLinks.

Go to **scilinks.org** to explore the topic.

Atmospheric Pressure and Winds **CODE: MDL010**

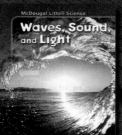

Waves, Sound, and Light
Contents Overview

Unit Features

(the **BIG** idea)

Waves transfer energy and
interact in predictable ways.

(the **BIG** idea)

Sound waves transfer energy
through vibrations.

(the **BIG** idea)

Electromagnetic waves transfer
energy through radiation.

(the **BIG** idea)

Optical tools depend on
the wave behavior of light.

FRONTIERS in Science

SCIENTIFIC AMERICAN FRONTIERS

EACH SOUND IS A PRESENT "Each Sound Is a Present" is a segment of the Scientific American Frontiers series that aired on PBS stations. This segment focuses on Kelley Flynn, who suffered severe hearing damage from an infection. At seven, Kelley's hearing is becoming worse, and she will have cochlear implant surgery. The goal of the implant is to compensate for hair cells destroyed by Kelley's infection.

The human ear contains the cochlea, a spiral bone lined with thousands of tiny hairs that bend from sound vibrations. This movement triggers electrical signals along nerves to the brain, which receives and processes these signals as sound. Damaged hairs do not recover.

Kelley's surgeons replace cochlear hairs with electrodes connected to a magnet and an antenna implanted in her skull. After surgery, a transmitter is attached to the magnet inside her head. The transmitter is attached to a small computer that converts sounds collected by a microphone to electrical signals that are sent to the brain.

National Science Education Standards

A.9.a–d Understandings About Scientific Inquiry

E.6.a–f Understandings About Science and Technology

F.5.a–e Science and Technology in Society

G.1.a–b Science as a Human Endeavor

G.2.a Nature of science

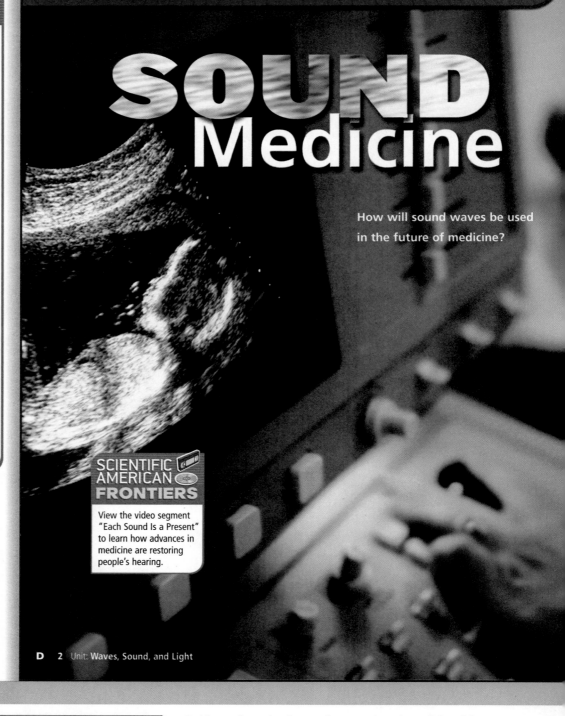

FRONTIERS in Science

SOUND Medicine

How will sound waves be used in the future of medicine?

SCIENTIFIC AMERICAN FRONTIERS

View the video segment "Each Sound Is a Present" to learn how advances in medicine are restoring people's hearing.

D 2 Unit: Waves, Sound, and Light

ADDITIONAL RESOURCES

Technology Resources

 Scientific American Frontiers Video: *Each Sound Is a Present:* 7-minute video segment introduces the unit.

 ClassZone.com
CAREER LINK, careers in audiology

Guide student viewing and comprehension of the video:

 Frontiers in Science Teaching Guide, pp. 1–2; Viewing Guide, p. 3, Video Wrap-Up, p. 4

Scientific American Frontiers Video Guide, pp. 51–54

Unit project procedures and rubrics:

 Unit Projects, pp. 5–10

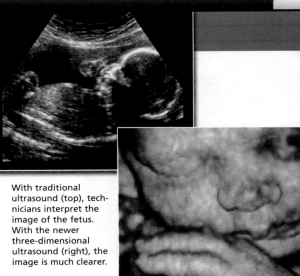

With traditional ultrasound (top), technicians interpret the image of the fetus. With the newer three-dimensional ultrasound (right), the image is much clearer.

Seeing Inside the Body

Have you ever wondered what the inside of your body looks like? Doctors have tried for many years to find ways of seeing what goes on inside a person's body that makes that person sick. Around 100 years ago, scientists found that a kind of wave called x-rays could be used to make images of the bones inside a person. This common method of seeing inside a body, is used mainly to show bones and teeth. However, repeated exposure to x-rays can be damaging to body cells. In the 1960s doctors started using a different kind of wave called ultrasound to make images of the organs inside the body.

Waves are now used in many medical applications. For example, cochlear implants use radio waves to help people hear. Ultrasound now has many new medical applications, from breaking up kidney stones to monitoring the flow of blood in the body.

?
A

Sound and Ultrasound

Sound is a type of wave, a vibration in the air. Humans can hear a wide range of different sounds, from very low pitches to very high. Sounds that are higher in pitch than humans can hear are referred to as ultrasound. They are no different from sounds we can hear, except they vibrate much faster than human ears can detect. Many animals can detect ultrasound; for example, dog whistles are in the ultrasound range.

?
B

◉ Set Learning Goals

Students will

- Observe practical uses of sound waves in medicine
- Compare technological uses of waves to similar uses in nature
- Design and produce a product that involves uses of waves

Remind students that frontiers are not totally unexplored areas; they are areas that are currently being explored and developed. Use the video segment "Each Sound Is a Present" to show that areas of science can have practical technologies yet still have new developments to come. Have students look at the ultrasound photographs and compare the effectiveness of the results shown.

INSTRUCT

Teach from Visuals

Have students examine the two ultrasound photographs and identify the heads and arms of the fetuses. Ask students if they know anyone who has had ultrasound tests or treatments. Emphasize that ultrasound is used for diagnosis of many diseases and disorders, including the presence and location of certain tumors.

Sharing Results

Ask: Why is it important that developments in medical research be shared among medical personnel? *Such developments are necessary for diagnosis and treatment of many diseases and disorders. The research that must be done is quite expensive and time-consuming, and specialized equipment is necessary. If the results of such research are not shared, the health of many people can be negatively affected.*

DIFFERENTIATE INSTRUCTION

❯ More Reading Support

A What type of wave is used to make images of bones inside a human body? *X-rays*

B Why can't humans hear ultrasound? *Its pitch is too high.*

Advanced Show students a copy of the electromagnetic spectrum. Point out the different types of electromagnetic waves. Ask: Why aren't ultrasound waves listed on the spectrum? *Ultrasound waves are sound waves, not electromagnetic waves.*

Teach from Visuals

Have students look at the figure of the dolphin and the sound waves reflecting off a fish, ask:

• Why do you think the process shown in this figure is called echolocation? *Sample answer: Dolphins use echoes to locate objects.*

• For what purposes other than finding food might a dolphin or bat use echolocation? *Sample answer: locate objects in its path*

Scientific Process

Emphasize to students that observations of wave behavior in nature were the bases for hypotheses regarding practical applications of waves. For example, observations of echolocation led to use of sound waves in ultrasound applications. By observing the results of these applications, further hypotheses can be made and conclusions can be drawn.

Asking a Question

Ask: In developing the technology that is used to help Kelley Flynn hear better, what questions might researchers have asked? *Sample answers: What part of the ear was damaged? What technology can duplicate the function of the damaged part of the ear?*

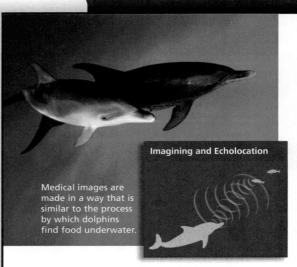

Imagining and Echolocation

Medical images are made in a way that is similar to the process by which dolphins find food underwater.

The technology of ultrasound in medicine is based upon a process similar to that used by bats and dolphins to find food, a process called echolocation. The animal emits an ultrasound click or chirp and then listens for an echo. The echo indicates that an object has reflected the sound back to the animal. Over time, these animals have evolved the ability to judge the distance of the object by noting the time required for the sound to travel to the object and return. Echolocation enables bats to capture flying insects at night and dolphins to catch fish in the ocean depths, where light doesn't penetrate.

Similarly, in ultrasound imaging, a machine sends a beam of ultrasound into a person's body and detects any echoes. The waves reflect whenever they strike a boundary between two objects with different densities. A computer measures the time required for the wave to travel to the boundary and reflect back; this information is used to determine the location and shape of the organ. The computer can then generate a live image of the organ inside the body.

Ultrasound imaging has been used most often to monitor the development of a fetus inside its mother and to observe the valves of the heart. Blood flow can be color coded with faster flow in one color and slower flow in another color. The colors make it easier to see the location of blockages affecting the rate of flow in the blood vessels. This helps doctors detect blockages and diagnose heart problems.

SCIENTIFIC AMERICAN FRONTIERS

View the "Each Sound Is a Present" segment of your *Scientific American Frontiers* video to learn how a cochlear implant restores hearing to a young girl.

IN THIS SCENE FROM THE VIDEO ▶
A young girl's cochlear implant is turned on for the first time.

HEARING IS A GIFT A recent development in technology is about to give seven-year-old Kelly Flynn something she has always wanted—better hearing. Kelly has been almost completely deaf since she was two years old, and now she is losing the little hearing she does have. The development is a device called a cochlear implant. Cochlear implants work inside the ear, stimulating the brain when a sound is detected.

Normally, sound travels as vibrations from the outer ear, through the middle ear to the inner ear, where thousands of tiny cells—called hair cells—register the quality of the sound and send a signal to the brain. In a cochlear implant, tiny electrical sensors, or electrodes, mimic the hair cells by registering the sound and sending a signal to the brain. The signals get to the electrodes through a system including a computer, microphone, and radio transmitter and receiver. Using this system, people with little or no hearing are able to sense sounds.

DIFFERENTIATE INSTRUCTION

❓ More Reading Support

C By what process do dolphins and bats locate food? *echolocation*

D What are two common medical uses of ultrasound? *studying a fetus and the heart*

Below Level Have each student write a short story about a bat on its nightly hunt for food. Stories should include how the bat uses echolocation to find food.

Recent advances in ultrasound technology include the development of portable devices that display images of the body, such as this hand-held device.

Advances in Ultrasound

Waves, including ultrasound, transfer energy. Physical therapists often use this fact when applying ultrasound to sore joints, heating the muscles and ligaments so they can move more freely. If the ultrasound waves are given stronger intensity and sharper focus, they can transfer enough energy to break up kidney stones in the body. The use of focused sound waves is now being tested for its ability to treat other problems, such as foot injuries.

Other recent advances in medical ultrasound include the development of devices that produce clearer images and use equipment that is smaller in size. In the late 1990's portable ultrasound devices were developed that allow the technology to be brought to the patient

? UNANSWERED Questions

As scientists learn more about the use of sound and other types of waves, new questions will arise.

- Will new methods of imaging the body change the way diseases are diagnosed?
- How closely do sounds heard using a cochlear implant resemble sounds heard by the ear?

UNIT PROJECTS

As you study this unit, work alone or with a group on one of these projects.

Magazine Article

Write a magazine article about the medical uses of ultrasound.

- Collect information about medical ultrasound and take notes about applications that interest you.
- If possible, conduct an interview with a medical practitioner who uses ultrasound.
- Read over all your notes and decide what information to include in your article.

Make a Music Video

Make a music video for a song of your choice, and explain how the video uses sound waves and light waves.

- Plan the sound portion of the video, including how the music will be played and amplified.
- For the lighting, use colored cellophane or gels to mix different colors of light. Explain your choices.
- Rehearse the video. Record the video and present it to the class.

Design a Demonstration

Design a hands-on demonstration of echolocation.

- Research the use of echolocation by animals.
- Design a demonstration of echolocation using a tennis ball and an obstacle.
- Present your demonstration to the class.

 **CAREER CENTER**
CLASSZONE.COM

Learn more about careers in audiology.

? UNANSWERED Questions

Have students read the questions and think of some of their own. Remind them that scientists always end up with more questions—that inquiry is the driving force of science.

- With the class, generate on the board a list of new questions.
- Students can add to the list after they watch the Scientific American Frontiers Video.
- Students can use the list as a springboard for choosing their Unit Projects.

UNIT PROJECTS

Encourage students to pick the project that most appeals to them. Point out that each is long-term and will take several weeks to complete. You might group or pair students to work on projects and in some cases guide student choice.

Each project has two worksheet pages, including a rubric. Use the pages to guide students through criteria, process, and schedule.

 Unit Projects, pp. 5–10

Technology Resources

Visit **ClassZone.com** for project procedures and for science career direction.

 RESOURCE CENTER, Unit Projects

REVISIT concepts introduced in this article:

Chapter 1
- Properties of waves, pp. 16–23

Chapter 2
- Sound is a wave, pp. 37–43
- Uses of sound, pp. 58–63

Chapter 3
- Uses of electromagnetic waves, pp. 79–86

DIFFERENTIATE INSTRUCTION

? More Reading Support

E Why can ultrasound be used to break kidney stones? *It transfers energy.*

F What are two recent advances in ultrasound? *clearer images and smaller equipment*

Differentiate Unit Projects Projects are appropriate for varying abilities. Allow students to choose the ones that interest them most. Encourage them to vary the products they produce throughout the year. Encourage below-level students to try "Design a Demonstration." Challenge advanced students to complete the "Magazine Article."

CHAPTER

1 Waves

Physical Science
UNIFYING PRINCIPLES

PRINCIPLE 1	PRINCIPLE 2	PRINCIPLE 3	PRINCIPLE 4
Matter is made of particles too small to see.	Matter changes form and moves from place to place.	Energy changes from one form to another, but it cannot be created or destroyed.	Physical forces affect the movement of all matter on Earth and throughout the universe.

Unit: Waves, Sound, and Light
BIG IDEAS

CHAPTER 1 Waves	CHAPTER 2 Sound	CHAPTER 3 Electromagnetic Waves	CHAPTER 4 Light and Optics
Waves transfer energy and interact in predictable ways.	Sound waves transfer energy through vibrations.	Electromagnetic waves transfer energy through radiation.	Optical tools depend on the wave behavior of light.

CHAPTER 1
KEY CONCEPTS

SECTION 1.1

Waves transfer energy.

1. A wave is a disturbance.

2. Waves can be classified by how they move.

SECTION 1.2

Waves have measurable properties.

1. Waves have amplitude, wavelength, and frequency.

2. Wave speed can be measured.

SECTION 1.3

Waves behave in predictable ways.

1. Waves interact with materials.

2. Waves interact with other waves.

T The Big Idea Flow Chart is available on p. T1 in the **UNIT TRANSPARENCY BOOK.**

Previewing Content

SECTION

 1.1 Waves transfer energy. pp. 9–15

1. A wave is a disturbance.

A **wave** is a disturbance that transfers energy from one place to another. **Mechanical waves** travel through a material, called a **medium,** transferring energy.

When a mechanical wave travels through a medium, such as water, ground, or air, the medium moves as the wave passes through it, but is not permanently moved. After the wave has passed, the medium returns to its former state.

2. Waves can be classified by how they move.

A **transverse wave** travels in the direction perpendicular to the disturbance that caused it. If you thrust your fist into a tub of water, waves travel out from the disturbance along the water's surface. These waves move at right angles to the downward force of your fist.

A **longitudinal wave** travels in the same direction as the disturbance that caused it. If you lay a spring toy on its side and push sharply on one end, waves will travel through the coils along the length of the spring toy. In this example, the disturbance—the push of your hand—is in the same direction in which the wave moves down the spring toy.

SECTION

1.2 Waves have measurable properties. pp. 16–23

1. Waves have amplitude, wavelength, and frequency.

A wave has repeating **crests** and **troughs,** which alternate in a wave. **Amplitude** is either how high a wave peaks above level (at its crests) or how low it dips below level (at its troughs). **Wavelength** is the distance from trough to trough or crest to crest. Wavelength can be measured from any part of one wave to the corresponding part on the next wave.

In a longitudinal wave such as in a spring toy, wavelength is the distance between compressions or rarefactions. Rarefactions are the spaces between compressions where the medium is spread out. Amplitude describes how tightly bunched the spring coils are in the wave.

Frequency is the number of wavelengths that pass a fixed point in a period of time—usually one second.

The graph below shows how wavelength, amplitude, and frequency are measured on a wave.

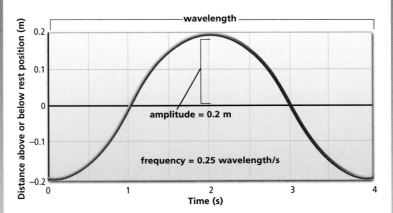

2. Wave speed can be measured.

The formula for calculating the speed of a wave is
$$S = \lambda \cdot f,$$
where S is speed, λ (lambda) is wavelength, and f is frequency.

 MISCONCEPTION DATABASE
CLASSZONE.COM **Background on student misconceptions**

Common Misconceptions

DISPLACEMENT OF MATTER Students may conceive of waves as the motion of matter from one place to another. Although matter can move in a wave, it is not permanently displaced. Rather, it is energy that is transferred in a wave.

 This misconception is addressed on p. 12.

Previewing Content

SECTION

1.3 **Waves behave in predictable ways.**
pp. 24–29

1. Waves interact with materials.

Waves behave predictably when they interact with barriers or other obstacles. Waves can undergo reflection, refraction, or diffraction.

- In **reflection,** waves meet a solid barrier and bounce back. For example, an echo occurs when a sound wave meets a wall and bounces back to the source of the sound.
- In **refraction,** waves move from one medium to another and bend, or refract. An example is a glass of water with a straw. Where the straw passes into the water, it looks broken. But the break is an illusion caused by refraction of light waves as they pass from air to water.
- In **diffraction,** waves interact with a partial barrier and a portion of the waves pass through and spread out. An example is the way sound waves spread around corners.

2. Waves interact with other waves.

In **constructive interference,** two waves combine in phase, so that a crest meets a crest or a trough meets a trough. In **destructive interference,** two waves of the same frequency meet up such that the trough of one wave joins with the crests of the other. If the amplitudes of the two original waves are equal, the two waves cancel each other out.

The following diagrams show how wave amplitudes can be added and subtracted as the waves interfere.

Constructive Interference   **Destructive Interference**

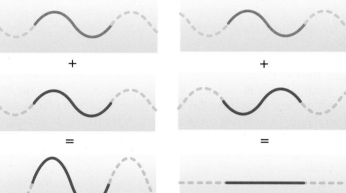

Previewing Labs

EXPLORE the BIG idea

How Can Energy Be Passed Along? p. 7
Students observe a domino effect and learn that energy can be transferred.

TIME 10 minutes
MATERIALS 4 video cassettes

How Can You Change a Wave? p. 7
Students make waves in a bowl and observe wave properties.

TIME 10 minutes
MATERIALS large bowl, half-full of water; pencil

Internet Activity: Waves, p. 7
Students find out how different forces start waves.

TIME 20 minutes
MATERIALS computer with Internet access

SECTION 1.1

EXPLORE Waves, p. 9
Students use a rope to find out how a wave travels.

TIME 5 minutes
MATERIALS 20 cm ribbon, 2 m rope, chair

INVESTIGATE Wave Types, p. 13
Students observe transverse and longitudinal waves in a spring toy.

TIME 10 minutes
MATERIALS spring toy

SECTION 1.2

INVESTIGATE Frequency, p. 20
Students vary the length of a pendulum to find out how length affects its frequency.

TIME 30 minutes
MATERIALS 3 metal washers, 60 cm string, 5 cm tape, stopwatch

CHAPTER INVESTIGATION Wavelength, pp. 22–23
Students vary the length of a pendulum to find out how it affects wavelength.

TIME 40 minutes
MATERIALS 1/2 sheet white paper, 16 cm tape, scissors, 80 cm string, meter stick, 40 mL fine sand, graduated cylinder, 2 sheets colored construction paper

---

SECTION 1.3

EXPLORE Reflection, p. 24
Students make ripples in water in a pan to find out how waves reflect.

TIME 10 minutes
MATERIALS wide, shallow pan, half-full of water; 3 drops food coloring; pencil

INVESTIGATE Diffraction, p. 26
Students manipulate water waves to go around an obstacle to learn about diffraction.

TIME 20 minutes
MATERIALS wide, shallow pan, half-full of water; 3 drops food coloring; plastic ruler; wooden block; sealable sandwich bag; 1–2 cups sand

---

R Additional **INVESTIGATION,** Tracking the Path of Light, A, B, & C, pp. 62–70; Teacher Instructions, pp. 284–285

Previewing Chapter Resources

INTEGRATED TECHNOLOGY	LABS AND ACTIVITIES

CHAPTER 1
Waves

 CLASSZONE.COM
- eEdition Plus
- EasyPlanner Plus
- Misconception Database
- Content Review
- Test Practice
- Visualization
- Simulation
- Resource Centers
- Internet Activity: Waves
- Math Tutorial

 SCILINKS.ORG
SCI LINKS

 CD-ROMS
- eEdition
- EasyPlanner
- Power Presentations
- Content Review
- Lab Generator
- Test Generator

 AUDIO CDS
- Audio Readings
- Audio Readings in Spanish

 EXPLORE the Big Idea, p. 7
- How Can Energy Be Passed Along?
- How Can You Change a Wave?
- Internet Activity: Waves

 UNIT RESOURCE BOOK
- Family Letter, p. vii
- Spanish Family Letter, p. viii
- Unit Projects, pp. 5–10

 Lab Generator CD-ROM
Generate customized labs.

SECTION
 1.1 **Waves transfer energy.**
pp. 9–15

Time: 2 periods (1 block)
 Lesson Plan, pp. 11–12

 • **RESOURCE CENTER,** Waves
• **MATH TUTORIAL**

 UNIT TRANSPARENCY BOOK
- Big Idea Flow Chart, p. T1
- Daily Vocabulary Scaffolding, p. T2
- Note-Taking Model, p. T3
- 3-Minute Warm-Up, p. T4

 • EXPLORE Waves, p. 9
• INVESTIGATE Wave Types, p. 13
• Math in Science, p. 15

UNIT RESOURCE BOOK
- Datasheet, Wave Types, p. 20
- Math Support, p. 49
- Math Practice, p. 50

SECTION
 1.2 **Waves have measurable properties.**
pp. 16–23

Time: 3 periods (1.5 blocks)
 Lesson Plan, pp. 22–23

 • **VISUALIZATION,** Wave Graphing
• **RESOURCE CENTER,** Wave Speed

 UNIT TRANSPARENCY BOOK
- Daily Vocabulary Scaffolding, p. T2
- 3-Minute Warm-Up, p. T4
- "Graphing a Wave" Visual, p. T6

 • INVESTIGATE Frequency, p. 20
• CHAPTER INVESTIGATION, Wavelength, pp. 22–23

UNIT RESOURCE BOOK
- Datasheet, Frequency, p. 31
- Math Support & Practice, pp. 51–52
- CHAPTER INVESTIGATION, Wavelength, A, B, & C, pp. 53–61

SECTION
 1.3 **Waves behave in predictable ways.**
pp. 24–29

Time: 3 periods (1.5 blocks)
 Lesson Plan, pp. 33–34

 UNIT TRANSPARENCY BOOK
- Big Idea Flow Chart, p. T1
- Daily Vocabulary Scaffolding, p. T2
- 3-Minute Warm-Up, p. T5
- Chapter Outline, pp. T7–T8

 • EXPLORE Reflection, p. 24
• INVESTIGATE Diffraction, p. 26
• Connecting Sciences, p. 29

UNIT RESOURCE BOOK
- Datasheet, Diffraction, p. 42
- Additional INVESTIGATION, Tracking the Path of Light, A, B, & C, pp. 62–70

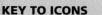
 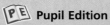
READING AND REINFORCEMENT

ASSESSMENT

STANDARDS

- Four Square, B22–23
- Combination Notes, C36
- Daily Vocabulary Scaffolding, H1–8

 UNIT RESOURCE BOOK
- Vocabulary Practice, pp. 46–47
- Decoding Support, p. 48
- Summarizing the Chapter, pp. 71–72

 UNIT RESOURCE BOOK
- Reading Study Guide, A & B, pp. 13–16
- Spanish Reading Study Guide, pp. 17–18
- Challenge and Extension, p. 19
- Reinforcing Key Concepts, p. 21
- Challenge Reading, pp. 44–45

 UNIT RESOURCE BOOK
- Reading Study Guide, A & B, pp. 24–27
- Spanish Reading Study Guide, pp. 28–29
- Challenge and Extension, p. 30
- Reinforcing Key Concepts, p. 32

 UNIT RESOURCE BOOK
- Reading Study Guide, A & B, pp. 35–38
- Spanish Reading Study Guide, pp. 39–40
- Challenge and Extension, p. 41
- Reinforcing Key Concepts, p. 43

- Chapter Review, pp. 31–32
- Standardized Test Practice, p. 33

 UNIT ASSESSMENT BOOK
- Diagnostic Test, pp. 1–2
- Chapter Test, A, B, & C, pp. 6–17
- Alternative Assessment, pp. 18–19

 Spanish Chapter Test, pp. 281–284

 **Test Generator CD-ROM**
Generate customized tests.

Lab Generator CD-ROM
Rubrics for Labs

 Ongoing Assessment, pp. 9–14

 Section 1.1 Review, p. 14

 UNIT ASSESSMENT BOOK
Section 1.1 Quiz, p. 3

 Ongoing Assessment, pp. 16–21

 Section 1.2 Review, p. 21

 UNIT ASSESSMENT BOOK
Section 1.2 Quiz, p. 4

 Ongoing Assessment, pp. 24–25, 27–28

 Section 1.3 Review, p. 28

UNIT ASSESSMENT BOOK
Section 1.3 Quiz, p. 5

National Standards
A.2–8, A.9.a–c, A.9.e–f, B.3.a, G.1.b

See p. 6 for the standards.

National Standards
A.2–8, A.9.a–c, A.9.e–f, B.3.a, G.1.b

National Standards
A.2–8, A.9.a–c, A.9.e–f, G.1.b

National Standards
A.2–8, A.9.a–c, A.9.e–f, G.1.b

Previewing Resources for Differentiated Instruction

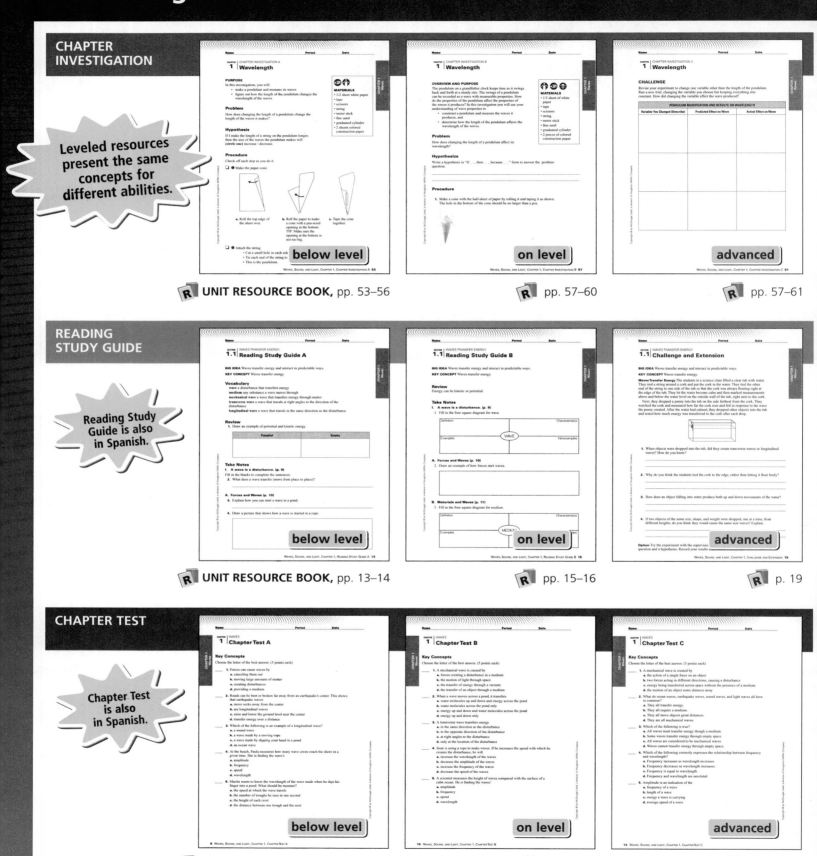

CHAPTER INVESTIGATION

Leveled resources present the same concepts for different abilities.

UNIT RESOURCE BOOK, pp. 53–56
pp. 57–60
pp. 57–61

below level
on level
advanced

READING STUDY GUIDE

Reading Study Guide is also in Spanish.

UNIT RESOURCE BOOK, pp. 13–14
pp. 15–16
p. 19

below level
on level
advanced

CHAPTER TEST

Chapter Test is also in Spanish.

UNIT ASSESSMENT BOOK, pp. 6–9
pp. 10–13
pp. 14–17

below level
on level
advanced

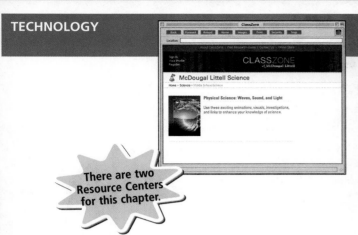

There are two Resource Centers for this chapter.

 CLASSZONE.COM

CD/CD-ROMS

CLASSZONE.COM

VISUAL CONTENT

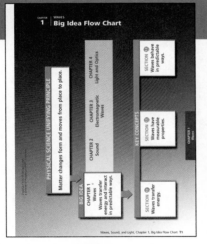

 **UNIT TRANSPARENCY BOOK,** p. T1

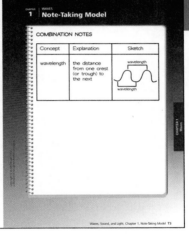

 p. T3

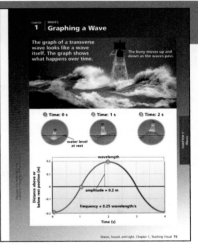

 p. T6

MORE SUPPORT

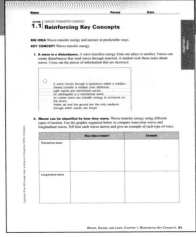

Reinforcing Key Concepts for each section

 **UNIT RESOURCE BOOK,** p. 21

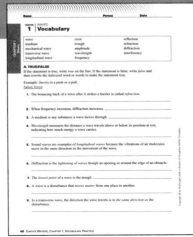

 pp. 46–47

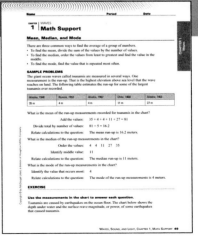

 p. 49

INTRODUCE

the **BIG** idea

Have students look at the photograph of the surfers riding the waves and discuss how the question in the box links to the Big Idea:

- What clues can you find that the wave carries energy?
- What parts of the wave can you see?
- What other kinds of waves can you think of?
- How do waves move from one place to another?

National Science Education Standards

Content

B.3.a Energy is a property of many substances and is associated with heat, light, electricity, mechanical motion, sound, nuclei, and the nature of a chemical. Energy is transferred in many ways.

Process

A.2–8 Design and conduct an investigation; use tools to gather and interpret data; use evidence to describe, predict, explain, model; think critically to make relationships between evidence and explanation; recognize different explanations and predictions; communicate scientific procedures and explanations; use mathematics.

A.9.a, A.9.c, A.9.e Understand scientific inquiry by using different investigations, methods, mathematics, and explanations based on logic, evidence, and skepticism.

G.1.b Science requires different abilities.

CHAPTER

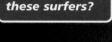 **Waves**

the **BIG** idea

Waves transfer energy and interact in predictable ways.

What is moving these surfers?

Key Concepts

SECTION
1.1 Waves transfer energy. Learn about forces and energy in wave motion.

SECTION
1.2 Waves have measurable properties. Learn how the amplitude, wavelength, and frequency of a wave are measured.

SECTION
1.3 Waves behave in predictable ways. Learn about reflection, refraction, diffraction, and interference.

 Internet Preview

CLASSZONE.COM
Chapter 1 online resources: Content Review, Simulation, Visualization, two Resource Centers, Math Tutorial, Test Practice

D 6 Unit: Waves, Sound, and Light

 **INTERNET PREVIEW**

CLASSZONE.COM For student use with the following pages:

Review and Practice
- Content Review, pp. 8, 30
- Math Tutorial: Finding the Mean, Median, and Mode, p. 15
- Test Practice, p. 33

Activities and Resources
- Internet Activity: Waves, p. 7
- Resource Centers: Waves, p. 14; Wave Speed, p. 21
- Visualization: Wave Graphing, p. 18

Seismic Waves **Code: MDL027**

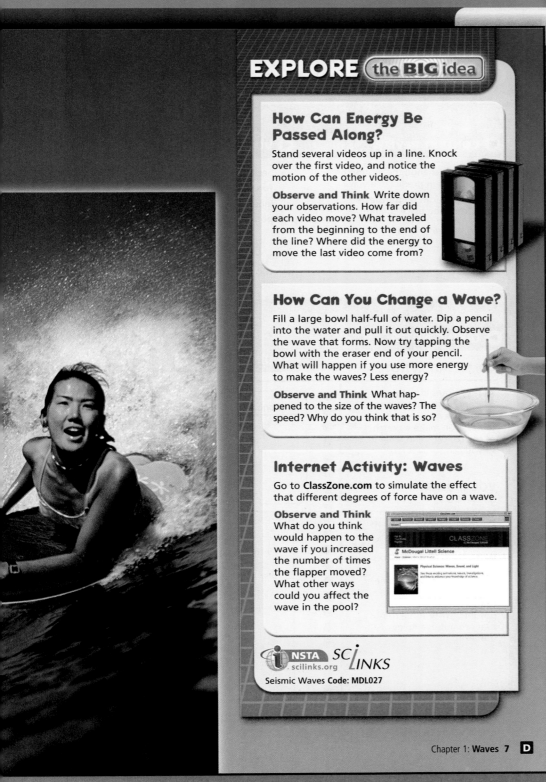

How Can Energy Be Passed Along?

Stand several videos up in a line. Knock over the first video, and notice the motion of the other videos.

Observe and Think Write down your observations. How far did each video move? What traveled from the beginning to the end of the line? Where did the energy to move the last video come from?

How Can You Change a Wave?

Fill a large bowl half-full of water. Dip a pencil into the water and pull it out quickly. Observe the wave that forms. Now try tapping the bowl with the eraser end of your pencil. What will happen if you use more energy to make the waves? Less energy?

Observe and Think What happened to the size of the waves? The speed? Why do you think that is so?

Internet Activity: Waves

Go to **ClassZone.com** to simulate the effect that different degrees of force have on a wave.

Observe and Think What do you think would happen to the wave if you increased the number of times the flapper moved? What other ways could you affect the wave in the pool?

NSTA
scilinks.org
SCiLINKS

Seismic Waves Code: MDL027

Chapter 1: Waves **7** **D**

EXPLORE the BIG idea

These inquiry-based activities are appropriate for use at home or as a supplement to classroom instruction.

How Can Energy Be Passed Along?

PURPOSE To introduce students to the idea that energy can be transferred. Students set up a row of video cassettes, push the first one, and observe as the whole row falls over.

TIP *10 min.* Students may use dominoes as an alternative to the video cassettes.

Answer: each video moved only to the next video; energy traveled from the beginning to the end of the line; the original source of energy could be the push, the first cassette falling, or the next to last cassette falling

REVISIT after p. 12.

How Can You Change a Wave?

PURPOSE To introduce students to wave properties such as amplitude and speed. Students generate and observe waves in a bowl.

TIP *10 min.* Only a small motion of the pencil is needed.

Answer: More energy produced larger waves; less energy produced smaller waves. The speed of the waves remained the same. Energy affects size of waves, but some other factor must affect speed of waves.

REVISIT after p. 21.

Internet Activity: Waves

PURPOSE To show how forces are involved in waves.

TIP *20 min.* Students observe waves produced by varying forces.

Answer: The waves would be produced at a higher rate if you increased the number of times the flapper moved. You could also affect the wave by moving the flapper slowly, or changing the amount of force applied by the flapper.

REVISIT after p. 10.

Chapter 1 **7** **D**

TEACHING WITH TECHNOLOGY

Computer Software and Overhead Projector Use a computer-generated wave to illustrate the form and motion of a transverse wave on p. 13. Help students identify crest, trough, amplitude, frequency, and wavelength in the wave.

Graphing Calculator Use a graphing calculator on p. 18 to graph sine and cosine functions. Graph $y = \sin x$ and $y = \cos x$, then point out to students that the graphs form a shape of a wave. Challenge students to alter the equations and report their findings. Ask them how the amplitude and wavelength vary.

PREPARE

◑ CONCEPT REVIEW

Activate Prior Knowledge

Put two marbles on a flat surface. Flick one marble so that it strikes the other one. The first marble must move forcefully enough to move the second marble as well.

- Ask: How did force change the motion of the first marble? *The first marble went from rest to motion.*
- Ask: How was energy transferred to the second marble? *Energy was transferred from the first marble to the second marble when they hit.*

◑ TAKING NOTES

Combination Notes

Combining pictures with notes will help students connect abstract concepts with concrete examples. The three-column format allows students to write their notes in the first and second columns and draw their pictures in the third column.

Vocabulary Strategy

The four square diagram helps students organize complex information in a user-friendly, visual way. The format also prompts students to process more information for a concept than they might otherwise do.

Vocabulary and Note-Taking Resources

- Vocabulary Practice, pp. 46–47
- Decoding Support, p. 48

- Daily Vocabulary Scaffolding, p. T2
- Note-Taking Model, p. T3

- Four Square, B22–23
- Combination Notes, C36
- Daily Vocabulary Scaffolding, H1–8

◑ CONCEPT REVIEW

- Forces change the motion of objects in predictable ways.
- Energy can be transferred from one place to another.

◑ VOCABULARY REVIEW

See Glossary for definitions.

force

kinetic energy

potential energy

 CONTENT REVIEW
CLASSZONE.COM
Review concepts and vocabulary.

▶ TAKING NOTES

COMBINATION NOTES

To take notes about a new concept, write an explanation of the concept in a table. Then make a sketch of the concept and label it so you can study it later.

VOCABULARY STRATEGY

Write each new vocabulary term in the center of a **four square** diagram. Write notes in the squares around each term. Include a definition, some characteristics, and some examples of the term. If possible, write some things that are not examples of the term.

See the Note-Taking Handbook on pages R45–R51.

SCIENCE NOTEBOOK

Concept	Explanation	Sketch
Forces and waves	Forces move a medium up and down or back and forth. A wave moves forward.	direction of force / direction of wave

Definition	Characteristics
A disturbance that transfers energy from one place to another	Matter moves in place. Energy travels entire distance.

WAVE

Examples	Nonexamples
Water wave	Ball rolling
Sound wave	Water rushing downstream

CHECK READINESS

Administer the Diagnostic Test to determine students' readiness for new science content and their mastery of requisite math skills.

 Diagnostic Test, pp. 1–2

Technology Resources

Students needing content and math skills should visit **ClassZone.com**.

 • CONTENT REVIEW
• MATH TUTORIAL

CONTENT REVIEW CD-ROM

KEY CONCEPT
Waves transfer energy.

 BEFORE, you learned

- Forces can change an object's motion
- Energy can be kinetic or potential

NOW, you will learn

- How forces cause waves
- How waves transfer energy
- How waves are classified

VOCABULARY

wave p. 9
medium p. 11
mechanical wave p. 11
transverse wave p. 13
longitudinal wave p. 14

EXPLORE Waves

How will the rope move?

PROCEDURE

1. Tie a ribbon in the middle of a rope. Then tie one end of the rope to a chair.

2. Holding the loose end of the rope in your hand, stand far enough away from the chair that the rope is fairly straight.

3. Flick the rope by moving your hand up and down quickly. Observe what happens.

MATERIALS
- ribbon
- rope
- chair

WHAT DO YOU THINK?
- How did the rope move? How did the ribbon move?
- What do you think starts a wave, and what keeps it going?

A wave is a disturbance.

You experience the effects of waves every day. Every sound you hear depends on sound waves. Every sight you see depends on light waves. A tiny wave can travel across the water in a glass, and a huge wave can travel across the ocean. Sound waves, light waves, and water waves seem very different from one another. So what, exactly, is a wave?

A **wave** is a disturbance that transfers energy from one place to another. Waves can transfer energy over distance without moving matter the entire distance. For example, an ocean wave can travel many kilometers without the water itself moving many kilometers. The water moves up and down—a motion known as a disturbance. It is the disturbance that travels in a wave, transferring energy.

 CHECK YOUR READING How does an ocean wave transfer energy across the ocean?

Chapter 1: **Waves 9**

1.1 FOCUS

◉ Set Learning Goals
Students will

- Explain how forces cause waves.
- Explain how waves transfer energy.
- Classify wave types.
- Compare and contrast different wave types in an experiment.

◐ 3-Minute Warm-Up

Display Transparency 4 or copy this exercise on the board:

1. Name two ways you can apply force to a soccer ball. *kicking with feet and hitting with hands*

2. Name two ways the force you apply can change the motion of the soccer ball. *The force changes the ball's speed and direction*

⊤ 3-Minute Warm-Up, p. T4

1.1 MOTIVATE

EXPLORE Waves

PURPOSE To think of how wave motion starts and how a wave moves through a medium

TIP *5 min.* Jump ropes work well.

WHAT DO YOU THINK? *A wave moved through the rope; the ribbon moved up and down. Forces start a wave and keep it going through the material.*

Ongoing Assessment

CHECK YOUR READING *Answer: through a moving disturbance, which we see as a moving wave*

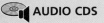
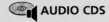

Teach from Visuals

To help students interpret the rope wave visual, ask:

- What represents the forces? *the up and down arrows*
- What produces the forces? *the person's hand and arm moving up and down*
- What is the result of the force? *a wave traveling down the rope*

Ask similar questions for the other visuals on the page.

Real World Example

A telephone cord wave can occur when you stretch out the cord and shake one end of the cord. The wave moves along the cord and reaches the handset.

EXPLORE (the BIG idea)

Revisit "Internet Activity: Waves" on p. 7. Have students explain the reasons for their results.

Ongoing Assessment

Explain how forces cause waves.

When a diver hits the smooth surface of water in a swimming pool, series of waves spread in all directions. Ask: How do forces make this happen? *The force of the diver striking the surface pushes water briefly out of the way; then the water rushes back in. These movements in the water set off the series of ripples, or waves, across the pool.*

Forces and Waves

READING TiP

As you read each example, think of how it is similar to and different from the other examples.

You know that a force is required to change the motion of an object. Forces can also start a disturbance, sending a wave through a material. The following examples describe how forces cause waves.

Example 1 Rope Wave Think of a rope that is tied to a doorknob. You apply one force to the rope by flicking it upward and an opposite force when you snap it back down. This sends a wave through the rope. Both forces—the one that moves the rope up and the one that moves the rope down—are required to start a wave.

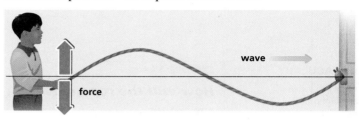

Example 2 Water Wave Forces are also required to start a wave in water. Think of a calm pool of water. What happens if you apply a force to the water by dipping your finger into it? The water rushes back after you remove your finger. The force of your finger and the force of the water rushing back send waves across the pool.

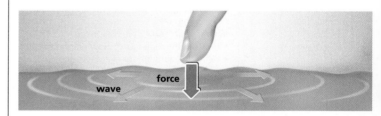

Example 3 Earthquake Wave An earthquake is a sudden release of energy that has built up in rock as a result of the surrounding rock pushing and pulling on it. When these two forces cause the rock to suddenly break, the energy is transferred as a wave through the ground.

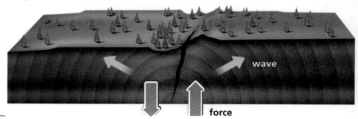

DIFFERENTIATE INSTRUCTION

 More Reading Support

A What is required to make a disturbance that will set off a wave? *forces*

English Learners English learners may not always understand certain abbreviated instructions. For example, when a student is asked to "study a sketch," the word *study* is vague to an English learner. Use precise instructions like *examine, analyze,* or *compare,* and make sure students know the exact meaning of these instructions.

Materials and Waves

A rope tied to a doorknob, water, and the ground all have something in common. They are all materials through which waves move. A **medium** is any substance that a wave moves through. Water is the medium for an ocean wave; the ground is the medium for an earthquake wave; the rope is the medium for the rope wave. In the next chapter, you will learn that sound waves can move through many mediums, including air.

Waves that transfer energy through matter are known as **mechanical waves**. All of the waves you have read about so far, even sound waves, are mechanical waves. Water, the ground, a rope, and the air are all made up of matter. Later, you will learn about waves that can transfer energy through empty space. Light is an example of a wave that transfers energy through empty space.

VOCABULARY
Add a four square for *medium* to your notebook.

CHECK YOUR READING How are all mechanical waves similar?

Energy and Waves

The waves caused by an earthquake are good examples of energy transfer. The disturbed ground shakes from side to side and up and down as the waves move through it. Such waves can travel kilometers away from their source. The ground does not travel kilometers away from where it began; it is the energy that travels in a wave. In the case of an earthquake, it is kinetic energy, or the energy of motion that is transferred.

This photograph was taken after a 1995 earthquake in Japan. A seismic wave transferred enough energy through the ground to bend the railroad tracks, leaving them in the shape of a wave.

Develop Critical Thinking

CLASSIFY Write the following choices on the board:

- air
- water
- rocks and soil

State that these can all be mediums for different kinds of waves. Have students identify the medium for each of the following waves:

- ocean wave *water*
- seismic wave in an earthquake *rocks and soil*
- pulsating sound from a bell *air*

Ongoing Assessment

CHECK YOUR READING *Answer: All mechanical waves transfer energy through matter.*

DIFFERENTIATE INSTRUCTION

More Reading Support

B What is the medium for an ocean wave? *water*

C Does a mechanical wave transfer material or energy? *energy*

Below Level Students may find the term *medium* confusing in the context of wave science. Students are likely to be familiar with *medium* meaning "halfway" or "in between."

Write the following sentences on the board:

- I like my steak medium rare.
- Water is the medium for an ocean wave . . . (p. 11).

Ask: Which of these sentences involves the scientific definition of *medium*? *the second sentence*

Address Misconceptions

IDENTIFY Ask students to imagine a wave in water. Ask: What moves from the beginning to the end of the wave? If students say water (or other matter), they may hold the misconception that matter is transferred in a wave.

CORRECT Have several students stand in a row. Have the first student tap the second on the shoulder, who in turn taps the third student, and so on. Notice that energy is transferred to the end of the row while each student stays in place.

REASSESS Ask: Why don't ocean waves that move toward the shore empty out the ocean? *Because waves transfer energy, not matter.*

Technology Resources

Visit **ClassZone.com** for background on common student misconceptions.

 MISCONCEPTION DATABASE

EXPLORE (the **BIG** idea)

Revisit "How Can Energy Be Passed Along?" on p. 7. Have students explain the reasons for their results.

Integrate the Sciences

Although ocean waves do not pile up water onshore, their energy dramatically shapes the coast. Examples of wave erosion on seacoasts include the formation of broad sandy beaches, sand spits and barrier islands, and stacks (rocky columns isolated in the water near the shoreline) or coves in stony coastlines.

Ongoing Assessment

Classify waves by how they move.

Ask: Name three directions waves can move. *up and down, side to side, or forward and backward*

READING VISUALS *Answer: The people move up and down. The wave moves around the stadium.*

CHECK YOUR READING *Answer: In an ocean wave, one part of the water pushes on the next. In the stadium wave, the people do not push on each other. Thus, the stadium wave is not a real wave because the energy is not transferred through a medium.*

A Wave Model

When these fans do "the wave" in a stadium, they are modeling the way a disturbance travels through a medium.

Each person only moves up and down.

The wave can move all the way around the stadium.

READING VISUALS In which direction do people move when doing the stadium wave? In which direction does the wave move?

Look at the illustration of people modeling a wave in a stadium. In this model, the crowd of people represents a wave medium. The people moving up and down represent the disturbance. The transfer of the disturbance around the stadium represents a wave. Each person only moves up and down, while the disturbance can move all the way around the stadium.

D

Ocean waves are another good example of energy transfer. Ocean waves travel to the shore, one after another. Instead of piling up all the ocean water on the shore, however, the waves transfer energy. A big ocean wave transfers enough kinetic energy to knock someone down.

CHECK YOUR READING How does the stadium wave differ from a real ocean wave?

Waves can be classified by how they move.

E

As you have seen, one way to classify waves is according to the medium through which they travel. Another way to classify waves is by how they move. You have read that some waves transfer an up-and-down or a side-to-side motion. Other waves transfer a forward-and-backward motion.

DIFFERENTIATE INSTRUCTION

? **More Reading Support**

D What does an ocean wave transfer? *energy*

E What are two ways to classify waves? *by medium and by how the wave moves*

Advanced Have students who are interested in learning about harnessing the energy of ocean waves read the following article:

 Challenge Reading, pp. 44–45

Transverse Waves

Think again about snapping the rope with your hand. The action of your hand causes a vertical, or up-and-down, disturbance in the rope. However, the wave it sets off is horizontal, or forward. This type of wave is known as a transverse wave. In a **transverse wave,** the direction in which the wave travels is perpendicular, or at right angles, to the direction of the disturbance. *Transverse* means "across" or "crosswise." The wave itself moves crosswise as compared with the vertical motion of the medium.

READING TiP

Perpendicular means at a 90° angle.

Transverse Wave

direction of disturbance   direction of wave

Water waves are also transverse. The up-and-down motion of the water is the disturbance. The wave travels in a direction that is perpendicular to the direction of the disturbance. The medium is the water, and energy is transferred outward in all directions from the source.

 CHECK YOUR READING What is a transverse wave? Find two examples in the paragraphs above.

INVESTIGATE Wave Types

How do waves compare?

PROCEDURE

① Place the spring toy on the floor on its side. Stretch out the spring. To start a disturbance in the spring, take one end and move it from side to side. Observe the movement in the spring. Remember that a transverse wave travels at right angles to the disturbance.

② Put the spring toy on the floor in the same position as before. Think about how you could make a different kind of disturbance to produce a different kind of wave. (**Hint:** Suppose you push the spring in the direction of the wave you expect to make.) Observe the movement in the spring.

WHAT DO YOU THINK?

• Compare the waves you made. How are they alike? How are they different?
• What kind of wave did you produce by moving the spring from side to side?

CHALLENGE Can you think of a third way to make a wave travel through a spring?

SKILL FOCUS
Comparing

MATERIALS
spring toy

TIME
10 minutes

DIFFERENTIATE INSTRUCTION

? More Reading Support

F What does transverse mean? *"across" or "crosswise"*

Inclusion Offer a wave model that is tailored to students with visual impairments. Ask students to imagine they are floating on an inner tube on a lake in which the water's surface is disturbed by waves. Ask how they would move with the waves. *They would bob up and down.* What types of waves are involved in this scenario? How do you know? *transverse; bobbing up and down is disturbance in a direction perpendicular to the direction of the wave.*

INVESTIGATE Wave Types

PURPOSE To observe a spring and explore the differences between transverse waves and longitudinal waves

TIPS *10 min.* Suggest the following:
• In step 1, make the shape of the spring toy look like a wavy line.
• In step 2, concentrate on what is happening inside the spring coils instead of the overall shape of the spring toy.

WHAT DO YOU THINK? *Both types travel from one end of the spring to the other. Transverse waves move the spring from side to side; longitudinal waves move the spring forward and backward.*

CHALLENGE *pinching together a bunch of coils and letting them go; moving the spring up and down*

 Datasheet, Wave Types, p. 20

Technology Resources

Customize this student lab as needed or look for an alternative. Print rubrics to assess student lab reports.

 Lab Generator CD-ROM

Teaching with Technology

Use a computer-generated sine wave to illustrate the form and motion of a transverse wave. Use an overhead projector and ask students to identify the parts and properties of the wave.

Teach Difficult Concepts

Some students may have difficulty with the concept of a right angle in a three-dimensional context. Place a box on a tabletop. Help students identify the right angle between the vertical side of the box and the table surface. Have them think of other familiar examples.

Ongoing Assessment

CHECK YOUR READING *Answer: a wave that moves perpendicular to the direction of the disturbance; water waves and rope waves*

Ongoing Assessment

Classify wave types.

Ask: How do a rope wave and the waves represented on this page compare and contrast? *Both start with a disturbance caused by a force; both transfer energy. A rope wave is a transverse wave; waves on this page are longitudinal waves.*

Teach from Visuals

Remind students that, in a transverse wave such as an ocean wave, a medium that alternates between high and low points forms the wave. Ask: What makes up the wave pattern in the spring toy? *alternating regions of bunched-up and spread-out spring coils*

Reinforce (the BIG idea)

Have students relate the section to the Big Idea.

 Reinforcing Key Concepts, p. 21

1.1 ASSESS & RETEACH

Assess

 Section 1.1 Quiz, p. 3

Reteach

Have students picture a tub with a toy boat floating on the water. Tell them to imagine plunging a fist into the water.

- Have students predict what would happen to the water's surface and the boat. *waves would move out from the disturbance; boat would bob up and down*

- Have students explain how waves start, move, and transfer energy.

Technology Resources

Have students visit **ClassZone.com** for reteaching of Key Concepts.

 CONTENT REVIEW

 CONTENT REVIEW CD-ROM

Longitudinal Waves

READING TiP
The word *long* can help you remember longitudinal waves. The disturbance moves along the length of the spring.

Another type of wave is a longitudinal wave. In a **longitudinal wave** (LAHN-jih-TOOD-n-uhl), the wave travels in the same direction as the disturbance. A longitudinal wave can be started in a spring by moving it forward and backward. The coils of the spring move forward and bunch up and then move backward and spread out. This forward and backward motion is the disturbance. Longitudinal waves are sometimes called compressional waves because the bunched-up area is known as a compression. How is a longitudinal wave similar to a transverse wave? How is it different?

Longitudinal Wave

direction of disturbance

Time 1

compression

Time 2

direction of wave

Time 3

RESOURCE CENTER
CLASSZONE.COM

Learn more about waves.

Sound waves are examples of longitudinal waves. Imagine a bell ringing. The clapper inside the bell strikes the side and makes it vibrate, or move back and forth rapidly. The vibrating bell pushes and pulls on nearby air molecules, causing them to move forward and backward. These air molecules, in turn, set more air molecules into motion. A sound wave pushes forward. In sound waves, the vibrations of the air molecules are in the same direction as the movement of the wave.

1.1 Review

KEY CONCEPTS

1. Describe how forces start waves.
2. Explain how a wave can travel through a medium and yet the medium stays in place. Use the term *energy* in your answer.
3. Describe two ways in which waves travel, and give an example of each.

CRITICAL THINKING

4. **Analyze** Does water moving through a hose qualify as a wave? Explain why or why not.
5. **Classify** Suppose you drop a cookie crumb in your milk. At once, you see ripples spreading across the surface of the milk. What type of waves are these? What is the disturbance?

CHALLENGE

6. **Predict** Suppose you had a rope long enough to extend several blocks down the street. If you were to start a wave in the rope, do you think it would continue all the way to the other end of the street? Explain why or why not.

ANSWERS

1. Forces start a disturbance that travels through a medium.

2. Energy travels through the medium, leaving the matter in place.

3. Transverse waves travel at right angles to the disturbance. Longitudinal waves travel in the same direction as the disturbance. Examples are ocean waves and sound waves.

4. No; the water travels from one end to the other end of the hose. In a wave, the disturbance moves, but the medium does not.

5. Transverse waves; the falling crumb

6. A rope wave would probably not continue all the way down the street. Gravity would be pulling down on the rope along the way, eventually stopping it.

Before going out on the water, boaters can check reports on wave conditions in their area.

MATH TUTORIAL
CLASSZONE.COM

Click on Math Tutorial for more help with finding the mean, median, and mode.

Wave Heights

Tracking stations throughout the world's oceans measure and record the height of water waves that pass beneath them. The data recorded by the stations can be summarized as average wave heights over one hour or one day.

How would you summarize the typical wave heights over one week? There are a few different ways in which data can be summarized. Three common ways are finding the mean, median, and mode.

Example

Wave height data for one week are shown below.

| 1.2 m | 1.5 m | 1.4 m | 1.7 m | 2.0 m | 1.4 m | 1.3 m |

(1) Mean To find the mean of the data, divide the sum of the values by the number of values.

$$\text{Mean} = \frac{1.2 + 1.5 + 1.4 + 1.7 + 2.0 + 1.4 + 1.3}{7} = 1.5 \text{ m}$$

ANSWER The mean wave height is 1.5 m.

(2) Median To find the median of the data, write the values in order from least to greatest. The value in the middle is the median.

1.2 m 1.3 m 1.4 m (1.4 m) 1.5 m 1.7 m 2.0 m

ANSWER The median wave height is 1.4 m.

(3) Mode The mode is the number that occurs most often.

ANSWER The mode for the data is also 1.4 m.

Use the data to answer the following questions.

The data below show wave heights taken from a station off the coast of Florida over two weeks.

| Wk 1 | 1.2 m | 1.1 m | 1.1 m | 1.5 m | 4.7 m | 1.2 m | 1.1 m |
| Wk 2 | 0.7 m | 0.8 m | 0.9 m | 0.8 m | 1.0 m | 1.1 m | 0.8 m |

1. Find the mean, median, and mode of the data for Week 1.

2. Find the mean, median, and mode of the data for Week 2.

CHALLENGE A storm carrying strong winds caused high waves on the fifth day of the data shown above for Week 1. Which of the following was most affected by the high value—the mean, median, or mode?

Chapter 1: Waves **15** **D**

Set Learning Goal

To learn how to calculate three kinds of averages: mean, median, and mode

Present the Science

Point out that wave height measures from trough to crest.

Develop Estimation Skills

Students may wonder how to calculate median for an even number of data values. Inform students that they would find the two center values and calculate their mean value.

Make up a row of data with an even number of values and have students calculate the median.

Close

Present the following scenario: You are collecting and analyzing test scores for all sixth-grade students in the Centerville school system. What special insight would each of the three kinds of averages give you about students' test performance? *Sample answer: The mean will give an idea of the overall success of all the classes; the median will show approximately what the average student's score was; the mode will show where scores clustered.*

• Math Support, p. 49
• Math Practice, p. 50

Technology Resources

Students can visit **ClassZone.com** for practice finding mean, median, and mode.

MATH TUTORIAL

ANSWERS

1. *mean (1.2 + 1.1 + 1.1 + 1.5 + 4.7 + 1.2 + 1.1) ÷ 7 = 1.7 m; mean = 1.7 m*
median 1.1 m; 1.1 m; 1.1 m; (1.2 m); 1.2 m; 1.5 m; 4.7 m; median = 1.2 m;
mode = 1.1 m

2. *mean (0.7 + 0.8 + 0.9 + 0.8 + 1.0 + 1.1 + 0.8) ÷ 7 = 0.9 m; mean = 0.9 m*
median 0.7 m; 0.8 m; 0.8 m; (0.8 m); 0.9 m; 1.0 m; 1.1 m
median = 0.8 m; mode = 0.8 m

CHALLENGE *the mean*

▶ Set Learning Goals
Students will

- Learn how to measure amplitude, wavelength, and frequency.
- Calculate a wave's speed.
- Collect data to investigate how to change frequency in an experiment.

◀ 3-Minute Warm-Up

Display Transparency 4 or copy this exercise on the board:

How could you measure each of these things? What problems might you encounter in doing so?

1. width of a door opening
2. volume of liquid
3. length of time

Sample answers:

1. *with measuring tape; keeping the tension in the tape across the door opening to get an accurate measurement*

2. *with a measuring cup; making sure liquid is stable*

3. *with a stopwatch; determining start and end time and variations in timing*

 3-Minute Warm-Up, p. T4

THINK ABOUT

PURPOSE To think about different ways a wave can be measured

DISCUSS Have students look at the photograph and describe the scene. Solicit ideas for ways in which waves can be measured.

Ongoing Assessment

CHECK YOUR READING *Answer: amplitude, frequency, and wavelength*

KEY CONCEPT

1.2 Waves have measurable properties.

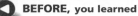

◀ BEFORE, you learned	▶ NOW, you will learn
• Forces cause waves • Waves transfer energy • Waves can be transverse or longitudinal	• How amplitude, wavelength, and frequency are measured • How to find a wave's speed

VOCABULARY

crest p. 17
trough p. 17
amplitude p. 17
wavelength p. 17
frequency p. 17

THINK ABOUT

How can a wave be measured?

This enormous wave moves the water high above sea level as it comes crashing through. How could you find out how high a water wave actually goes? How could you find out how fast it is traveling? In what other ways do you think a wave can be measured? Read on to find out.

Waves have amplitude, wavelength, and frequency.

COMBINATION NOTES
Use combination notes in your notebook to describe how waves can be measured.

The tallest ocean wave ever recorded was measured from the deck of a ship during a storm. An officer on the ship saw a wave reach a height that was level with a point high on the ship, more than 30 meters (100 ft)! Height is a property of all waves—from ripples in a glass of water to gigantic waves at surfing beaches—and it can be measured.

The speed of a water wave is another property that can be measured—by finding the time it takes for one wave peak to travel a set distance. Other properties of a wave that can be measured include the time between waves and the length of a single wave. Scientists use the terms *amplitude*, *wavelength*, and *frequency* to refer to some commonly measured properties of all waves.

 What are three properties of a wave that can be measured?

RESOURCES FOR DIFFERENTIATED INSTRUCTION

Below Level
UNIT RESOURCE BOOK
- Reading Study Guide A, pp. 24–25
- Decoding Support, p. 48

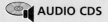 **AUDIO CDS**

Advanced
UNIT RESOURCE BOOK
Challenge and Extension, p. 30

English Learners
UNIT RESOURCE BOOK
Spanish Reading Study Guide, pp. 28–29

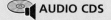 **AUDIO CDS**

- Audio Readings in Spanish
- Audio Readings (English)

Measuring Wave Properties

A **crest** is the highest point, or peak, of a wave. A **trough** is the lowest point, or valley, of a wave. Suppose you are riding on a boat in rough water. When the boat points upward and rises, it is climbing to the crest of a wave. When it points downward and sinks, the boat is falling to the trough of the wave.

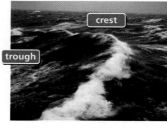

① **Amplitude** is the distance between a line through the middle of a wave and a crest or trough. In an ocean wave, amplitude measures how far the wave rises above, or dips below, its original position, or rest position.

Amplitude is an important measurement, because it indicates how much energy a wave is carrying. The bigger the amplitude, the more energy the wave has. Find amplitude on the diagram below.

② The distance from one wave crest to the very next crest is called the **wavelength.** Wavelength can also be measured from trough to trough. Find wavelength on the diagram below.

③ The number of wavelengths passing a fixed point in a certain amount of time is called the **frequency.** The word *frequent* means "often," so frequency measures how often one wavelength occurs. Frequency is often measured by counting the number of crests or troughs that pass by a given point in one second. Find frequency on the diagram below.

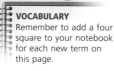

VOCABULARY
Remember to add a four square to your notebook for each new term on this page.

CHECK YOUR READING How is amplitude related to energy?

Wave Properties

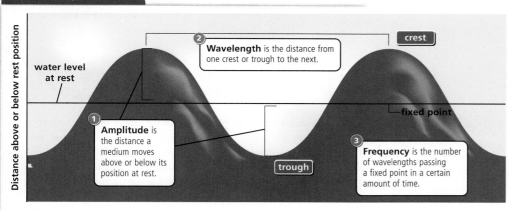

water level at rest

Distance above or below rest position

Wavelength is the distance from one crest or trough to the next.

crest

fixed point

① **Amplitude** is the distance a medium moves above or below its position at rest.

trough

③ **Frequency** is the number of wavelengths passing a fixed point in a certain amount of time.

READING VISUALS How many wavelengths are shown in this diagram? How do you know?

Develop Critical Thinking

EVALUATE Instruct students to think about two waves having equal frequencies and wavelengths. Ask them if that means the waves are identical. Have them draw a diagram to explain their answer. *Two waves can have the same frequency and wavelength without being identical if they have different amplitudes.*

Ongoing Assessment

Learn how to measure frequency.

Ask: How is frequency measured? *Frequency is measured by counting the number of crests or troughs that pass by a fixed point in one second.*

CHECK YOUR READING *Answer: The larger the amplitude, the more energy a wave carries.*

READING VISUALS *Answer: two wavelengths, shown from trough to trough*

DIFFERENTIATE INSTRUCTION

More Reading Support

A What is the crest of a wave? *the highest point, or peak*

B What is the trough of a wave? *the lowest point, or valley*

English Learners Be aware that English learners do not always have the same background knowledge as the rest of the class. For example, this section's Investigate refers to small, metal rings called *washers.* English learners most likely won't connect the term *washer* to the metal rings in front of them. You may also want to point out the comparative structure of sentences such as: The bigger the amplitude, the more energy the wave has (p. 17).

Teach Difficult Concepts

Students may have trouble with the concept of wave frequency. Ask students to brainstorm examples of repetitive events. Students might think of natural cycles such as the sun rising and setting each day. Point out that the patterns change repeatedly with a particular frequency, or repeating time period. For example, the sun rises every 24 hours.

Relate examples to frequency in wave motion: in a transverse wave, the pattern is repeating crests or troughs; in a longitudinal wave, the pattern is repeating bunched-up areas. Emphasize that frequency describes *regular* repeating patterns.

Teach from Visuals

Draw attention to the visual of the spring toy. Ask:

- What repeating pattern do you observe? *bunching up/ spreading out*
- What part of the spring coil corresponds to the wave crest? *where the coils are spread out*
- What part corresponds to the wave trough? *where the coils are bunched up*

Teaching with Technology

On a graphing calculator, have students graph $y = \sin x$ and $y = \cos x$. Point out that the graphs form a shape of a wave. Ask students to alter the equations (for example, $y = 3 \sin x$, $y = \sin 2x$) and have them compare the different wavelengths and amplitudes.

Ongoing Assessment

CHECK YOUR READING *Answer: Amplitude is a measure of how compressed the medium gets. Wavelength is a measure of the distance between compressions. Frequency is the number of compressions that occur within a certain time.*

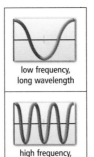

low frequency, long wavelength

high frequency, short wavelength

▼ REMINDER

Frequency is the number of wavelengths that pass a given point in a certain amount of time.

🌐 VISUALIZATION
CLASSZONE.COM
Watch the graph of a wave form.

How Frequency and Wavelength Are Related

The frequency and wavelength of a wave are related. When frequency increases more wave crests pass a fixed point each second. That means the wavelength shortens. So, as frequency increases, wavelength decreases. The opposite is also true—as frequency decreases, wavelength increases.

Suppose you are making waves in a rope. If you make one wave crest every second, the frequency is one wavelength per second. Now suppose you want to increase the frequency to more than one wavelength per second. You flick the rope up and down faster. The wave crests are now closer together. In other words, their wavelengths have decreased.

Graphing Wave Properties

The graph of a transverse wave looks much like a wave itself. The illustration on page 19 shows the graph of an ocean wave. The measurements for the graph come from a float, or buoy (BOO-ee), that keeps track of how high or low the water goes. The graph shows the positioning of the buoy at three different points in time. These points are numbered. Since the graph shows what happens over time, you can see the frequency of the waves.

Unlike transverse waves, longitudinal waves look different from their graphs. The graph of a longitudinal wave in a spring is drawn below. The coils of the spring get closer and then farther apart as the wave moves through them.

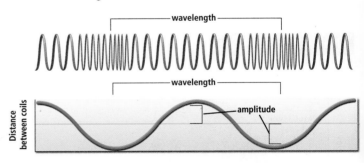

The shape of the graph resembles the shape of a transverse wave. The wavelength on a longitudinal wave is the distance from one compression to the next. The amplitude of a longitudinal wave measures how compressed the medium gets. Just as in a transverse wave, frequency in a longitudinal wave is the number of wavelengths passing a fixed point in a certain amount of time.

▲ CHECK YOUR READING How are longitudinal waves measured?

DIFFERENTIATE INSTRUCTION

❓ **More Reading Support**

C Explain how frequency and wavelength of a wave are related. *If frequency increases, wavelength decreases. If frequency decreases, wavelength increases.*

Below Level Draw two identical rectangular boxes far apart on the board, each about 65 cm wide and 25 cm deep. Ask two volunteers to each draw waves in one of the boxes. Have one student draw lazy, slow waves and the other fast-moving waves.

Ask: Which box has the most waves? Which box has the longest wavelengths? *Students should see that more waves mean shorter wavelengths (as in the box with fast-moving waves) and that fewer waves mean longer wavelengths (as in the box with slow waves).*

Graphing a Wave

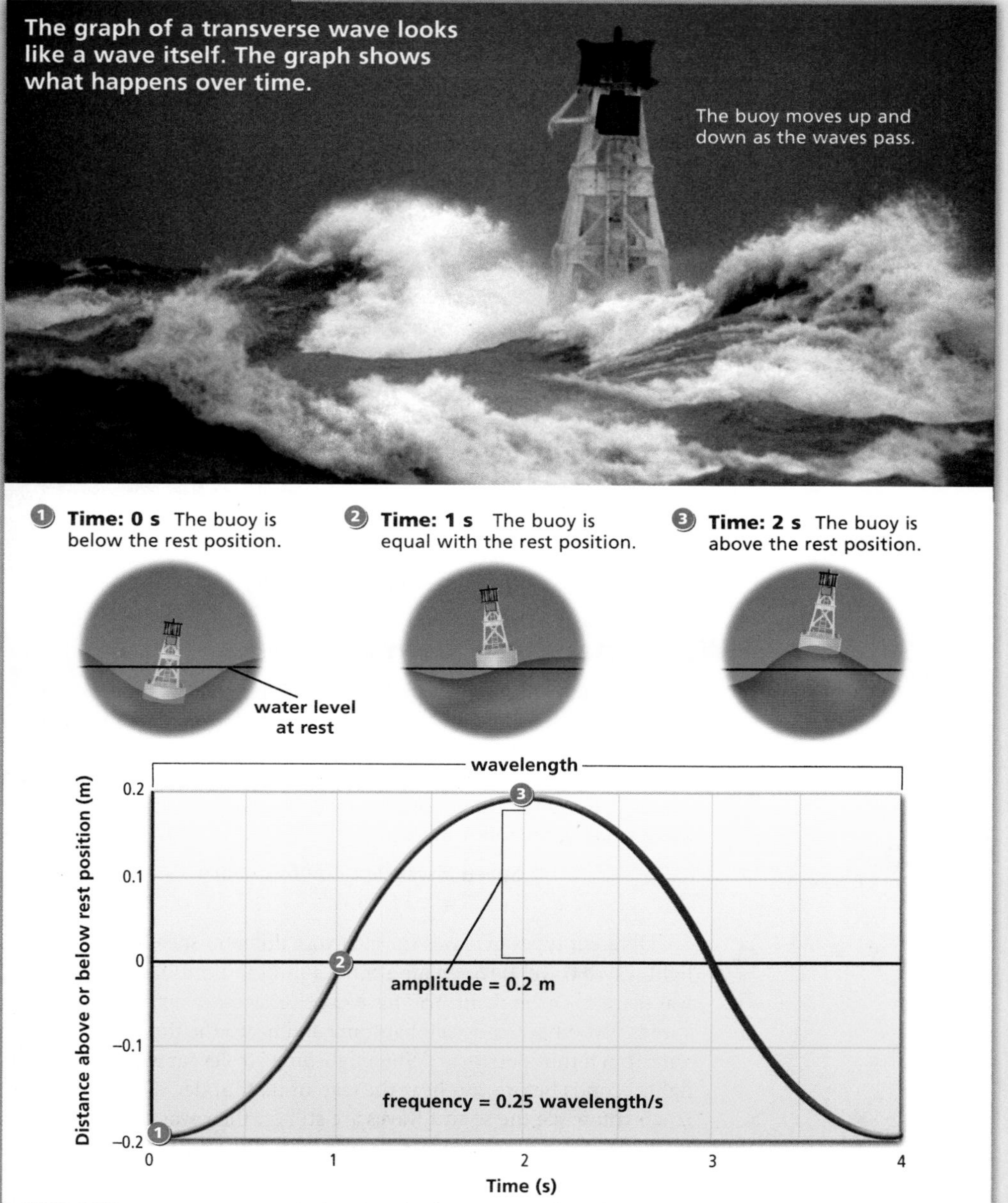

The graph of a transverse wave looks like a wave itself. The graph shows what happens over time.

The buoy moves up and down as the waves pass.

① **Time: 0 s** The buoy is below the rest position.

② **Time: 1 s** The buoy is equal with the rest position.

③ **Time: 2 s** The buoy is above the rest position.

water level at rest

Distance above or below rest position (m)

wavelength

amplitude = 0.2 m

frequency = 0.25 wavelength/s

Time (s)

READING VISUALS How many seconds does it take for one wavelength to pass? How much of the wave passes in one second?

Teach from Visuals

To help students interpret the "Graphing a Wave" visual, ask:

- What are the different parts of the visual on this page? *a photo, three diagrams, and a graph*
- What does the curved line on the graph represent? *the wave in the photo*
- What do the *x*-axis and the *y*-axis represent? *time in seconds, and distance (amplitude) in meters*
- What do the bubble pictures above the graph show? Why are they important? *The buoy bobs up and down as a wave passes underneath; it is the fixed point from which frequency and other wave measurements can be taken.*

T The visual "Graphing a Wave" is available as T6 in the Unit Transparency Book.

Teacher Demo

Demonstrate wavelength and frequency in a wave. Tie a cord to a fixed object such as a doorknob; holding the other end, move back until the cord is taut. Pull up and down on the cord to set off waves, and then have students identify crests and troughs. Ask students how they could measure wavelength. Measure distance from one crest or trough to the next.

Place an object under the cord between you and the tied end. Prompt students to count waves passing this reference point as you call time. Say "start" when you set off the rope wave, time three seconds, and then say "stop." Get students' wave counts. Have students discuss how they counted the waves and how they could use their data to determine frequency.

DIFFERENTIATE INSTRUCTION

 More Reading Support

D Which type of wave has a graph that looks like the real wave—transverse or longitudinal? *transverse*

Advanced Instruct students to use the data in the graph on this page. Have them make a graph in which the frequency is doubled (0.5 wavelengths/second).

Have students write a sentence or two comparing their modified graph to the printed one.

R Challenge and Extension, p. 30

Ongoing Assessment

 Answer: 4 seconds; one-fourth of the wave

INVESTIGATE Frequency

PURPOSE To investigate how changing the length of a pendulum affects wave frequency

TIPS *30 min.* If students have trouble with the technique, here are some suggestions.

- Tie the washers tightly together so that they will not shift weight.
- Make sure the washers swing freely, without bumping anything.
- Count one full swing each time the pendulum returns to the start position.

WHAT DO YOU THINK? *The longer the string, the lower the frequency. The swinging string is similar because it has frequency and amplitude. It is different because it does not transfer energy.*

CHALLENGE *Pulling the washers back farther would change the amplitude. Changing the amplitude does not change the frequency.*

 Datasheet, Frequency, p. 31

History of Science

Scientists are continually refining measurement of the speed of sound. The standard speed of sound in air is measured at sea level at 0°C. In 1738, French scientists determined the speed of sound in air to be 332 m/s. In 1942, scientists made a more precise measurement: 331.45 m/s. As recently as 1986, scientists further refined the measurement to 331.29 m/s.

Ongoing Assessment

Calculate a wave's speed.

Ask: What measurements do you need to calculate a wave's speed? *wavelength and frequency*

INVESTIGATE Frequency

How can you change frequency?

PROCEDURE

① Tie 3 washers to a string. Tape the string to the side of your desk so that it can swing freely. The swinging washers can model wave action.

② Pull the washers slightly to the side and let go. Find the frequency by counting the number of complete swings that occur in 1 minute.

③ Make a table in your notebook to record both the length of the string and the frequency.

④ Shorten the string by moving and retaping it. Repeat for 5 different lengths. Keep the distance you pull the washers the same each time.

WHAT DO YOU THINK?

- How did changing the length of the string affect the frequency?
- How does this model represent a wave? How does it differ from a wave?

CHALLENGE How could you vary the amplitude of this model? Predict how changing the amplitude would affect the frequency.

SKILL FOCUS
Collecting data

MATERIALS
- 3 metal washers
- piece of string
- tape
- stopwatch

TIME
30 minutes

Wave speed can be measured.

In addition to amplitude, wavelength, and frequency, a wave's speed can be measured. One way to find the speed of a wave is to time how long it takes for a wave crest to get from one point to another. Another way to find the speed of a wave is to calculate it. The speed of any wave can be determined when both the frequency and the wavelength are known, using the following equation:

$$\text{Speed} = \text{wavelength} \cdot \text{frequency}$$
$$S = \lambda \cdot f$$

Different types of waves travel at very different speeds. For example, light waves travel through air almost a million times faster than sound waves travel through air. You have experienced the difference in wave speeds if you have ever seen lightning and heard the thunder that comes with it in a thunderstorm. When lightning strikes far away, you see the light seconds before you hear the clap of its thunder. The light waves reach you while the sound waves are still on their way.

How fast do you think water waves can travel? Water waves travel at different speeds. Use the equation for wave speed on the next page to find out how fast an ocean wave can travel.

REMINDER
The symbol λ represents wavelength.

 E

 F

DIFFERENTIATE INSTRUCTION

? More Reading Support

E Wavelength times frequency is the formula for which wave measurement? *wave speed*

F Do all kinds of waves travel at the same speed? *no*

Below Level Give this analogy for the formula for wave speed: To find the frequency of cars on a moving train, you could watch a moving train and count the number of railroad cars that pass in a certain length of time, such as 10 seconds, using a stopwatch. Wavelength is like the length of a railroad car. Frequency is the number of cars that pass in a certain length of time.

Calculating Wave Speed

▶ Sample Problem

An ocean wave has a wavelength of 16 meters and a frequency of 0.31 wavelengths per second. What is the speed of the wave?

What do you know?	1 wavelength = 16 m, frequency = 3 wavelengths/s
What do you want to find out?	Speed
Write the formula:	$S = \lambda \cdot f$
Substitute into the formula:	$S = 16 \dfrac{m}{wavelength} \cdot 0.31 \dfrac{wavelengths}{s}$
Calculate and simplify:	$16 \dfrac{m}{\cancel{wavelength}} \cdot 0.31 \dfrac{\cancel{wavelengths}}{s} = 5 \dfrac{m}{s}$
Check that your units agree:	Unit is m/s. Unit for speed is m/s. Units agree.
Answer:	$S = 5$ m/s

▶ Practice the Math

1. In a stormy sea, 2 waves pass a fixed point every second, and the waves are 10 m apart. What is the speed of the waves?
2. In a ripple tank, the wavelength is 0.1 cm, and 10 waves occur each second. What is the speed of the waves (in cm/s)?

RESOURCE CENTER
CLASSZONE.COM

Find out more about wave speed.

Geologists use the wave speed equation to find oil deep underground. They send sound waves into the ground. The speed of the waves that bounce back indicates what the ground is made up of. This is possible because sound travels at different speeds through different kinds of rock. The geologists then use this information to determine where to drill for oil and how far down they will need to go.

CHECK YOUR READING How do geologists use sound waves to find oil in the ground?

1.2 Review

KEY CONCEPTS

1. Make a simple diagram of a wave, labeling amplitude, frequency, and wavelength. For frequency, you will need to indicate a span of time, such as one second.

2. What two measurements of a wave do you need to calculate its speed?

CRITICAL THINKING

3. **Observe** Suppose you are watching water waves pass under the end of a pier. How can you figure out their frequency?

4. **Calculate** A wave has a speed of 3 m/s and a frequency of 6 wavelengths/s. What is its wavelength?

⚠ CHALLENGE

5. **Apply** Imagine you are on a boat in the middle of the sea. You are in charge of recording the properties of passing ocean waves into the ship's logbook. What types of information could you record? How would this information be useful? Explain your answer.

Chapter 1: **Waves** 21 **D**

ANSWERS

1. See students' diagrams.

2. wavelength and frequency

3. Sample answer: Look at a fixed point below, and time how many waves pass that point within a unit of time.

4. $S = \lambda \cdot f$
$\lambda = f \div s$
$\lambda = 6 \dfrac{wavelength}{s} \div 3 \text{ m/s}$
$= 0.5$ m

5. Sample answer: You could record the amplitude, wavelength, and frequency of waves at different times during the day. Changes in these measures could help indicate the approach of a storm.

Ongoing Assessment

CHECK YOUR READING *Answer: measure the speed of sound traveling through rock*

▶ Practice the Math

1. $S = \lambda \cdot f$
$S = 10 \dfrac{m}{wavelength} \cdot 2 \dfrac{wavelength}{s}$
$= 20$ m/s

2. $S = \lambda \cdot f$
$S = 0.1 \dfrac{cm}{wavelength} \cdot 10 \dfrac{wavelength}{s}$
$= 1$ cm/s

 Math Support, p. 51
Math Practice, p. 52

EXPLORE (the **BIG** idea)

Revisit "How Can You Change Waves?" on p. 7. Have students explain the reasons for their results.

Reinforce (the **BIG** idea)

Have students relate the section to the Big Idea.

Reinforcing Key Concepts, p. 32

1.2 ASSESS & RETEACH

Assess

A Section 1.2 Quiz, p. 4

Reteach

Have students make the following drawings to illustrate the main concepts.

1. A sketch of an ocean wave (transverse), with wavelength and amplitude clearly marked, and a dotted or dashed line to show level.

2. A sketch of a wave in a spring toy (longitudinal wave), illustrating regions of compression and wavelength.

Technology Resources

Have students visit **ClassZone.com** for reteaching of Key Concepts.

 CONTENT REVIEW

 CONTENT REVIEW CD-ROM

CHAPTER INVESTIGATION

Focus

PURPOSE Students will learn how to measure waves produced by a pendulum and will investigate how the length of the pendulum affects wavelength.

OVERVIEW Students will:

- Make a pendulum from which sand drains.
- Record the pendulum's wave pattern with sand on construction paper.
- Repeat the activity with varying pendulum lengths.
- Evaluate the different waves produced.

Lab Preparation

- Students can prepare the paper cones ahead of time. Be sure students make the hole no larger than a pea. Have them make extra cones in case some perform poorly.
- For homework the night before, have students read the investigation, write their hypothesis, and draw their data table. Or you may wish to copy and distribute datasheets and rubrics.

 UNIT RESOURCE BOOK, pp. 53–61

SCIENCE TOOLKIT, F14

Lab Management

- Pair students. Have students choose roles: one partner will manipulate the pendulum, while the other will fill the pendulum with sand and then pull the paper under the swinging pendulum.
- The waves may not look like those in the text, but should have clear crests and troughs.

INCLUSION Encourage students working together to discuss in advance which tasks each will perform. Help students identify tasks that they can do well or alternate approaches that suit their abilities.

CHAPTER INVESTIGATION

Wavelength

OVERVIEW AND PURPOSE The pendulum on a grandfather clock keeps time as it swings back and forth at a steady rate. The swings of a pendulum can be recorded as a wave with measurable properties. How do the properties of the pendulum affect the properties of the waves it produces? In this investigation you will use your understanding of wave properties to

- construct a pendulum and measure the waves it produces, and
- determine how the length of the pendulum affects the wavelength of the waves.

Problem

How does changing the length of a pendulum affect the wavelength?

Hypothesize

Write a hypothesis in "If . . . , then . . . , because . . ." form to answer the problem question.

Procedure

MATERIALS
- 1/2 sheet white paper
- tape
- scissors
- string
- meter stick
- fine sand
- graduated cylinder
- 2 sheets colored construction paper

1. Make a data table like the one shown on the sample notebook page.

2. Make a cone with the half-sheet of paper by rolling it and taping it as shown. The hole in the bottom of the cone should be no larger than a pea.

3. Cut a hole in each side of the cone and tie the ends of the string to the cone to make a pendulum.

4. Hold the string on the pendulum so that the distance from your fingers holding the string to the bottom of the cone is 20 cm.

5. Cover the bottom of the cone with your fingertip. While you hold the cone, have your partner pour about 40 ml of sand into the cone.

 D 22 Unit: Waves, Sound, and Light

INVESTIGATION RESOURCES

 CHAPTER INVESTIGATION, Wavelength
- Level A, pp. 53–56
- Level B, pp. 57–60
- Level C, p. 61

Advanced students should complete Levels B & C.

 Writing a Lab Report, D12–13

Technology Resources

Customize this student lab as needed or look for an alternative. Print rubrics to assess student lab reports.

Lab Generator CD-ROM

6. Hold the pendulum about 5 cm above the construction paper as shown. Pull the pendulum from the bottom to one side of the construction paper. Be careful not to move the pendulum at the top, or to pull the pendulum over the edge of the paper.

7. Let the pendulum go while your partner gently pulls the paper forward so that the sand makes waves on the paper. Be sure to pull the paper at a steady rate. Let the remaining sand pile up on the end of the paper.

8. Measure the wavelength from crest to crest or trough to trough. Record the wavelength in your table.

9. Run two more trials, repeating steps 5–8. Be sure to pull the paper at the same speed for each trial. Calculate the average wavelength over all three trials, and record it in your table.

10. Repeat steps 4–8, changing the length of the pendulum to 30 cm and then to 40 cm.

▶ Observe and Analyze Write It Up

1. **RECORD OBSERVATIONS** Draw the setup of your procedure. Be sure your data table is complete.

2. **IDENTIFY VARIABLES AND CONSTANTS** Identify the variables and constants that affected the wave produced by the moving pendulum. List them in your notebook.

3. **ANALYZE** What patterns can you find in your data? For example, do the numbers increase or decrease as you read down each column?

▶ Conclude Write It Up

1. **INFER** Answer your problem question.

2. **INTERPRET** Compare your results with your hypothesis. Do your data support your hypothesis?

3. **IDENTIFY LIMITS** What possible limitations or sources of error could have affected your results?

4. **APPLY** Suppose you were examining the tracing made by a seismograph, a machine that records an earthquake wave. What would happen if you increased the speed at which the paper ran through the machine? What do you think the amplitude of the tracing represents?

▶ INVESTIGATE Further

CHALLENGE Revise your experiment to change one variable other than the length of the pendulum. Run a new trial, changing the variable you choose but keeping everything else constant. How did changing the variable affect the wave produced?

Wavelength
Problem How does changing the length of a pendulum affect the wavelength?

Hypothesize

Observe and Analyze
Table 1. Wavelengths Produced by Pendulums

Pendulum Length (cm)	Trial 1	Trial 2	Trial 3	Average Wavelength (cm)
20				
30				
40				

Conclude

▶ Observe and Analyze Write It Up

SAMPLE DATA wavelength with 20-cm string: 3.5 cm; wavelength with 30-cm string: 5 cm; wavelength with 40-cm string: 8 cm

1. See students' diagrams. See students' data tables.

2. The variable is the length of the pendulum. Ideally, constants will include the speed of the paper as the partner pulls it forward, the height of the pendulum above the table, the amount the pendulum is pulled back before letting go, and the amount of sand used in each trial.

3. Sample answer: Most of the numbers increase as you read down the columns.

▶ Conclude Write It Up

1. Changing the length of the pendulum changes the wavelength of waves it produces.

2. Student answers will vary, depending on students' original hypotheses.

3. Sample answer: The speed at which the paper is pulled each time may have varied, affecting the results.

4. If you increase the speed of paper that runs through a seismograph, you would increase the wavelength. The amplitude represents the amount of energy carried by a seismic wave.

▶ INVESTIGATE Further

CHALLENGE Students could choose to vary the weight of the pendulum, the size of the cone, the speed at which the paper is pulled under the pendulum, or the height of the pendulum above the paper. Any of these changes could affect the wave produced by the pendulum.

Post-Lab Discussion

Discuss with students how the activity enabled them to graph the movement of a pendulum.

Point out that almost any motion that is regular and repeating can be graphed as a wave. Challenge students to describe how the movement of a clock's minute hand could be interpreted and graphed as wave motion. You could graph the 12 o'clock position as the crest of the wave and the 6 o'clock position as the trough.

◉ Set Learning Goals

Students will

- Describe how waves change as they encounter a barrier.
- Explain what happens when waves enter a new medium.
- Identify ways in which waves interact with one another.
- Investigate in an experiment how waves behave when they meet a barrier.

◐ 3-Minute Warm-Up

Display Transparency 5 or copy this exercise on the board:

Draw a diagram that shows how a duck floats on waves. Use arrows to show how the waves move and how the duck moves. *Arrows for the duck should show it staying in the same position but bobbing up and down vertically; arrows for waves should point in a horizontal direction.*

 3-Minute Warm-Up, p. T5

1.3 MOTIVATE

EXPLORE Reflection

PURPOSE To introduce the concept that waves reflect off a barrier

TIP *10 min.* The waves will be circular.

WHAT DO YOU THINK? *The ripples travel back in the opposite direction after they encounter the side of the pan. Waves encounter a medium that they cannot travel through.*

Ongoing Assessment

CHECK YOUR READING *Answer: reflection, refraction, diffraction*

KEY CONCEPT

1.3 Waves behave in predictable ways.

◀ **BEFORE, you learned**

- Waves transfer energy
- Amplitude, wavelength, and frequency can be measured

▶ **NOW, you will learn**

- How waves change as they encounter a barrier
- What happens when waves enter a new medium
- How waves interact with other waves

VOCABULARY

reflection p. 25
refraction p. 25
diffraction p. 26
interference p. 27

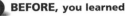

EXPLORE Reflection

How do ripples reflect?

PROCEDURE

1. Put a few drops of food coloring into the pan of water.
2. Dip the pencil in the water at one end of the pan to make ripples in the water.
3. Observe the ripples as they reflect off the side of the pan. Draw a sketch of the waves reflecting.

WHAT DO YOU THINK?
- What happens when the waves reach the side of the pan?
- Why do you think the waves behave as they do?

MATERIALS
- wide pan, half full of water
- food coloring
- pencil

COMBINATION NOTES
Use combination notes in your notebook to describe how waves interact with materials.

Waves interact with materials.

You have read that mechanical waves travel through a medium like air, water, or the ground. In this section, you will read how the motion of waves changes when they encounter a new medium. For instance, when an ocean wave rolls into a ship or a sound wave strikes a solid wall, the wave encounters a new medium.

When waves interact with materials in these ways, they behave predictably. All waves, from water waves to sound waves and even light waves, show the behaviors that you will learn about next. Scientists call these behaviors reflection, refraction, and diffraction.

CHECK YOUR READING What behaviors do all waves have in common?

RESOURCES FOR DIFFERENTIATED INSTRUCTION

Below Level
UNIT RESOURCE BOOK
- Reading Study Guide A, pp. 35–36
- Decoding Support, p. 48

 AUDIO CDS

 Additional INVESTIGATION,
Tracking the Path of Light, A, B, & C, pp. 62–70;
Teacher Instructions, pp. 284–285

Advanced
UNIT RESOURCE BOOK
Challenge and Extension, p. 41

English Learners
UNIT RESOURCE BOOK
Spanish Reading Study Guide, pp. 39–40

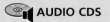 AUDIO CDS

- Audio Readings in Spanish
- Audio Readings (English)

Reflection

What happens to water waves at the end of a swimming pool? The waves cannot travel through the wall of the pool. Instead, the waves bounce off the pool wall. The bouncing back of a wave after it strikes a barrier is called **reflection.**

Remember what you have learned about forces. A water wave, like all waves, transfers energy. When the water wave meets the wall of the pool, it pushes against the wall. The wall applies an equal and opposite force on the water, sending the wave back in another direction. In the illustration on the right, you can see water waves reflecting off a barrier.

Sound and light waves reflect too. Sound waves reflecting off the walls of a canyon produce an echo. Light waves reflecting off smooth metal behind glass let you see an image of yourself in the mirror. The light waves bounce off the metal just as the water waves bounce off the pool wall. You will learn more about how sound and light waves reflect in the next chapters.

Reflection Water waves move in predictable ways. Here waves are shown from above as they reflect off a barrier.

 CHECK YOUR READING How would you define *reflection* in your own words?

Refraction

Sometimes, a wave does not bounce back when it encounters a new medium. Instead, the wave continues moving forward. When a wave enters a new medium at an angle, it bends, or refracts. **Refraction** is the bending of a wave as it enters a new medium at an angle other than 90 degrees. Refraction occurs because waves travel at different speeds in different mediums. Because the wave enters the new medium at an angle, one side of the wave enters the new medium before the rest of the wave. When one side of a wave speeds up or slows down before the other side, it causes the wave to bend.

You have probably noticed the refraction of light waves in water. Objects half-in and half-out of water look broken or split. Look at the photograph of the straw in the glass. What your eyes suggest—that the straw is split—is not real, is it? You are seeing the refraction of light waves caused by the change of medium from air to water. You will learn more about the refraction of light waves in Chapter 4.

Refraction The light waves refract as they pass from air to water, making this straw look split.

1.3 INSTRUCT

Real World Example

Bands sometimes play on stages surrounded by a large curved backdrop. This curved backdrop, or shell, reflects sound waves out to the audience. Without the band shell, sound waves produced by the performers would shoot out in all directions in the open air, lost to listeners, and the music would sound faint and dull.

Integrate the Sciences

Seismologists—scientists who study earthquakes—measure seismic waves all over Earth. These waves, set off by earthquakes in Earth's crust, travel in all directions and eventually reach the surface. Surface stations make precise measurements of the seismic waves. By studying how these waves travel and are refracted by different materials, seismologists have learned much about the density and composition of Earth's core and mantle.

Ongoing Assessment

Describe how waves change as they encounter a barrier.

Ask: What happens when sound waves strike solid rock? Give an example. *They reflect from the rock, as in an echo.*

Explain what happens when waves enter a new medium.

Ask: Why do you see distorted shapes when you look into the water in a swimming pool? *Light waves bend, or undergo refraction, when they pass from the air into the water. This refraction distorts underwater objects.*

CHECK YOUR READING *Sample answer: waves bouncing off a barrier*

INVESTIGATE Diffraction

PURPOSE To investigate how waves behave when they meet a partial barrier

TIPS *20 min.* Allow students a few minutes to explore, then suggest the following:

• Make sure the block sticks out of the water.

• Use a gentle push on the ruler to create waves.

WHAT DO YOU THINK? *by putting a partial barrier in their path; students' answers will vary depending on their predictions.*

CHALLENGE *Sample answer: You could try blocks with different sizes and shapes to better diffract the waves.*

 Datasheet, Diffraction, p. 42

Metacognitive Strategy

Ask students to write a paragraph describing the reasoning behind their prediction about what would happen when the waves hit the barrier in the pan.

Teach from Visuals

To help students interpret the visual of diffraction, ask:

• Through what medium are the waves moving? *water*

• From what perspective are you viewing the waves? *from above*

• What happens to the waves as they go through the opening? *The waves spread out.*

Diffraction

 C

You have seen how waves reflect off a barrier. For example, water waves bounce off the side of a pool. But what if the side of the pool had an opening in it? Sometimes, waves interact with a partial barrier, such as a wall with an opening. As the waves pass through the opening, they spread out, or diffract. **Diffraction** is the spreading out of waves through an opening or around the edge of an obstacle. Diffraction occurs with all types of waves.

Look at the photograph on the right. It shows water waves diffracting as they pass through a small gap in a barrier. In the real world, ocean waves diffract through openings in cliffs or rock formations.

Similarly, sound waves diffract as they pass through an open doorway. Turn on a TV or stereo, and walk into another room. Listen to the sound with the door closed and then open. Then try moving around the room. You can hear the sound wherever you stand because the waves spread out, or diffract, through the doorway and reflect from the walls.

Diffraction through an opening

INVESTIGATE Diffraction

How can you make a wave diffract?

SKILL FOCUS
Predicting

PROCEDURE

① Put a few drops of food coloring into the container of water.

② Experiment with quick motions of the ruler to set off waves in the container.

③ Place the block on its side in the center of the container. Set the bag of sand on the block to hold it down. Predict how the waves will interact with the barrier you have added.

④ Make another set of waves, and observe how they interact with the barrier.

WHAT DO YOU THINK?

• How did you make the waves diffract?

• How did your observations compare with your prediction?

CHALLENGE How could you change the experiment to make the effect of the diffraction more obvious?

MATERIALS
• wide pan of water
• food coloring
• ruler
• wooden block
• bag of sand

TIME
20 minutes

DIFFERENTIATE INSTRUCTION

(?) More Reading Support

C What do waves do when they pass through an opening in a barrier? *They diffract.*

Below Level Students may have difficulty with the similar-sounding vocabulary in this lesson. Write "reflection," "refraction," and "diffraction" on the board, dividing them into syllables. Pronounce each word. Ask students to visually scan each word and find similarities and differences among them. *Reflection and refraction have identical prefixes and suffixes but different roots. Refraction and diffraction have identical roots and suffixes but different prefixes.*

Diffraction also occurs as waves pass the edge of an obstacle. The photograph at the right shows water waves diffracting as they pass an obstacle. Ocean waves also diffract in this way as they pass large rocks in the water.

Light waves diffract around the edge of an obstacle too. The edges of a shadow appear fuzzy because of diffraction. The light waves spread out, or diffract, around the object that is making the shadow.

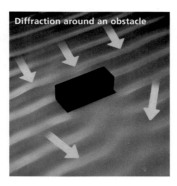
Diffraction around an obstacle

 CHECK YOUR READING Describe what happens when waves diffract.

Waves interact with other waves.

Just as waves sometimes interact with new mediums, they can also interact with other waves. Two waves can add energy to or take away energy from each other in the place where they meet. **Interference** is the meeting and combining of waves.

Waves Adding Together

Suppose two identical waves coming from opposite directions come together at one point. The waves' crests and troughs are aligned, which means they join up exactly. When the two waves merge, their amplitudes are added together, and the result is a bigger wave. After the waves have merged, they return to their original amplitudes and continue in their original directions.

The adding of two waves is called constructive interference. It builds up, or constructs, a larger wave out of two smaller ones. Look at the diagram at the right to see what happens in constructive interference.

Because the waves in the example joined together perfectly, the amplitude of the new wave equals the combined amplitudes of the 2 original waves. For example, if the crest of a water wave with an amplitude of 1 meter (3.3 ft) met up with the crest of another wave with an amplitude of 1 meter (3.3 ft), there would be a 2 meter (6.6 ft) crest in the spot where they met.

Constructive Interference

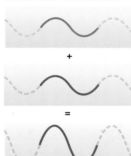

When two wave crests with amplitudes of 1 m each combine, a wave with an amplitude of 2 m is formed.

Chapter 1: **Waves** 27 **D**

Real World Example

Piano tuners use interference in sound waves to tune pianos. When two piano strings go out of tune with each other, they have slightly different frequencies, or pitches. When their sound waves collide, the troughs and crests join out of step. The result is a throbbing sound, called beats. When two strings are exactly in tune, the crests and troughs of the sound waves join completely in step, and the beats disappear.

History of Science

West Coast tsunamis have been recorded since 1737, from Alaska to California. Because of a particularly damaging tsunami in 1946, the Seismic Sea Wave Warning System was started in 1948. It predicted the arrival of tsunamis and warned coastal dwellers. This system is now the Pacific Tsunami Warning System, which tracks tsunamis and warns nations around the Pacific Basin.

Ongoing Assessment

Identify ways in which waves interact with one another.

Ask: What must happen for two waves to interact in constructive interference? *The waves must be aligned so that their crests and troughs match.*

CHECK YOUR READING *Answer: Waves spread out as they diffract around edges of openings and obstacles.*

DIFFERENTIATE INSTRUCTION

? More Reading Support

D What can happen to the energy in two waves when they meet? *Two waves can add to or subtract energy from each other.*

Advanced Interference among light waves is responsible for some commonly seen color effects, such as rainbow colors in oil slicks on pavement or on bubbles. Challenge students to find detailed explanations for these interference effects. Have them demonstrate the effects with oil or bubbles and share their scientific findings in class.

 Challenge and Extension, p. 41

Ongoing Assessment

Reinforce  (the **BIG** idea)

Have students relate the section to the Big Idea.

 Reinforcing Key Concepts, p. 43

1.3 ASSESS & RETEACH

Assess

 Section 1.3 Quiz, p. 5

Reteach

Present these four wave scenarios:

- You shout into a cave and hear an echo.

- Two people in a choir are singing slightly off-key, and their voices combine to produce loud, unpleasant beats.

- From shore, it looks like all of a bridge's pilings are broken at the water level, but trucks are safely driving over it.

- Before you reach the corner of a huge stone building, you hear a noisy parade approaching down the other street.

Have each student choose a scenario and find the appropriate presentation for that topic in the lesson. Have students diagram the wave scenario and write a scientific explanation for it.

Technology Resources

Have students visit **ClassZone.com** for reteaching of Key Concepts.

 CONTENT REVIEW

CONTENT REVIEW CD-ROM

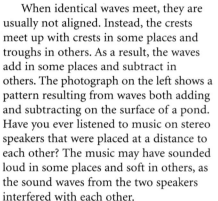

Wave interference produces this pattern on a pond as two sets of waves interact.

Waves Canceling Each Other Out

Imagine again that two very similar waves come together. This time, however, the crests of one wave join with the troughs of the other. The energy of one wave is subtracted from the energy of the other. The new wave is therefore smaller than the original wave. This process is called destructive interference. Look at the diagram below to see what happens in destructive interference.

For example, if a 2 meter (6.6 ft) crest met up with a 1 meter (3.3 ft) trough, there would be a crest of only 1 meter (3.3 ft) where they met. If the amplitudes of the 2 original waves are identical, the 2 waves can cancel each other out completely!

When identical waves meet, they are usually not aligned. Instead, the crests meet up with crests in some places and troughs in others. As a result, the waves add in some places and subtract in others. The photograph on the left shows a pattern resulting from waves both adding and subtracting on the surface of a pond. Have you ever listened to music on stereo speakers that were placed at a distance to each other? The music may have sounded loud in some places and soft in others, as the sound waves from the two speakers interfered with each other.

 CHECK YOUR READING Summarize in your own words what happens during interference.

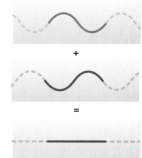

Destructive Interference

+

=

When a 1 m wave crest meets a 1 m wave trough, the amplitudes cancel each other out. A wave with an amplitude of 0 m is formed where they meet.

1.3 Review

KEY CONCEPTS

1. Explain what happens when waves encounter a medium that they cannot travel through.

2. Describe a situation in which waves would diffract.

3. Describe two ways that waves are affected by interference.

CRITICAL THINKING

4. **Synthesize** Explain how reflection and diffraction can happen at the same time in a wave.

5. **Compare** How is interference similar to net force? How do you think the two concepts might be related? **Hint:** Think about how forces are involved in wave motion.

⚠ CHALLENGE

6. **Predict** Imagine that you make gelatin in a long, shallow pan. Then you scoop the gelatin out of one end of the pan and add icy cold water to the exact same depth as the gelatin. Now suppose you set off waves at the water end. What do you think will happen when the waves meet the gelatin?

ANSWERS

1. Waves reflect.

2. Sample answer: Ocean waves hit a spot of land.

3. Amplitudes can add up or cancel each other out.

4. When a wave hits a partial barrier, the part of the wave that hits the barrier reflects

into the oncoming waves. The part of the wave that passes through the gap in the barrier diffracts.

5. Forces are responsible for moving a wave medium up and down as a wave passes through it. Forces can be

added and subtracted on a wave medium just as they can on any object.

6. Answers could describe refraction, as waves meet and enter a new medium, or reflection, as waves encounter a barrier.

CONNECTING SCIENCES

Tsunamis!

Tsunamis (tsu-NAH-mees) are among the most powerful waves on Earth. They can travel fast enough to cross the Pacific Ocean in less than a day! When they reach shore, these powerful waves strike with enough force to destroy whole communities.

What Causes Tsunamis?

Tsunamis are caused by an undersea volcanic eruption, an earthquake, or even a landslide. This deep-sea event sends out a series of waves. Surprisingly, if you were out at sea, you would not even notice these powerful waves. The reason has to do with the physics of waves—their velocity, wavelength, and amplitude.

A tsunami generated by a powerful earthquake struck Japan in 1983. The photograph above shows a scene before the tsunami struck. What changes do you see in the picture below showing the scene after the tsunami struck?

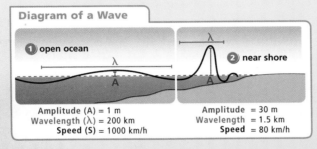

Diagram of a Wave

1 open ocean

2 near shore

Amplitude (A) = 1 m
Wavelength (λ) = 200 km
Speed (S) = 1000 km/h

Amplitude = 30 m
Wavelength = 1.5 km
Speed = 80 km/h

The Changing Wave

1 On the open ocean, the waves of a tsunami are barely visible. The amplitude of the waves is less than a few meters, but the energy of the waves extends to the sea floor. The tsunami's wavelength is extremely long—up to 200 kilometers (120 mi). These long, low waves can travel as fast as a jet—almost 1000 kilometers per hour (600 mi/h).

2 Near shore, the waves slow down as they approach shallow water. As their velocity drops, their wavelengths get shorter, but their amplitude gets bigger. All the energy that was spread out over a long wave in deep water is now compressed into a huge wave that can reach a height of more than 30 meters (100 ft).

Individual tsunami waves may arrive more than an hour apart. Many people have lost their lives returning home between waves, making the fatal mistake of thinking the danger was over.

EXPLORE

1. **VISUALIZE** Look at 2 on the diagram. How tall is 30 meters (100 ft)? Find a 100-foot building or structure near you to visualize the shore height of a tsunami.

2. **CHALLENGE** Use library or Internet resources to prepare a chart on the causes and effects of a major tsunami event.

CONNECTING SCIENCES
Integration of Sciences

Set Learning Goal

To understand how the physics of waves applies to tsunamis

Present the Science

Share the following facts with students:

- The word *tsunami* is a Japanese word meaning "harbor wave."
- The most devastating tsunami occurred in Papua New Guinea in 1998.
- The states of Alaska and Hawaii have frequently felt tsunamis hitting their shores.
- Tsunamis are often confused with tidal waves, which are caused by the gravitational attraction of the Sun and Moon.
- Tsunamis are often called "seismic sea waves" because they are usually the result of "underwater earthquakes."

Discussion Questions

Ask: Why does the amplitude of the wave go up in shallow water? *It has nowhere else to go.*

Ask: Where will the energy of the wave go when it hits shore? *It will strike the shoreline with the force of the wave.*

Ask: How is the cause of a tsunami different from the cause of regular waves? *Regular waves are formed from wind and weather, while tsunamis develop from underwater volcanic eruptions or underwater earthquakes.*

Ask: Why do you think the waves slow down when they reach water that is more shallow? *They are slowed down by friction with the ocean floor.*

Close

Ask: How do tsunamis show that waves carry energy? *by the destruction they cause*

EXPLORE

1. **VISUALIZE** Discuss with students the impact that 30 meters of water might have.

2. **CHALLENGE** While many pictures and Internet sites show the devastation caused by tsunamis, there are very few pictures of a tsunami in action. Artists have attempted to depict the phenomenon. Students should have no trouble finding information on tsunamis.

PHOTO CAPTION Answer: There is a significant amount of water on the shore, the wall surrounding the water is destroyed, and the objects near the shore are pulled into the water by the force of the tsunami hitting the land.

BACK TO

the BIG idea

Ask students to describe a demonstration they could present to show that waves transfer energy. *Sample answer: You could go to the beach and put a plastic bucket in the sand a few centimeters out in the surf. The incoming waves would push the bucket, showing the transfer of energy.*

◯ KEY CONCEPTS SUMMARY

SECTION 1.1

For the picture on the left, have students describe the direction of the disturbance compared to the direction of the wave. *It is at right angles with the direction of the wave.* For the picture on the right, have students describe the direction of the disturbance compared to the direction of the wave. *They are both in the same direction.*

SECTION 1.2

Ask: What wave properties are identified in the picture? Which property indicates the amount of energy the wave has? *The wave properties are amplitude, wavelength, and frequency. Amplitude indicates the energy in the wave.*

SECTION 1.3

Have students describe the three wave behaviors pictured here. *In reflection, waves bounce off a solid barrier. In refraction, waves bend as they pass from one medium into another. In diffraction, waves pass by a barrier and spread out.*

Review Concepts

- Big Idea Flow Chart, p. T1
- Chapter Outline, pp. T7–T8

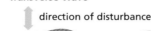 **Chapter Review**

the BIG idea

Waves transfer energy and interact in predictable ways.

 **CONTENT REVIEW**
CLASSZONE.COM

◯ KEY CONCEPTS SUMMARY

1.1 Waves transfer energy.

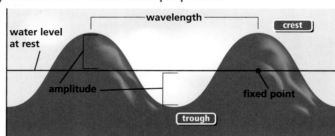

Transverse Wave

direction of disturbance

direction of wave
transfer of energy

Longitudinal Wave

direction of disturbance

direction of wave
transfer of energy

VOCABULARY
wave p. 9
medium p. 11
mechanical wave p. 11
transverse wave p. 13
longitudinal wave p. 14

1.2 Waves have measurable properties.

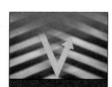

water level at rest
wavelength
crest
amplitude
fixed point
trough

Frequency is the number of wavelengths passing a fixed point in a certain amount of time.

VOCABULARY
crest p. 17
trough p. 17
amplitude p. 17
wavelength p. 17
frequency p. 17

1.3 Waves behave in predictable ways.

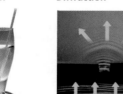

Reflection   Refraction   Diffraction

VOCABULARY
reflection p. 25
refraction p. 25
diffraction p. 26
interference p. 27

Technology Resources

Have students visit **ClassZone.com** or use the CD-ROM for a cumulative review of concepts.

 CONTENT REVIEW

CONTENT REVIEW CD-ROM

Engage students in a whole-class interactive review of Key Concepts. Edit content as you wish.

 POWER PRESENTATIONS

Reviewing Vocabulary

Draw a triangle for each of the terms below. On the wide bottom of the triangle, write the term and your own definition of it. Above that, write a sentence in which you use the term correctly. At the top of the triangle, draw a small picture to show what the term looks like. The first triangle is completed for you.

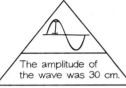

The amplitude of the wave was 30 cm.

Amplitude is the distance between a line through the middle of a wave and a crest or trough.

1. amplitude
2. diffraction
3. frequency
4. medium
5. crest
6. interference
7. reflection
8. trough
9. refraction
10. wavelength

Reviewing Key Concepts

Multiple Choice *Choose the letter of the best answer.*

11. The direction in which a transverse wave travels is
    a. in the same direction as the disturbance
    b. toward the disturbance
    c. from the disturbance downward
    d. at right angles to the disturbance

12. An example of a longitudinal wave is a
    a. water wave
    b. stadium wave
    c. sound wave
    d. rope wave

13. Which statement best defines a wave medium?
    a. the material through which a wave travels
    b. a point half-way between the crest and trough of a wave
    c. the distance from one wave crest to the next
    d. the speed at which waves travel in water

14. As you increase the amplitude of a wave, you also increase the
    a. frequency
    b. wavelength
    c. speed
    d. energy

15. To identify the amplitude in a longitudinal wave, you would look for areas of
    a. reflection
    b. compression
    c. crests
    d. refraction

16. Which statement describes the relationship between frequency and wavelength?
    a. When frequency increases, wavelength increases.
    b. When frequency increases, wavelength decreases.
    c. When frequency increases, wavelength remains constant.
    d. When frequency increases, wavelength varies unpredictably.

17. For wave refraction to take place, a wave must
    a. increase in velocity
    b. enter a new medium
    c. increase in frequency
    d. merge with another wave

18. Which setup in a wave tank would enable you to demonstrate diffraction?
    a. water only
    b. water and sand
    c. water and food coloring
    d. water and a barrier with a small gap

19. Two waves come together and interact to form a new, smaller wave. This process is called
    a. destructive interference
    b. constructive interference
    c. reflective interference
    d. positive interference

Reviewing Vocabulary

Sample definitions are given below.

1. Amplitude is how high or low a wave is from its test position.
2. Diffraction is the spreading of waves through a hole.
3. Frequency is the number of waves passing a fixed point in a certain time.
4. A medium is anything a wave travels through.
5. A crest is the highest point of a wave.
6. Interference is the meeting of two waves.
7. Reflection is the bouncing back of a wave after it meets a barrier.
8. A trough is the lowest point of a wave.
9. Refraction is the bending of a wave as it moves from one medium into another.
10. Wavelength is the distance from crest to crest.

Reviewing Key Concepts

11. d
12. c
13. a
14. d
15. b
16. b
17. b
18. d
19. a

ASSESSMENT RESOURCES

UNIT ASSESSMENT BOOK
- Chapter Test A, pp. 6–9
- Chapter Test B, pp. 10–13
- Chapter Test C, pp. 14–17
- Alternative Assessment, pp. 18–19

SP

SPANISH ASSESSMENT BOOK
Spanish Chapter Test, pp. 281–284

Technology Resources

Edit test items and answer choices.

 Test Generator CD-ROM

Visit **ClassZone.com** to extend test practice.

 Test Practice

Thinking Critically

20. a and b measure amplitude

21. wavelength

22. 3 waves

23. frequency

24. 3 wavelengths/s

25. No, it could be a liquid or gas.

26. By drawing the pendulum back farther; The swinging motion has amplitude, frequency, and wavelength.

27. It might diffract through the gap.

28. Sample answer: Yes, in each instance one wave meets another and changes it in some way.

Using Math in Science

29. $S = \lambda \cdot f$

$S = 1.2 \; \dfrac{m}{\text{wavelength}} \cdot 2 \; \dfrac{\text{wavelength}}{s}$

$= 2.4 \; m/s$

30. $S = \lambda \cdot f$

$S = 9 \; \dfrac{m}{\text{wavelength}} \cdot 0.42 \; \dfrac{\text{wavelength}}{s}$

$= 3.78 \; m/s$

31. $S = \lambda \cdot f; \; \lambda = s \div f$

$\lambda = 340 \; m/s \div 10{,}000 \; \dfrac{\text{wavelength}}{s}$

$= 0.034 \; m, \; or \; 3.4 \; cm$

32. $S = \lambda \cdot f; \; f = s \div \lambda$

$f = 2.5 \; m/s \div 4 \; \dfrac{m}{\text{wavelength}}$

$= 0.625, \; round \; to \; 0.63 \; \dfrac{\text{wavelengths}}{s}$

the BIG idea

33. Sample answer: The wave has transferred energy to a surfer.

34. Sample answer: (1) Sound in an empty room. (2) Ripples in a pond. (3) Waves in a swimming pool.

35. Students' paragraphs should demonstrate an understanding of the Big Idea.

UNIT PROJECTS

Give students the appropriate Unit Project worksheets from the URB for their projects. Both directions and rubrics can be used as a guide.

 Unit Projects, pp. 5–10

Thinking Critically

Use the diagram below to answer the next two questions.

20. What two letters in the diagram measure the same thing? What do they both measure?

21. In the diagram above, what does the letter c measure?

Use the diagram below to answer the next three questions. The diagram shows waves passing a fixed point.

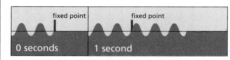

22. At 0 seconds, no waves have passed. How many waves have passed after 1 second?

23. What is the measurement shown in the diagram?

24. How would you write the measurement taken in the diagram?

25. **EVALUATE** Do you think the following is an accurate definition of medium? Explain your answer.

A **medium** is any solid through which waves travel.

26. **APPLY** Picture a pendulum. The pendulum is swinging back and forth at a steady rate. How could you make it swing higher? How is swinging a pendulum like making a wave?

27. **PREDICT** What might happen to an ocean wave that encounters a gap or hole in a cliff along the shore?

28. **EVALUATE** Do you think *interference* is an appropriate name for the types of wave interaction you read about in Section 1.3? Explain your answer.

Using Math in Science

29. At what speed is the wave below traveling if it has a frequency of 2 wavelengths/s?

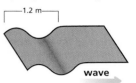

wave

30. An ocean wave has a wavelength of 9 m and a frequency of 0.42 wavelengths/s. What is the wave's speed?

31. Suppose a sound wave has a frequency of 10,000 wavelengths/s. The wave's speed is 340 m/s. Calculate the wavelength of this sound wave.

32. A water wave is traveling at a speed of 2.5 m/s. The wave has a wavelength of 4 m. Calculate the frequency of this water wave.

the BIG idea

33. **INTERPRET** Look back at the photograph at the start of the chapter on pages 6–7. How does this photograph illustrate a transfer of energy?

34. **SYNTHESIZE** Describe three situations in which you can predict the behavior of waves.

35. **SUMMARIZE** Write a paragraph summarizing this chapter. Use the big idea from page 6 as the topic sentence. Then write an example from each of the key concepts listed under the big idea.

UNIT PROJECTS

If you are doing a unit project, make a folder for your project. Include in your folder a list of the resources you will need, the date on which the project is due, and a schedule to track your progress. Begin gathering data.

MONITOR AND RETEACH

If students have difficulty answering questions 20 and 21, make flash cards for all the vocabulary words in the chapter. Draw a wave diagram on the board. Have students play a flash-card game. When a word and definition apply to some part of the diagram, have students identify it. For vocabulary words that are illustrated in the chapter, have students find a visual that depicts the vocabulary word.
Students may benefit from summarizing one or more sections of the chapter.

Summarizing the Chapter, pp. 71–72

Standardized Test Practice

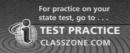 For practice on your state test, go to . . .
TEST PRACTICE
CLASSZONE.COM

Interpreting Diagrams

Study the illustration below and then answer the questions.

The illustration below shows a wave channel, a way of making and studying water waves. The motor moves the rod, which moves the paddle back and forth. The movement of the paddle makes waves, which move down the length of the channel. The material behind the paddle absorbs the waves generated in that direction.

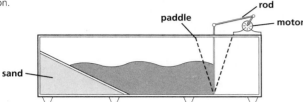

1. An experimenter can adjust the position of the rod on the arm of the motor. Placing it closer to the motor makes shallower waves. Placing it farther from the motor makes deeper waves. What property of waves does this affect?
   a. amplitude
   b. direction
   c. frequency
   d. wavelength

2. By changing motor speeds, an experimenter can make the paddle move faster or slower. What property of waves does this affect?
   a. amplitude
   b. direction
   c. trough depth
   d. wavelength

3. Sand is piled up in the channel at the end of the tank opposite the motor. When waves pass over this sand, their wavelengths shorten. Assuming that the speed of the waves stays the same, their frequency
   a. stays the same
   b. increases
   c. decreases
   d. cannot be predicted

4. Suppose there was no sand at the end of the tank opposite the paddle. In that case, the waves would hit the glass wall. What would they do then?
   a. stop
   b. reflect
   c. refract
   d. diffract

Extended Response

Answer the two questions below in detail.

5. Suppose temperatures in one 10-day period were as follows: 94°, 96°, 95°, 97°, 95°, 98°, 99°, 97°, 99°, and 98°. Make a simple line graph of the data. In what ways is the series of temperatures similar to a wave, and in what ways does it differ?

6. Lydia and Bill each drop a ball of the same size into the same tank of water but at two different spots. Both balls produce waves that spread across the surface of the water. As the two sets of waves cross each other, the water forms high crests in some places. What can you say about both waves? Explain your answer.

Interpreting Diagrams

1. a   3. c
2. d   4. b

Extended Response

5. RUBRIC
4 points for a response that correctly answers the question and uses the following terms accurately:
- amplitude
- frequency
- wavelength

Sample: The series of high temperatures is similar to a wave in that its graph has <u>amplitude</u>, <u>frequency</u>, *and* <u>wavelength</u>, *but it is different in that energy is not transferred from one place to another.*

3 points for a response that correctly answers the question and uses two terms accurately
2 points for a response that correctly uses one term accurately
1 point for a response that correctly answers the question but doesn't use the terms

6. RUBRIC
4 points for a response that correctly answers the question and uses the following terms accurately:
- amplitude
- crests
- trough

Sample: Both waves have the same <u>amplitude</u> *and cancel each other out because the* <u>crests</u> *and troughs were matched exactly. This is an example of destructive interference showing that when the two waves meet, the* <u>trough</u> *of one wave joins with the crest of the other.*

3 points for a response that correctly answers the question and uses two terms accurately
2 points for a response that correctly uses one term accurately
1 point for a response that correctly answers the question but doesn't use the terms

METACOGNITIVE ACTIVITY

Have students answer the following questions in their **Science Notebook:**

1. What did you find most surprising about how waves behave?

2. What questions do you still have about the ways waves interact?

3. What did you learn about waves that can be applied to your Unit Project?

Physical Science
UNIFYING PRINCIPLES

PRINCIPLE 1

Matter is made of particles too small to see.

PRINCIPLE 2

Matter changes form and moves from place to place.

PRINCIPLE 3

Energy changes from one form to another, but it cannot be created or destroyed.

PRINCIPLE 4

Physical forces affect the movement of all matter on Earth and throughout the universe.

Unit: Waves, Sound, and Light
BIG IDEAS

CHAPTER 1
Waves

Waves transfer energy and interact in predictable ways.

CHAPTER 2
Sound

Sound waves transfer energy through vibrations.

CHAPTER 3
Electromagnetic Waves

Electromagnetic waves transfer energy through radiation.

CHAPTER 4
Light and Optics

Optical tools depend on the wave behavior of light.

CHAPTER 2
KEY CONCEPTS

SECTION 2.1

Sound is a wave.

1. Sound is a type of mechanical wave.

2. Sound waves vibrate particles.

3. The speed of sound depends on its medium.

SECTION 2.2

Frequency determines pitch.

1. Pitch depends on the frequency of a sound wave.

2. The motion of the source of a sound affects its pitch.

SECTION 2.3

Intensity determines loudness.

1. Intensity depends on the amplitude of a sound wave.

2. The intensity of sound can be controlled.

3. Intense sound can damage hearing.

SECTION 2.4

Sound has many uses.

1. Ultrasound waves are used to detect objects.

2. Sound waves can produce music.

3. Sound can be recorded and reproduced.

The Big Idea Flow Chart is available on p. T9 in the **UNIT TRANSPARENCY BOOK**.

Previewing Content

SECTION

 2.1 **Sound is a wave.** pp. 37–44

1. Sound is a type of mechanical wave.

Sound is a longitudinal wave. **Vibrations** in the wave move in the same direction as the wave. Because sound is a mechanical wave, it must travel through a medium.

Humans detect sound because of vibrations in the ear caused by sound waves.

2. Sound waves vibrate particles.

As sound waves push against molecules in the medium, they compress the molecules, creating bands of high and low pressure. These bands of pressure push and pull on the surrounding air, which then pushes and pulls on the air around that and so on. This creates a sound wave travelling through the air.

Materials and Sound Speeds		
Medium	**State**	**Speed of Sound**
Air (20°C)	Gas	344 m/s (769 mi/h)
Water (20°C)	Liquid	1,400 m/s (3,130 mi/h)
Steel (20°C)	Solid	5,000 m/s (11,200 mi/h)

Temperature and Sound Speeds		
Medium	**Temperature**	**Speed of Sound**
Air	0°C (32°F)	331 m/s (741 mi/h)
Air	100°C (212°F)	386 m/s (864 mi/h)

Sound waves can travel through mediums that are made up of particles, but sound waves cannot travel through a vacuum.

3. The speed of sound depends on its medium.

The speed of sound depends on the state and the temperature of the medium.
- Sound usually travels most quickly through a solid and most slowly through a gas.
- Sound travels more quickly through a specific medium at higher temperatures.

SECTION

2.2 **Frequency determines pitch.** pp. 45–51

1. Pitch depends on the frequency of a sound wave.

Pitch is an indication of how high or how low a sound is. A high-frequency wave has a short wavelength and produces a high pitch. A low-frequency wave has a long wavelength and produces a low pitch.

Hertz is the unit used to measure frequency and therefore pitch. One hertz is one wavelength per second.

All objects have a natural frequency of vibrations. When a sound wave is produced that matches an object's natural frequency, its waves combine to create sound with a larger amplitude. This increase is **resonance.**

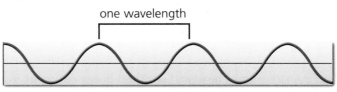

Low-frequency, low-pitched sound wave

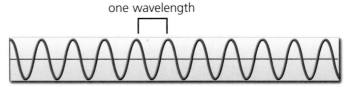

High-frequency, high-pitched sound wave

Timbre, or sound quality, is affected by
- the combination of waves produced by an object and
- how the sound starts and stops.

2. The motion of the source of a sound affects its pitch.

The **Doppler effect** is a change in pitch that occurs because the source or receiver of a sound is moving. Because the sound source is a little closer to the receiver each time it vibrates, it takes less time for the compression to reach the receiver. The decrease in distance makes the wavelength shorter, and the pitch rises.

Common Misconceptions

TYPES OF WAVES Students might think that all waves are the same. They should be aware that waves are classified according to whether they need a medium to travel or not. Electromagnetic waves do not need a medium, and mechanical waves do.

 This misconception is addressed on p. 38.

MISCONCEPTION DATABASE
CLASSZONE.COM Background on student misconceptions

SPEED OF SOUND The phrase "the speed of sound" might lead students to assume that sound always travels at the same speed. Mechanical waves, such as sound, travel at a speed that depends on the nature and temperature of the medium.

 This misconception is addressed on p. 42.

Previewing Content

SECTION

 2.3 **Intensity determines loudness.**
pp. 52–57

1. Intensity depends on the amplitude of a sound wave.
Intensity is the amount of energy a sound wave has, measured in **decibels** (dB). Low-intensity sound waves are heard as quiet sounds. Louder sounds are produced by high-intensity sound waves.

2. The intensity of sound can be controlled.
Changing the amount of energy in a sound wave changes the sound's intensity without changing its pitch or quality.
- A muffler decreases intensity.
- An **amplifier** increases intensity.

3. Intense sound can damage hearing.
The hair cells in the cochlea in the ear are easily damaged by loud sounds. Long-term exposure to sounds of 90 dB or more can damage human hearing. Even short bursts of very intense sound can deafen a person.

SECTION

 2.4 **Sound has many uses.** pp. 58–65

1. Ultrasound waves are used to detect objects.
Reflected ultrasound waves are used to detect the presence and location of objects.
- Some animals, such as bats, use **echolocation,** which involves sending out ultrasound signals and interpreting the returning sound echoes.
- Humans use **sonar,** a form of echolocation, to locate objects underwater.
- In medicine, **ultrasound** is used to treat stones that form in the human body and to scan internal organs.

2. Sound waves can produce music.
Noise is sound with no pattern. Music is sound with clear patterns of pitch and rhythm. Stringed, wind, and percussion instruments all produce vibrations in different ways, which accounts for their distinctive timbres.

3. Sound can be recorded and reproduced.
Vibrations can be changed to other types of signals or stored as reproducible information.
- Some methods of communication, such as the telephone, change sound waves into electrical signals. These signals travel to a receiver that changes them back into sound.
- Sound can be recorded as physical grooves (records) or pits (CDs) or as magnetic information (tapes) that can be changed back to sound waves.

Common Misconceptions

HEARING DAMAGE Students commonly assume that any hearing damage resulting from loud noises is temporary. Noises greater than 90 dB can cause permanent hearing damage.

T E This misconception is addressed on p. 55.

 MISCONCEPTION DATABASE
CLASSZONE.COM Background on student misconceptions

Previewing Labs

Lab Generator CD-ROM
Edit these Pupil Edition labs and generate alternative labs.

EXPLORE the BIG idea

What Gives a Sound Its Qualities? p. 35
Students will explore sound to discover its properties.

TIME 5 minutes
MATERIALS table

How Does Size Affect Sound? p. 35
Students test three nails to find that the size of a sample affects its pitch.

TIME 10 minutes
MATERIALS 3 different-sized nails of the same material, string (at least 1 m), scissors, metal spoon

---

Internet Activity: Sound, p. 35
Students discover how air particles move as sound travels through air.

TIME 20 minutes
MATERIALS computer with Internet access

SECTION 2.1

EXPLORE Sound, p. 37
Students listen to sounds traveling via string to understand that sound travels through a medium.

TIME 10 minutes
MATERIALS 75 cm of string; large, metal spoon

INVESTIGATE Sound Energy, p. 41
Students observe that sound waves move salt, to find out that sound transfers energy.

TIME 10 minutes
MATERIALS clean jar, a pinch of table salt, balloon, scissors, rubber band, pencil with good eraser end

---

SECTION 2.2

EXPLORE Pitch, p. 45
Students observe sounds to find out what affects pitch.

TIME 5 minutes
MATERIALS ruler

INVESTIGATE Sound Frequency, p. 48
Students listen to rubber bands at differing tensions and infer the relationship between frequency and pitch.

TIME 20 minutes
MATERIALS two rubber bands of different sizes; small, open box; 16 cm tape

---

SECTION 2.3

INVESTIGATE Loudness, p. 53
Students pluck a rubber band to observe the relationship between amplitude and loudness.

TIME 15 minutes
MATERIALS piece of cardboard, 25 cm long; scissors; large rubber band; two pencils; ruler

---

SECTION 2.4

EXPLORE Echoes, p. 58
Students use sound to detect an object.

TIME 10 minutes
MATERIALS two cardboard tubes, 10 cm tape, book

CHAPTER INVESTIGATION
Build a Stringed Instrument, pp. 64–65
Students make a simple stringed instrument and find out how it can produce different pitches.

TIME 40 minutes
MATERIALS book, 3–5 rubber bands, two pencils, ruler, shoebox, scissors

---

R **Additional INVESTIGATION,** Exploring Resonance, A, B, & C, pp. 132–140; Teacher Instructions, pp. 284–285

Previewing Chapter Resources

	INTEGRATED TECHNOLOGY	LABS AND ACTIVITIES

CHAPTER 2
Sound

 CLASSZONE.COM
- eEdition Plus
- EasyPlanner Plus
- Misconception Database
- Content Review
- Test Practice
- Visualizations
- Resource Centers
- Internet Activity: Sound
- Math Tutorial

 SCILINKS.ORG
 SCI LINKS

 CD-ROMS
- eEdition
- EasyPlanner
- Power Presentations
- Content Review
- Lab Generator
- Test Generator

 AUDIO CDS
- Audio Readings
- Audio Readings in Spanish

EXPLORE the Big Idea, p. 35
- What Gives a Sound Its Qualities?
- How Does Size Affect Sound?
- Internet Activity: Sound

UNIT RESOURCE BOOK
Unit Projects, pp. 5–10

 Lab Generator CD-ROM
Generate customized labs.

SECTION
2.1 Sound is a wave.
pp. 37–44

Time: 2 periods (1 block)
 Lesson Plan, pp. 73–74

 RESOURCE CENTER, Supersonic Aircraft

 UNIT TRANSPARENCY BOOK
- Big Idea Flow Chart, p. T9
- Daily Vocabulary Scaffolding, p. T10
- Note-Taking Model, p. T11
- 3-Minute Warm-Up, p. T12

 • EXPLORE Sound, p. 37
• INVESTIGATE Sound Energy, p. 41

 UNIT RESOURCE BOOK
Datasheet, Sound Energy, p. 82

SECTION
2.2 Frequency determines pitch.
pp. 45–51

Time: 2 periods (1 block)
 Lesson Plan, pp. 84–85

 VISUALIZATION, Doppler Effect

 UNIT TRANSPARENCY BOOK
- Daily Vocabulary Scaffolding, p. T10
- 3-Minute Warm-Up, p. T12
- "Sound Frequencies Heard by Animals" Visual, p. T14

 • EXPLORE Pitch, p. 45
• INVESTIGATE Sound Frequency, p. 48

 UNIT RESOURCE BOOK
- Datasheet, Sound Frequency, p. 93
- Additional INVESTIGATION, Exploring Resonance, A, B, & C, pp. 132–140

SECTION
2.3 Intensity determines loudness.
pp. 52–57

Time: 2 periods (1 block)
 Lesson Plan, pp. 95–96

 RESOURCE CENTER, Sound Safety
MATH TUTORIAL

 UNIT TRANSPARENCY BOOK
- Daily Vocabulary Scaffolding, p. T10
- 3-Minute Warm-Up, p. T13

 • INVESTIGATE Loudness, p. 53
• Math in Science, p. 57

 UNIT RESOURCE BOOK
- Datasheet, Loudness, p. 104
- Math Support, p. 121
- Math Practice, p. 122

SECTION
2.4 Sound has many uses.
pp. 58–65

Time: 4 periods (2 blocks)
 Lesson Plan, pp. 106–107

 RESOURCE CENTER, Musical Instruments

 UNIT TRANSPARENCY BOOK
- Big Idea Flow Chart, p. T9
- Daily Vocabulary Scaffolding, p. T10
- 3-Minute Warm-Up, p. T13
- Chapter Outline, pp. T15–16

 • EXPLORE Echoes, p. 58
• CHAPTER INVESTIGATION, Build a Stringed Instrument, pp. 64–65

 UNIT RESOURCE BOOK
CHAPTER INVESTIGATION, Build a Stringed Instrument, A, B, & C, pp. 123–131

READING AND REINFORCEMENT

- Description Wheel, B20–21
- Outline, C43
- Daily Vocabulary Scaffolding, H1–8

 UNIT RESOURCE BOOK
- Vocabulary Practice, pp. 118–119
- Decoding Support, p. 120
- Summarizing the Chapter, pp. 141–142

 Audio Readings CD
Listen to Pupil Edition.

 Audio Readings in Spanish CD
Listen to Pupil Edition in Spanish

 UNIT RESOURCE BOOK
- Reading Study Guide, A & B, pp. 75–78
- Spanish Reading Study Guide, pp. 79–80
- Challenge and Extension, p. 81
- Reinforcing Key Concepts, p. 83

 UNIT RESOURCE BOOK
- Reading Study Guide, A & B, pp. 86–89
- Spanish Reading Study Guide, pp. 90–91
- Challenge and Extension, p. 92
- Reinforcing Key Concepts, p. 94
- Challenge Reading, pp. 116–117

UNIT RESOURCE BOOK
- Reading Study Guide, A & B, pp. 97–100
- Spanish Reading Study Guide, pp. 101–102
- Challenge and Extension, p. 103
- Reinforcing Key Concepts, p. 105

 UNIT RESOURCE BOOK
- Reading Study Guide, A & B, pp. 108–111
- Spanish Reading Study Guide, pp. 112–113
- Challenge and Extension, p. 114
- Reinforcing Key Concepts, p. 115

ASSESSMENT

- Chapter Review, pp. 67–68
- Standardized Test Practice, p. 69

 UNIT ASSESSMENT BOOK
- Diagnostic Test, pp. 20–21
- Chapter Test, A, B, & C, pp. 26–37
- Alternative Assessment, pp. 38–39

 Spanish Chapter Test, pp. 285–288

 Test Generator CD-ROM
Generate customized tests.

 Lab Generator CD-ROM
Rubrics for Labs

 Ongoing Assessment, pp. 37–43

 Section 2.1 Review, p. 43

A **UNIT ASSESSMENT BOOK**
Section 2.1 Quiz, p. 22

 Ongoing Assessment, pp. 45–51

 Section 2.2 Review, p. 51

A **UNIT ASSESSMENT BOOK**
Section 2.2 Quiz, p. 23

 Ongoing Assessment, pp. 52–56

 Section 2.3 Review, p. 56

A **UNIT ASSESSMENT BOOK**
Section 2.3 Quiz, p. 24

 Ongoing Assessment, pp. 58, 60–63

 Section 2.4 Review, p. 63

 UNIT ASSESSMENT BOOK
Section 2.4 Quiz, p. 25

STANDARDS

National Standards
A.1–8, A.9.a–g, E.2–5, F.5.a–d

See p. 34 for the standards.

National Standards
A.2–8, A.9.a–c, A.9.e–f, F.5.d

National Standards
A.2–8, A.9.a–f, F.5.b

National Standards
A.2–8, A.9.a–f, F.5.a, F.5.c

National Standards
A.1–7, A.9.a–b, A.9.d–g, E.2–5, F.5.a–c

Previewing Resources for Differentiated Instruction

CHAPTER INVESTIGATION

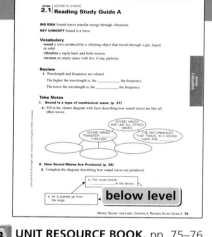

CHAPTER INVESTIGATION A
2 Build a Stringed Instrument

PURPOSE
In this lab, you will:
- make a simple instrument with strings and see how it makes sounds
- change the design to change the sound of your instrument

MATERIALS
- book
- 3–5 rubber bands
- 2 pencils
- ruler
- shoe box
- scissors

EXTRA MATERIALS
- stapler
- cardboard

Problem
How does the sound made by a plucked string change when the length of the string is changed?

Hypothesis
After completing step 3, complete the sentence below. If the length of the string on a stringed instrument is increased, then the sound will (circle one) be higher/be lower/stay the same because

Procedure
Check off each step as you do it.
□ Build a simple stringed instrument.
  - Stretch a rubber band around a textbook. The rubber band is the string.
  - Place two pencils under the rubber band. These pencils are the bridges.
□ Adjust the bridges. Play the instrument.

a. Move the bridges so they are far apart, at either end of the book
b. Use a ruler to measure the distance between the two bridges. Record the
c. Pluck the string to make it vibrate. Watch the

below level

WAVES, SOUND, AND LIGHT, CHAPTER 2, CHAPTER INVESTIGATION A **123**

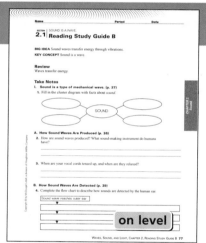

CHAPTER INVESTIGATION B
2 Build a Stringed Instrument

OVERVIEW AND PURPOSE
People make music by plucking strings, blowing through tubes, and striking things. Part of each musical instrument vibrates to produce sounds that form the building blocks of music. In this lab, you will see what you have learned about sound to
- make a simple stringed instrument and see how the vibrating string produces sounds and
- change the design so that your stringed instrument produces more than one pitch.

MATERIALS
- book
- 3–5 rubber bands
- two pencils
- ruler
- shoebox
- scissors

Problem
How does the length of a string affect the pitch of the sound it produces when plucked?

Hypothesize
Write a hypothesis to explain how changing the length of the string affects the pitch of sound that is produced. Your hypothesis should take the form of an "If ..., then ..., because ..." statement. Complete steps 1–3 before writing your hypothesis.

Procedure
1. Try out the following idea for a simple stringed instrument. Stretch a rubber band around a textbook. Put two pencils under the rubber band to serve as bridges.

2. Put the bridges far apart at either end of the book. Find the string length by measuring the distance between the two bridges. Record this measurement. Pluck the rubber band to make it vibrate. Watch it vibrate and listen to the sound it makes.
   Simple instrument: initial string length _____

on level

WAVES, SOUND, AND LIGHT, CHAPTER 2, CHAPTER INVESTIGATION B **127**

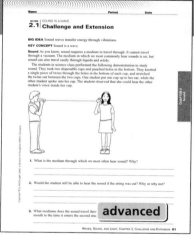

CHAPTER INVESTIGATION C
2 Build a Stringed Instrument

Challenge
Stringed instruments vary the pitch of musical sounds in several other ways. In addition to the length of the string, pitch depends on the tension, weight, and thickness of the string. Design an experiment to test one of these variables. How does it alter the range of sounds produced by your stringed instrument?

My experiment changes (circle one) tension / weight / thickness

Describe the steps of your experiment:

Record your results.

String Modification (Record weight, tension, or thickness of each trial)	Predicted Effect on Pitch	Actual Effect on Pitch

- What conclusion can you draw about how your variable affected pitch?

advanced

WAVES, SOUND, AND LIGHT, CHAPTER 2, CHAPTER INVESTIGATION C **131**

R **UNIT RESOURCE BOOK,** pp. 123–126

R pp. 127–130

R pp. 127–131

Leveled resources present the same concepts for different abilities.

READING STUDY GUIDE

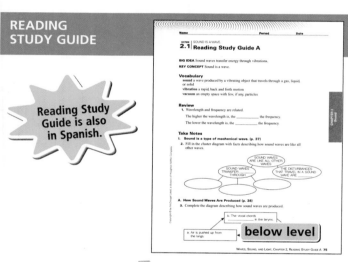

SOUND IS A WAVE
2.1 Reading Study Guide A

BIG IDEA Sound waves transfer energy through vibrations.

KEY CONCEPT Sound is a wave.

Vocabulary
sound a wave produced by a vibrating object that travels through a gas, liquid, or solid
vibration a rapid, back and forth motion
vacuum an empty space with few, if any, particles

Review
1. Wavelength and frequency are related.
   The higher the wavelength is, the _____ the frequency.
   The lower the wavelength is, the _____ the frequency

Take Notes
I. Sound is a type of mechanical wave. (p. 37)
1. Fill in the cluster diagram with facts describing how sound waves are like all other waves.

A. How Sound Waves Are Produced (p. 38)
2. Complete the diagram describing how sound waves are produced.

a. Air is pushed up from the lungs.
b. The vocal chords _____ in the larynx.

below level

WAVES, SOUND, AND LIGHT, CHAPTER 2, READING STUDY GUIDE A **75**

SOUND IS A WAVE
2.1 Reading Study Guide B

BIG IDEA Sound waves transfer energy through vibrations.

KEY CONCEPT Sound is a wave.

Review
Waves transfer energy.

Take Notes
I. Sound is a type of mechanical wave. (p. 37)
1. Fill in the cluster diagram with facts about sound.

SOUND

A. How Sound Waves Are Produced (p. 38)
2. How are sound waves produced? What sound-making instrument do humans have?

3. When are your vocal cords tensed up, and when are they relaxed?

B. How Sound Waves Are Detected (p. 39)
4. Complete the flow chart to describe how sounds are detected by the human ear.

Sound wave reaches outer ear

on level

WAVES, SOUND, AND LIGHT, CHAPTER 2, READING STUDY GUIDE B **77**

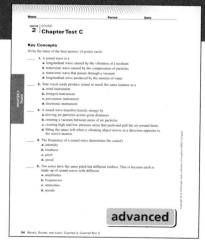

SOUND IS A WAVE
2.1 Challenge and Extension

BIG IDEA Sound waves transfer energy through vibrations.

KEY CONCEPT Sound is a wave.

Sound As you know, sound requires a medium to travel through. It cannot travel through a vacuum. The medium in which we most commonly hear sounds is air, but sound can also travel easily through liquids and solids.
The students in science class performed the following demonstration to study sound. They took two disposable cups and punched holes in the bottom. They knotted a single piece of twine through the holes in the bottom of each cup, and stretched the twine out between the two cups. One student put one cup up to her ear, while the other student spoke into his cup. The student observed that she could hear the other student's voice inside her cup.

1. What is the medium through which we most often hear sound? Why?

2. Would the student still be able to hear the sound if the string was cut? Why or why not?

3. What mediums does the sound travel through from the mouth to the time it enters the second stu...

advanced

WAVES, SOUND, AND LIGHT, CHAPTER 2, CHALLENGE AND EXTENSION **81**

R **UNIT RESOURCE BOOK,** pp. 75–76

R pp. 77–78

R p. 81

Reading Study Guide is also in Spanish.

CHAPTER TEST

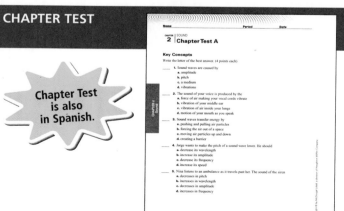

SOUND
2 Chapter Test A

Key Concepts
Write the letter of the best answer. (4 points each)

____ 1. Sound waves are caused by
   a. amplitude
   b. pitch
   c. a medium
   d. vibrations

____ 2. The sound of your voice is produced by the
   a. force of air making your vocal cords vibrate
   b. vibration of your middle ear
   c. vibration of air inside your lungs
   d. motion of your mouth as you speak

____ 3. Sound waves transfer energy by
   a. pushing and pulling air particles
   b. forcing the air out of a space
   c. moving air particles up and down
   d. creating a barrier

____ 4. Jorge wants to make the pitch of a sound wave lower. He should
   a. decrease its wavelength
   b. increase its amplitude
   c. decrease its frequency
   d. increase its speed

____ 5. Nina listens to an ambulance as it travels past her. The sound of the siren
   a. decreases in pitch
   b. increases in wavelength
   c. decreases in amplitude
   d. increases in frequency

below level

28 WAVES, SOUND, AND LIGHT, CHAPTER 2, CHAPTER TEST A

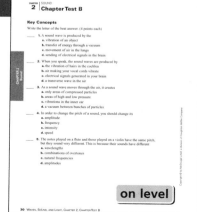

SOUND
2 Chapter Test B

Key Concepts
Write the letter of the best answer. (4 points each)

____ 1. A sound wave is produced by the
   a. vibration of an object
   b. transfer of energy through a vacuum
   c. movement of air in the lungs
   d. sending of electrical signals in the brain

____ 2. When you speak, the sound waves are produced by
   a. the vibration of hairs in the cochlea
   b. air making your vocal cords vibrate
   c. electrical signals generated in your brain
   d. a transverse wave in the air

____ 3. As a sound wave moves through the air, it creates
   a. only areas of compressed particles
   b. areas of high and low pressure
   c. vibrations in the inner ear
   d. a vacuum between bunches of particles

____ 4. In order to change the pitch of a sound, you should change its
   a. amplitude
   b. frequency
   c. intensity
   d. speed

____ 5. The notes played on a flute and those played on a violin have the same pitch, but they sound very different. This is because their sounds have different
   a. wavelengths
   b. combinations of overtones
   c. natural frequencies
   d. amplitudes

on level

30 WAVES, SOUND, AND LIGHT, CHAPTER 2, CHAPTER TEST B

SOUND
2 Chapter Test C

Key Concepts
Write the letter of the best answer. (4 points each)

____ 1. A sound wave is a
   a. longitudinal wave caused by the vibration of a medium
   b. transverse wave caused by the compression of particles
   c. transverse wave that passes through a vacuum
   d. longitudinal wave produced by the motion of water

____ 2. Your vocal cords produce sound in much the same manner as a
   a. wind instrument
   b. stringed instrument
   c. percussion instrument
   d. electronic instrument

____ 3. A sound wave transfers kinetic energy by
   a. moving air particles across great distances
   b. creating a vacuum between areas of air particles
   c. creating high and low pressure areas that push and pull the air around them
   d. filling the space left when a vibrating object moves in a direction opposite to the wave's motion

____ 4. The frequency of a sound wave determines the sound's
   a. intensity
   b. loudness
   c. pitch
   d. speed

____ 5. Two notes have the same pitch but different timbres. This is because each is made up of sound waves with different
   a. amplitudes
   b. frequencies
   c. intensities
   d. speeds

advanced

34 WAVES, SOUND, AND LIGHT, CHAPTER 2, CHAPTER TEST C

A **UNIT ASSESSMENT BOOK,** pp. 26–29

A pp. 30–33

A pp. 34–37

Chapter Test is also in Spanish.

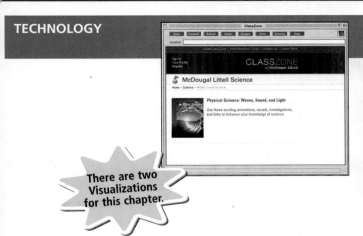

There are two Visualizations for this chapter.

 CLASSZONE.COM

 CD/CD-ROMS

 CLASSZONE.COM

VISUAL CONTENT

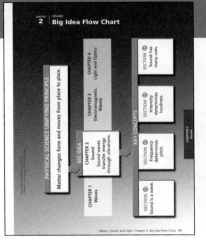

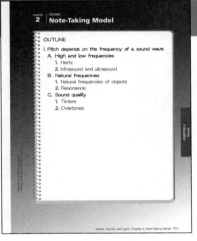

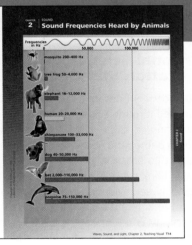

T **UNIT TRANSPARENCY BOOK,** p. T9

T p. T11

T p. T14

MORE SUPPORT

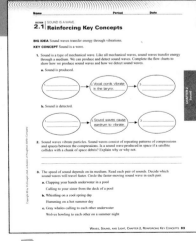

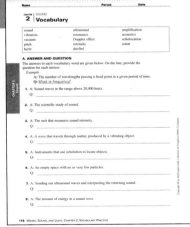

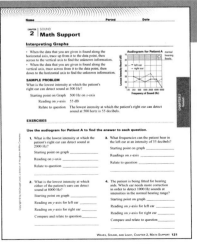

Reinforcing Key Concepts for each section

R **UNIT RESOURCE BOOK,** p. 83

R pp. 118–119

R p. 121

CHAPTER

2 Sound

INTRODUCE

the **BIG** idea

Have students look at the photograph of the guitar player and discuss how the question in the box links to the Big Idea:

- What role do guitar strings play in producing vibrations?
- What role does the body of the guitar play in producing sound?
- What part of the human body receives the vibrations from the guitar?
- How are the vibrations from the guitar perceived as sound?

National Science Education Standards

A.1–8 Identify questions that can be answered through scientific investigations; design and conduct an investigation; use tools to gather and interpret data; use evidence to describe, predict, explain, model; think critically to make relationships between evidence and explanation; recognize different explanations and predictions; communicate scientific procedures and explanations; use mathematics.

A.9.a–g Understand scientific inquiry by using different investigations, methods, mathematics, technology, and explanations based on logic, evidence, and skepticism. Data often results in new investigations.

E.2-5 Design, implement, evaluate a solution or product; communicate technological design.

F.5.a–d Science and technology in society

the **BIG** idea

Sound waves transfer energy through vibrations.

Key Concepts

SECTION
2.1 Sound is a wave. Learn how sound waves are produced and detected.

SECTION
2.2 Frequency determines pitch. Learn about the relationship between the frequency of a sound wave and its pitch.

SECTION
2.3 Intensity determines loudness. Learn how the energy of a sound wave relates to its loudness.

SECTION
2.4 Sound has many uses. Learn how sound waves are used to detect objects and to make music.

Internet Preview

CLASSZONE.COM
Chapter 2 online resources: Content Review, two Visualizations, three Resource Centers, Math Tutorial, Test Practice

D 34 Unit: Waves, Sound, and Light

How is this guitar player producing sound?

INTERNET PREVIEW

CLASSZONE.COM For student use with the following pages:

Review and Practice
- Content Review, pp. 36, 66
- Math Tutorial: Interpreting Line Graphs, p. 57
- Test Practice, p. 69

Activities and Resources
- Internet Activity, p. 35
- Visualization: Doppler Effect, p. 51
- Resource Centers: Supersonic Aircraft, p. 44; Sound Safety, p. 56; Musical Instruments, p. 60

NSTA SC*L*INKS
scilinks.org
What is Sound?
Code: MDL028

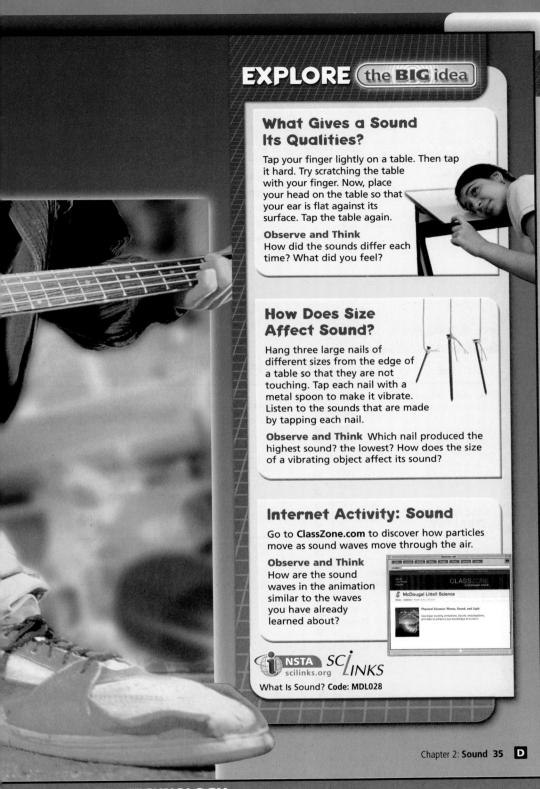

EXPLORE (the BIG idea)

What Gives a Sound Its Qualities?

Tap your finger lightly on a table. Then tap it hard. Try scratching the table with your finger. Now, place your head on the table so that your ear is flat against its surface. Tap the table again.

Observe and Think
How did the sounds differ each time? What did you feel?

How Does Size Affect Sound?

Hang three large nails of different sizes from the edge of a table so that they are not touching. Tap each nail with a metal spoon to make it vibrate. Listen to the sounds that are made by tapping each nail.

Observe and Think Which nail produced the highest sound? the lowest? How does the size of a vibrating object affect its sound?

Internet Activity: Sound

Go to **ClassZone.com** to discover how particles move as sound waves move through the air.

Observe and Think
How are the sound waves in the animation similar to the waves you have already learned about?

NSTA
scilinks.org
SCiLINKS
What Is Sound? Code: MDL028

TEACHING WITH TECHNOLOGY

Tape Recorder Use a tape recorder to record the sounds of the stringed instruments the students build in the Chapter Investigation on pp. 64–65. Have students listen to the sounds so they can hear the differences in pitch between the different instrument designs.

CBL and Probeware If students have probeware, you might encourage them to use a microphone during activities and demonstrations throughout the chapter.

EXPLORE (the BIG idea)

These inquiry-based activities are appropriate for use at home or as a supplement to classroom instruction.

What Gives a Sound Its Qualities?

PURPOSE To introduce students to sound properties such as loudness, pitch, and general sound quality.

TIP *10 min.* Adapt the activity for students who might be hearing impaired. Depending on the degree of impairment, students might tap with an object that makes more noise than their fingernail.

Answer: The sounds each had a different quality. Students should feel the vibrations in the table.

REVISIT after p. 49.

How Does Size Affect Sound?

PURPOSE To show that the size of a nail affects the pitch produced by striking it. Students observe that the size and shape of an object affects the frequency of the vibration produced by the object.

TIP *10 min.* All nails must be made of the same material, so that comparison involves only the size and not the type of material.

Answer: The smallest nail produced the highest sound. The largest nail produced the lowest sound. The size of an object affects how high or low its sound is.

REVISIT after p. 61.

Internet Activity: Sound

PURPOSE To help students visualize sound moving through air as longitudinal waves.

TIP *20 min.* Ask students to predict the difference in the sound waves produced by a jet engine and those produced by a whisper. Have students keep their predictions for evaluation at the end of Section 3.

Answer: Accept any of the following: The sound waves in the animation have amplitude, frequency, and wavelength; transfer energy; are longitudinal waves.

REVISIT after p. 41.

CONCEPT REVIEW

Activate Prior Knowledge

- Ask students to list examples of how waves transfer energy from one place to another. Examples include electromagnetic waves from the Sun warming Earth, and water waves moving a boat.
- Ask students to explain what a medium is.
- Review the concept of matter and that matter is present in a medium.

TAKING NOTES

Outline

Outlining the chapter will help students pull together their ideas about sound and waves. Using the headings as a skeleton helps them identify the important points in the chapter.

Vocabulary Strategy

Students can use the description wheels not only to describe terms but also to differentiate between them.

Vocabulary and Note-Taking Resources

- Vocabulary Practice, pp. 118–119
- Decoding Support, p. 120

- Daily Vocabulary Scaffolding, p. T10
- Note-Taking Model, p. T11

- Description Wheel, B20–21
- Outline, C43
- Daily Vocabulary Scaffolding, H1–8

CHAPTER 2
Getting Ready to Learn

CONCEPT REVIEW

- A wave is a disturbance that transfers energy from one place to another.
- Mechanical waves are waves that travel through matter.

VOCABULARY REVIEW

medium p. 11
longitudinal wave p. 14
amplitude p. 17
wavelength p. 17
frequency p. 17

CONTENT REVIEW
CLASSZONE.COM
Review concepts and vocabulary.

TAKING NOTES

OUTLINE

As you read, copy the headings on your paper in the form of an outline. Then add notes in your own words that summarize what you have read.

VOCABULARY STRATEGY

Place each vocabulary term at the center of a **description wheel** diagram. Write some words on the spokes describing it.

See the Note-Taking Handbook on pages R45–R51.

SCIENCE NOTEBOOK

I. Sound is a type of mechanical wave
  A. How sound waves are produced
    1.
    2.
    3.
  B. How sound waves are detected
    1.
    2.
    3.

rapid back-and-forth motion / can produce a sound / **VIBRATION** / usually too small to see / can make with vocal cords

---

CHECK READINESS

Administer the Diagnostic Test to determine students' readiness for new science content and their mastery of requisite math skills.

 Diagnostic Test, pp. 20–21

Technology Resources

Students needing content and math skills should visit **ClassZone.com**.

- CONTENT REVIEW
- MATH TUTORIAL
- CONTENT REVIEW CD-ROM

2.1 Sound is a wave.

◀ **BEFORE, you learned**
- Waves transfer energy
- Waves have wavelength, amplitude, and frequency

▶ **NOW, you will learn**
- How sound waves are produced and detected
- How sound waves transfer energy
- What affects the speed of sound waves

VOCABULARY

sound p. 37
vibration p. 37
vacuum p. 41

EXPLORE Sound

What is sound?

PROCEDURE

1. Tie the middle of the string to the spoon handle.

2. Wrap the string ends around your left and right index fingers. Put the tips of these fingers gently in your ears and hold them there.

3. Stand over your desk so that the spoon dangles without touching your body or the desk. Then move a little to make the spoon tap the desk lightly. Listen to the sound.

WHAT DO YOU THINK?
- What did you hear when the spoon tapped the desk?
- How did sound travel from the spoon to your ears?

MATERIALS
- piece of string
- large metal spoon

OUTLINE
Start an outline for this heading. Remember to leave room for details.

I. Main idea
  A. Supporting idea
    1. Detail
    2. Detail
  B. Supporting idea

Sound is a type of mechanical wave.

In the last chapter, you read that a mechanical wave travels through a material medium. Such mediums include air, water, and solid materials. Sound is an example of a mechanical wave. **Sound** is a wave that is produced by a vibrating object and travels through matter.

The disturbances that travel in a sound wave are vibrations. A **vibration** is a rapid, back-and-forth motion. Because the medium vibrates back and forth in the same direction as the wave travels, sound is a longitudinal wave. Like all mechanical waves, sound waves transfer energy through a medium.

CHECK YOUR READING What do sound waves have in common with other mechanical waves? Your answer should include the word *energy*.

RESOURCES FOR DIFFERENTIATED INSTRUCTION

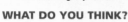

Below Level

UNIT RESOURCE BOOK
- Reading Study Guide A, pp. 75–76
- Decoding Support, p. 120

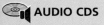 **AUDIO CDS**

Advanced

UNIT RESOURCE BOOK
Challenge and Extension, p. 81

English Learners

UNIT RESOURCE BOOK
Spanish Reading Study Guide, pp. 79–80

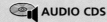 **AUDIO CDS**

- Audio Readings in Spanish
- Audio Readings (English)

2.1 FOCUS

▶ **Set Learning Goals**

Students will
- Explain how sound waves are produced and detected.
- Explain how sound waves transfer energy.
- Describe what affects the speed of sound waves.
- Observe through experimentation that sound transfers energy.

◀ **3-Minute Warm-Up**

Display Transparency 12 or copy this exercise on the board:

In each situation below, what might happen to show that the waves transfer energy?

1. a seaside beach in a storm *waves in the water eroding the beach*

2. an earthquake *waves in the earth shaking buildings*

3. movement of coils *a wave travelling from one end to the other*

T 3-Minute Warm-Up, p. T12

2.1 MOTIVATE

EXPLORE Sound

PURPOSE To introduce the concept that sound travels through a medium

TIP *10 min.* If metal spoons are not available, use another metal object, such as a metal coat hanger.

WHAT DO YOU THINK? *A loud, ringing sound; the sound traveled through the string*

Ongoing Assessment

CHECK YOUR READING *Answer: Sound waves transfer energy.*

2.1 INSTRUCT

Address Misconceptions

IDENTIFY Ask: Do all waves have the same traits, or are there different types of waves? If students answer that all waves have the same traits, they hold the misconception that all waves are the same.

CORRECT Place a small amount of water in a flask. Attach a bell to a solid stopper with a wire and a thumbtack, so that it will hang in the middle of the flask when you insert the stopper. Shake the flask and listen to the bell. Shine a light through the flask against a dark surface.

Remove the stopper and heat the water to boiling. Immediately replace the stopper and allow the flask to cool. Shake the flask again, and note that the sound is much less. Shine a light through the flask again, and note that the intensity of the light is the same.

REASSESS Ask students to explain why sound decreased in intensity and light did not. *Sound is a mechanical wave and needs a medium. Heating the water causes the gases in the flask to expand. Cooling creates a partial vacuum, decreasing the amount of medium. Light is an electromagnetic wave that does not need a medium, so the partial vacuum does not affect its intensity.*

Technology Resources

Visit **ClassZone.com** for background on common student misconceptions.

 MISCONCEPTION DATABASE

Ongoing Assessment

CHECK YOUR READING *Answer: The vocal cords vibrate, producing sound waves.*

READING VISUALS *Answer: The vibrations start when air from the lungs passes through the vocal cords.*

How Sound Waves Are Produced

READING TiP
When you see the words *push* or *pull,* think of force.

The disturbances in a sound wave are vibrations that are usually too small to see. Vibrations are also required to start sound waves. A vibrating object pushes and pulls on the medium around it and sends out waves in all directions.

You have a sound-making instrument within your own body. It is the set of vocal cords within the voice box, or larynx, in your throat. Put several of your fingers against the front of your throat. Now hum. Do you feel the vibrations of your vocal cords?

Your vocal cords relax when you breathe to allow air to pass in and out of your windpipe. Your vocal cords tense up and draw close together when you are about to speak or sing. The illustration below shows how sound waves are produced by the human vocal cords.

1. Your muscles push air up from your lungs and through the narrow opening between the vocal cords.

2. The force of the air causes the vocal cords to vibrate.

3. The vibrating vocal cords produce sound waves.

CHECK YOUR READING How do human vocal cords produce sound waves?

How Vocal Cords Produce Sound

Sound waves are produced by vibrations.

Sound waves are produced.

The **vocal cords** vibrate in the larynx.

enlargement of vocal cords

Air is pushed up from the lungs.

READING VISUALS What starts the vibrations in the vocal cords?

D 38 Unit: Waves, Sound, and Light

DIFFERENTIATE INSTRUCTION

? More Reading Support

A What makes sounds in your throat? *the vibrations of vocal cords*

English Learners Students new to English may be confused about what is required in review items, such as those on p. 43. For instructions such as "describe and explain," tell students exactly what each direction asks them to do.

• *describe:* Write as if drawing a picture. Give physical details.
• *explain:* Write as if listing steps in a process.

Students can practice both types of answers with simple questions, such as, "Describe and explain what happens in your classroom." It is also helpful to provide model answers.

How Sound Waves Are Detected

The shape of a human ear helps it collect sound waves. Picture a satellite dish. It collects radio waves from satellites. Your ear works in much the same way. Actually, what we typically call the ear is only the outer section of the ear. The illustration below shows the main parts of the human ear.

① Your outer ear collects sound waves and reflects them into a tiny tube called the ear canal. At the end of the ear canal is a thin, skin-like membrane stretched tightly over the opening, called the eardrum. When sound waves strike the eardrum, they make it vibrate.

② The middle ear contains three tiny, connected bones called the hammer, anvil, and stirrup. These bones carry vibrations from the eardrum to the inner ear.

③ One of the main parts of the inner ear, the cochlea, contains about 30,000 hair cells. Each of these cells has tiny hairs on its surface. The hairs bend as a result of the vibrations. This movement triggers changes that cause the cell to send electrical signals along nerves to your brain. Only when your brain receives and processes these signals do you actually hear a sound.

> **READING TiP**
> As you read each numbered description here, match it to the number on the illustration below.

How the Ear Detects Sound

Sound waves are detected in the human ear, beginning with vibrations of the eardrum.

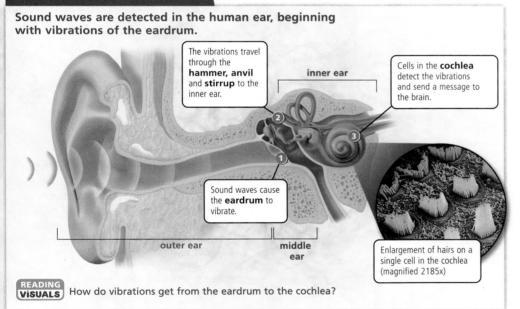

The vibrations travel through the **hammer, anvil** and **stirrup** to the inner ear.

Cells in the **cochlea** detect the vibrations and send a message to the brain.

inner ear

Sound waves cause the **eardrum** to vibrate.

outer ear

middle ear

Enlargement of hairs on a single cell in the cochlea (magnified 2185x)

> **READING ViSUALS** How do vibrations get from the eardrum to the cochlea?

Integrate the Sciences

Although the ears detect sound and change it to electrical signals, the brain interprets the signals sent to it. One important function of the brain is to sort out unneeded noise, focusing on important or unusual sounds. This filtering occurs in a network of nerve cells in the brain stem, known as the reticular activating system (RAS). The action of the RAS enables you to study with the radio on or hear the sound of an announcer over the noise of a crowd.

Teach from Visuals

To help students interpret the visual of the ear, ask:

• What causes the eardrum to vibrate? *sound waves striking the eardrum*

• What effect would an inner ear infection, when the inner ear fills up with fluid, have on hearing? What might happen? *Students may infer that sound is blocked or it changes the perception of sounds. In fact, most ear infections block some vibrations from passing through the inner ear.*

Ongoing Assessment

Explain how sound waves are produced and detected.

Ask: If two people talk, how is sound produced and detected? *Air pushes between vocal cords to produce sound. The ear detects vibrations that enter it and interprets them as sound.*

> **READING ViSUALS** *Answer: The vibrations get to the cochlea by passing through the hammer, anvil, and stirrup.*

DIFFERENTIATE INSTRUCTION

More Reading Support

B What happens when sound hits the eardrum? *The eardrum vibrates.*

Advanced

R Challenge and Extension, p. 81

Explain how sound waves transfer energy.

Ask: What happens to the particles in a medium when sound waves pass through it? *The particles vibrate.*

CHECK YOUR READING *Sample answer: Sound travels as compressions in the particles that make up the air.*

Sound waves vibrate particles.

You can see the motion of waves in water. You can even ride them with a surfboard. But you cannot see air. How, then, can you picture sound waves moving through air? Sound waves transfer the motion of particles too small to see from one place to another.

For example, think about a drum that has been struck. What happens between the time the drum is struck and the sound is heard?

- The drum skin vibrates rapidly. It pushes out and then in, over and over again. Of course, this happens very, very fast. The vibrating drum skin pushes against nearby particles in the air. The particles in the air become bunched together, or compressed.
- When the drum skin pushes the opposite way, a space opens up between the drum's surface and the particles. The particles rush back in to fill the space.
- The back-and-forth movement, or vibration, of the particles is the disturbance that travels to the listener. Both the bunched up areas, or compressions, and the spaces between the compressions are parts of the wave.

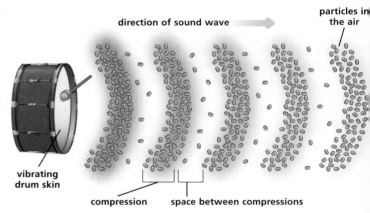

direction of sound wave

particles in the air

vibrating drum skin

compression   space between compressions

Notice that the waves consist of repeating patterns of compressions and spaces between the compressions. The compressions are areas of high air pressure. The spaces between the compressions are areas of low air pressure. The high- and low-pressure air pushes and pulls on the surrounding air, which then pushes and pulls on the air around that. Soon a sound wave has traveled through the air and has transferred kinetic energy from one place to another.

▼ **REMINDER**
Kinetic energy is the energy of motion.

CHECK YOUR READING Summarize in your own words how sound travels through air.

DIFFERENTIATE INSTRUCTION

(?) More Reading Support

C What is the name for air particles bunching up in a sound wave? *compression*

Below Level Have two students model how a sound wave travels. Have them place a spring toy on the floor and stretch it between them until it is approximately 2 meters long. Have one student squeeze together about 20 coils on his or her end. They should see that the other coils become farther apart. When they release the coils, the compression travels down the spring.

In the middle 1600s, scientists began to do experiments to learn more about air. They used pumps to force the air out of enclosed spaces to produce a vacuum. A **vacuum** is empty space. It has no particles—or very, very few of them. Robert Boyle, a British scientist, designed an experiment to find out if sound moves through a vacuum.

Boyle put a ticking clock in a sealed jar. He pumped some air out of the jar and still heard the clock ticking. Then he pumped more air out. The ticking grew quieter. Finally, when Boyle had pumped out almost all the air, he could hear no ticking at all. Boyle's experiment demonstrated that sound does not travel through a vacuum.

The photograph at the right shows equipment that is set up to perform an experiment similar to Boyle's. A bell is placed in a sealed jar and powered through the electrical connections at the top. The sound of the loudly ringing bell becomes quieter as air is pumped out through the vacuum plate.

Sound is a mechanical wave. It can move only through a medium that is made up of matter. Sound waves can travel through air, solid materials, and liquids, such as water because all of these mediums are made up of particles. Sound waves cannot travel through a vacuum.

CHECK YOUR READING How did Boyle's experiment show that sound cannot travel through a vacuum?

Sound Experiment

connections

sealed jar

bell

vacuum plate

INFER As air is pumped out of the jar, the sound of the bell becomes quieter. Why do you think the bell is suspended?

INVESTIGATE Sound Energy

How does sound transfer energy?
PROCEDURE

1. Sprinkle a few grains of salt into the jar. Put the jar on a flat surface in a well-lit place.

2. Cut off the neck of the balloon with the scissors.

3. Stretch the balloon over the mouth of the jar and pull the sides down past the rim of the jar's mouth. Use a rubber band to make a tight fit.

4. Tap the balloon with the eraser end of the pencil. Observe what happens to the salt on the bottom of the jar.

WHAT DO YOU THINK?
• What happens to the salt?
• How can you explain what you observed?

CHALLENGE Suppose you could pump all the air out of the jar and could leave the salt grains in the jar and the tight rubber cover on top. If you repeated the experiment, do you think the results would be different? Explain your answer.

SKILL FOCUS
Observing

MATERIALS
• clean jar
• table salt
• balloon
• scissors
• rubber band
• pencil with good eraser end

TIME
10 minutes

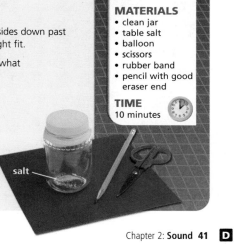

salt

Chapter 2: **Sound 41** D

EXPLORE (the BIG idea)
Revisit "Internet Activity: Sound" on p. 35. Have students explain their results.

INVESTIGATE
Sound Energy

PURPOSE To observe how sound transfers energy from one place to another

TIPS *10 min.*
• Fine sand can be used instead of salt.
• Be sure all jars are completely dry.

WHAT DO YOU THINK? *The salt vibrates. The energy of vibration has traveled from the balloon to the salt.*

CHALLENGE *The salt would not vibrate because there would be no air to transfer the wave. Another acceptable answer is that the salt would still vibrate because the sound would travel through the glass.*

 Datasheet, Sound Energy, p. 82

Technology Resources

Customize this student lab as needed or look for an alternative. Print rubrics to assess student lab reports.

Lab Generator CD-ROM

Ongoing Assessment
Observe that sound transfers energy.

Ask: Where did the energy that moved the salt come from? *from the sound waves*

CHECK YOUR READING *Answer: When air was removed from the jar, the sound was no longer heard, showing that sound cannot travel through a vacuum.*

DIFFERENTIATE INSTRUCTION

More Reading Support

D Can sound travel through a vacuum? *No, it must have a medium.*

E What makes up mediums such as air and water? *particles or matter*

Alternative Assessment Have students perform the following activity and write a paragraph or draw a diagram that answers the question.
• Seal a bowl with plastic wrap and tape the wrap to the bowl.
• Tape a rubber band to the center of the plastic wrap.
• Snap the rubber band. Tape an identical rubber band to a table, and snap it against the table. Ask: Why do the sounds differ? *The surface, as well as particles in the air, vibrate. However, the wrap vibrates much more easily, so the sound is louder.*

Address Misconceptions

IDENTIFY Ask: How fast does sound travel? If students answer with a specific speed, they may hold the misconception that sound always travels at the same speed.

CORRECT Elicit from students that sound will travel faster when particles are able to quickly pass a vibration to another particle. Have students draw molecular models of warm air and cold air. Also have them draw molecular models of solids and gases.

REASSESS Ask students to use their models to explain why sound travels more quickly at increased temperature in the same medium and more quickly in a solid than in a gas. *Particles in warm gas move more quickly than particles in a cooler gas, so they collide more often and with more energy, quickly moving the sound wave. Particles in a solid are closer than in a gas; thus, they pass vibrations more quickly from one particle to another.*

Ongoing Assessment

Describe what affects the speed of sound waves.

Ask: How does increasing temperature affect the speed of sound? *Sound travels more quickly as temperature rises.*

CHECK YOUR READING *Answer: The material and temperature of the medium affect the speed of sound.*

The speed of sound depends on its medium.

Suppose you are in the baseball stands during an exciting game. A pitch flies from the mound toward home plate, and you see the batter draw back, swing, and hit the ball high. A split second later you hear the crack of the bat meeting the ball. You notice that the sound of the hit comes later than the sight. Just how fast does sound travel?

 Sound travels more slowly than light, and it does not always travel at the same speed. Two main factors affect the speed of sound: the material that makes up the medium—such as air or water—and the temperature. If we know the medium and the temperature, however, we can predict the speed of sound.

CHECK YOUR READING Which two factors affect the speed of sound?

The Effect of the Material

You have probably heard sounds in more than one medium. Think about the medium in which you most often hear sound—air. You listen to a radio or a compact disk player. You hear the siren of a fire truck. These sound waves travel through air, a mixture of gases.

Now think about going swimming. You dip below the water's surface briefly. Someone jumps into the water nearby and splashes water against the pool wall. You hear strange underwater sounds. These sound waves travel through water, a liquid.

Sound travels faster through liquids than it does through gases because liquids are denser than gases. That means that the particles are packed closer together. It takes less time for a water particle to push on the water particles around it because the particles are already closer together than are the particles in air. As a result, divers underwater would hear a sound sooner than people above water would.

Sound can also travel through solid materials that are elastic, which means they can vibrate back and forth. In solid materials, the particles are packed even closer together than they are in liquids or gases. Steel is an example of an elastic material that is very dense. Sound travels very rapidly through steel. Look at the chart on the left. Compare the speed of sound in air with the speed of sound in steel.

These divers can hear the motor of a distant boat before their friends above water hear it.

Materials and Sound Speeds

Medium	State	Speed of Sound
Air (20°C)	Gas	344 m/s (769 mi/h)
Water (20°C)	Liquid	1,400 m/s (3,130 mi/h)
Steel (20°C)	Solid	5,000 m/s (11,200 mi/h)

DIFFERENTIATE INSTRUCTION

? More Reading Support

F Which moves faster, light waves or sound waves? *light*

G Does sound travel faster through a gas or a liquid? *a liquid*

Advanced

- Ask: what might prevent you from performing an experiment in class to measure the speed of sound in a medium? *The speed is too great to measure without specialized equipment.*

- Ask: What requirements do you think are needed for the equipment that is used to measure the speed of sound in a medium? *Answers might include the ability to detect small vibrations and measure extremely small periods of time.*

The Effect of Temperature

Sound also travels faster through a medium at higher temperatures than at lower ones. Consider the medium of air, a mixture of gases. Gas particles are not held tightly together as are particles in solids. Instead, the gas particles bounce all around. The higher the temperature, the more the gas particles wiggle and bounce. It takes less time for particles that are already moving quickly to push against the particles around them than it takes particles that are moving slowly. Sound, therefore, travels faster in hot air than in cold air.

Look at the picture of the snowboarders. The sound waves they make by yelling will travel more slowly through air than similar sounds made on a hot day. If you could bear to stand in air at a temperature of 100°C (212°F—the boiling point of water) and listen to the same person yelling, you might notice that the sound of the person's voice reaches you faster.

The chart on the right shows the speed of sound in air at two different temperatures. Compare the speed of sound at the temperature at which water freezes with the speed of sound at the temperature at which water boils. Sound travels about 17 percent faster in air at 100°C than in air at 0°C.

These snowboarders' shouts reach their friends more slowly in this cold air than they would in hot air.

Temperature and Sound Speeds

Medium	Temperature	Speed of Sound
Air	0°C (32°F)	331 m/s (741 mi/h)
Air	100°C (212°F)	386 m/s (864 mi/h)

 CHECK YOUR READING What is the difference between the speed of sound in air at 0°C and at 100°C?

2.1 Review

KEY CONCEPTS

1. Describe how sound waves are produced.
2. Describe how particles move as energy is transferred through a sound wave.
3. Explain how temperature affects the speed of sound.

CRITICAL THINKING

4. **Predict** Would the sound from a distant train travel faster through air or through steel train tracks? Explain.
5. **Evaluate** Suppose an audience watching a science fiction movie hears a loud roar as a spaceship explodes in outer space. Why is this scene unrealistic?

CHALLENGE

6. **Evaluate** A famous riddle asks this question: If a tree falls in the forest and there is no one there to hear it, is there any sound? What do you think? Give reasons for your answer.

Reinforce (the **BIG** idea)

Have students relate the section to the Big Idea.

R Reinforcing Key Concepts, p. 83

2.1 ASSESS & RETEACH

Assess

A Section 2.1 Quiz, p. 22

Reteach

Have students perform the following activity:

- Remove both the top and the bottom of a coffee can or an oatmeal box.
- Stretch a balloon tightly over one end and secure it with a rubber band.
- Glue a small mirror to the center of the balloon.
- Lay the can on its side on a desk.
- Shine a flashlight on the mirror at a 45° angle so that the reflection is on a white wall or screen.
- Yell into the open end and explain what happens. *The sound waves cause the balloon to vibrate, which causes the mirror to vibrate. These vibrations can be seen in the reflection of the flashlight on the wall or screen.*

Technology Resources

Have students visit **ClassZone.com** for reaching of Key Concepts.

 CONTENT REVIEW

 CONTENT REVIEW CD-ROM

ANSWERS

Sound waves are produced when an object vibrates.

Particles compress and then spread out as a sound wave passes through them.

the higher the temperature, the faster the speed of sound

4. The sound would travel faster through steel train tracks than through air because steel is a denser material than air.

5. Sound must travel through a material medium such as a gas, liquid, or solid. Outer space is a vacuum, and so the sound of the explosion would not be heard.

6. Accept either of these answers if well-defended: The sole criterion for sound is the production of sound waves. The sound waves must be detected by some receiver to qualify as sound.

Set Learning Goal

To find out more about supersonic aircraft and the sonic booms they produce

Present the Science

A sonic boom is caused by an extremely high-pressure wave that is produced as an object accelerates to a speed faster than its sound waves and breaks through the pressure barrier.

The pressure from a sonic boom normally causes no damage on Earth. The energy range of a sonic boom is less than that of most industrial noise. Occasionally, minor damage such as broken glass results from a sonic boom.

Discussion Questions

Ask: How is the atmosphere different at the height of an airplane flying compared to close to Earth? *Air at the height of the plane is colder and less dense.*

Ask: Based on your answer to the first question, would the speed of sound be greater or less at Earth's surface? *greater*

Close

Ask: How do the sound waves produced by an airplane compare to the waves produced by the front of a ship as it moves through water? *The waves are similar because the moving object generates them. They collect at the front of the object, and the object passes through them.*

Technology Resources

Students can visit **ClassZone.com** to find out more about supersonic aircraft.

 RESOURCE CENTER

EXTREME SCIENCE

SURPASSING THE SPEED OF SOUND

RESOURCE CENTER
CLASSZONE.COM
Find out more about supersonic aircraft.

This photograph may actually show the wake of a sonic boom. It was taken on a very humid day, and water vapor may have condensed in the low-pressure part of the sound wave.

Boom Notes

- The pilot of an airplane cannot hear the sonic boom because the sound waves are behind the plane.

- Lightning heats particles in the air so rapidly that they move faster than the speed of sound and cause a shock wave, which is what makes the boom of thunder. If a lightning strike is very close, you will hear a sharp crack.

- Large meteors enter the atmosphere fast enough to make a sonic boom.

Sonic Booms

Airplanes traveling faster than the speed of sound can produce an incredibly loud sound called a sonic boom. The sonic boom from a low-flying airplane can rattle and even break windows!

How It Works

Breaking the Barrier

The sound waves produced by this airplane begin to pile up and produce a pressure barrier.

This airplane has broken through the pressure barrier and has produced a loud boom.

When an airplane reaches extremely high speeds, it actually catches up to its own sound waves. The waves start to pile up and form a high-pressure area in front of the plane. If the airplane has enough acceleration, it breaks through the barrier, making a sonic boom. The airplane gets ahead of both the pressure barrier and the sound waves and is said to be traveling at supersonic speeds—speeds faster than the speed of sound.

Boom and It's Gone

If an airplane that produces a boom is flying very high, it may be out of sight by the time the sonic boom reaches a hearer on the ground. To make a sonic boom, a plane must be traveling faster than about 1240 kilometers per hour (769 mi/h)! The sound does not last very long—about one-tenth of a second for a small fighter plane to one-half second for a supersonic passenger plane.

EXPLORE

1. **PREDICT** Specially designed cars have traveled faster than the speed of sound. Would you expect them to produce a sonic boom?
2. **CHALLENGE** The space shuttles produce sonic booms when they are taking off and landing, but not while they are orbiting Earth, even though they are moving much faster than 1240 km/h. Can you explain why?

EXPLORE

1. **PREDICT** *Yes, a car traveling faster than the speed of sound would produce a sonic boom.*

2. **CHALLENGE** *Sonic waves would not be produced in space because there is no medium in which the sound can travel.*

2.2 Frequency determines pitch.

◀ **BEFORE,** you learned

- Sound waves are produced by vibrations
- Frequency measures the number of wavelengths passing a fixed point per second

▶ **NOW,** you will learn

- How the frequency of a wave affects the way it sounds
- How sound quality differs from pitch
- How the Doppler effect works

VOCABULARY

pitch p. 45
hertz p. 46
ultrasound p. 46
resonance p. 48
Doppler effect p. 50

EXPLORE Pitch

Why does the sound change?

PROCEDURE

1. Hold the ruler flat on the edge of a desk so that it sticks out about 25 centimeters beyond the edge.

2. With your free hand, push the tip of the ruler down and then let it go. As the ruler vibrates, slide it back onto the desk. Listen to the sounds the ruler makes.

WHAT DO YOU THINK?

- What happened to the sound as you slid the ruler back onto the desk?
- Describe the motion of the ruler.

MATERIALS
ruler

VOCABULARY
Remember to add a description wheel in your notebook for each new term.

Pitch depends on the frequency of a sound wave.

When you listen to music, you hear both high and low sounds. The quality of highness or lowness of a sound is called **pitch.** The frequency of a sound wave determines the pitch of the sound you hear. Remember that frequency is the number of wavelengths passing a fixed point in a given period of time. A high-frequency wave with short wavelengths, such as that produced by a tiny flute, makes a high-pitched sound. A low-frequency wave with long wavelengths, such as the one produced by the deep croak of a tuba, makes a low-pitched sound. An object vibrating very fast produces a high-pitched sound, while an object vibrating slower produces a lower-pitched sound.

CHECK YOUR READING How is frequency related to pitch?

⭕ **Set Learning Goals**
Students will

- Describe how the frequency of a wave affects the way it sounds.
- Describe how sound quality differs from pitch.
- Learn about the Doppler effect.
- Discover through an experiment how frequency and pitch are related.

◀ **3-Minute Warm-Up**

Display Transparency 12 or copy this exercise on the board:

Decide if these statements are true. If not true, correct them.

1. In a longitudinal wave, the vibrations move perpendicular to the direction of the wave. *In a longitudinal wave, the vibrations move in the same direction as the wave.*

2. Sound is a type of electromagnetic wave. *Sound is a type of mechanical wave.*

3. Vibrations pass through many parts of the ear, not just the eardrum. *True*

▮T▮ 3-Minute Warm-Up, p. T12

2.2 MOTIVATE

EXPLORE Pitch

PURPOSE To examine factors that affect the pitch of a sound

TIP *10 min.* Experiment with available rulers before class begins. Use those that produce the best results. Do not use plastic rulers that might break.

WHAT DO YOU THINK? *The sound became higher. The ruler moved up and down a greater distance when the ruler extended farther than when it was pulled back.*

Ongoing Assessment

CHECK YOUR READING *Answer: The higher the frequency of the sound wave, the higher pitched the sound.*

RESOURCES FOR DIFFERENTIATED INSTRUCTION

Below Level

UNIT RESOURCE BOOK
Reading Study Guide A, pp. 86–87
Decoding Support, p. 120

💿 **AUDIO CDS**

▮R▮ **Additional INVESTIGATION,**
Exploring Resonance, A, B, & C, pp. 132–140;
Teacher Instructions, pp. 284–285

Advanced

UNIT RESOURCE BOOK
- Challenge and Extension, p. 92
- Challenge Reading, pp. 116–117

English Learners

UNIT RESOURCE BOOK
Spanish Reading Study Guide, pp. 90–91

💿 **AUDIO CDS**

- Audio Readings in Spanish
- Audio Readings (English)

Teach from Visuals

To help students interpret the visual on frequency and pitch, have pairs of students prepare a series of flash cards, each of which contains a wave of a different wavelength. Have students work together, choosing two cards at a time and deciding which wave has a lower pitch. Have them identify the wavelength of each wave and compare their frequencies. You may want to remind students that wavelength and frequency are inversely related—the higher the frequency, the shorter the wavelength.

Develop Critical Thinking

APPLY Have students apply their knowledge of pitch by having them compare the length of vocal cords, based on the pitch of voices. Have them use the categories of male child, female child, male adult, and female adult. *In general, male adult vocal cords are longest, as indicated by the lower pitches of adult male voices. Female adult vocal cords are the next longest. The lengths of the vocal cords of a male and female child are about the same and are shorter than the vocal cords of a female adult.*

Ongoing Assessment

Describe the effect of frequency on how a wave sounds.

Ask: Does a wave with a high frequency have a high pitch or a low pitch? *a high pitch*

CHECK YOUR READING *Answer: Humans can hear frequencies between 20 hertz and 20,000 hertz.*

High and Low Frequencies

Frequency is a measure of how often a wavelength passes a fixed point. The unit for measuring frequency, and also pitch, is the hertz. A **hertz** (Hz) is one complete wavelength per second. For example, a wave with a frequency of 20 hertz has 20 wavelengths per second. In a wave with a frequency of 100 hertz, 100 wavelengths pass a given point every second. One complete wavelength can also be called a cycle. The diagram below shows how frequency and pitch are related.

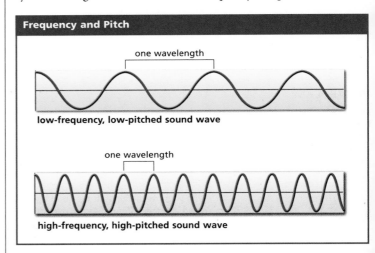

Frequency and Pitch

one wavelength

low-frequency, low-pitched sound wave

one wavelength

high-frequency, high-pitched sound wave

Human ears can hear a wide range of pitches. Most people with good hearing can hear sounds in the range of 20 hertz to 20,000 hertz. The note of middle C on a piano, for example, has a frequency of 262 hertz.

READING TiP
The prefix *infra* means "below," and the prefix *ultra* means "beyond."

Sound waves with wavelengths below 20 hertz are called infrasound. People cannot hear sounds in this range. Infrasound waves have a very long wavelength and can travel great distances without losing much energy. Elephants may use infrasound to communicate over long distances. Some of the waves that elephants use travel through the ground instead of the air, and they may be detected by another elephant up to 32 kilometers (about 20 miles) away.

The highest frequency that humans can hear is 20,000 hertz. Sound waves in the range above 20,000 hertz are called **ultrasound.** Though people cannot hear ultrasound, it is very useful. Later in this chapter, you will learn about some of the uses of ultrasound. Many animals can hear sound waves in the ultrasound range. The chart on the next page shows the hearing ranges of some animals.

CHECK YOUR READING What is the range of frequencies that humans can hear?

DIFFERENTIATE INSTRUCTION

?) More Reading Support

A If a wave has a frequency of 16 Hz, how many wavelengths per second does it have? *16*

B What are sound waves above 20,000 Hz called? *ultrasound*

Inclusion If you have students in class who have visual impairments, sketch the waves in the figure on this page and glue yarn on the waves. Have students feel the difference in the high-frequency wave and the low-frequency wave.

English Learners Be aware that English learners do not always have necessary background knowledge. For example, English learners may not be familiar with some musical terminology, such as *cymbals* and *clarinet* (p. 49), and *middle C* and *piano* (above).

Sound Frequencies Heard by Animals

Frequencies in Hz

0   50,000   100,000

 mosquito 200–400 Hz

tree frog 50–4,000 Hz

elephant 16–12,000 Hz

human 20–20,000 Hz

chimpanzee 100–33,000 Hz

dog 40–50,000 Hz

bat 2,000–110,000 Hz

porpoise 75–150,000 Hz

Although people can hear a wide range of frequencies, there are many sounds that people cannot hear.

Some animals can hear frequencies that are higher than those that people can hear. Dog whistles produce ultrasound.

READING VISUALS Which animals on this chart can hear frequencies above those that humans can hear?

Teach from Visuals

To help students interpret the sound frequency visual, ask:

- Why are devices that send out high-frequency sounds sometimes used as repellents for pests such as mice? (A mouse can hear frequencies from 1,000 to 90,000 hertz.) *Frequencies too high for human hearing can be heard by mice. Broadcasting upper frequencies that are uncomfortable for mice could repel them.*

- Why can't people hear the sound produced by a dog whistle? *The whistle sends out sound between 20,000 and 50,000 hertz. Dogs hear it, but it is barely audible to humans.*

- Cats have a hearing range from approximately 100 to 60,000 hertz. Where would cats be placed on the diagram on this page? *between dogs and bats*

  T This visual is also available as T14 in the Unit Transparency Book.

Real World Example

Hearing aids amplify vibrations so that people with hearing loss can detect sound. However, hearing aids amplify all frequencies of sound. Very few people have equal hearing loss at all frequency levels. Some people have difficulty wearing hearing aids because, to make certain frequencies loud enough to hear, other frequencies become too loud.

Ongoing Assessment

READING VISUALS *Answer: chimpanzee, dog, bat, and porpoise*

DIFFERENTIATE INSTRUCTION

More Reading Support

C Which animals on the chart hear frequencies below those that humans can hear? *elephant*

Inclusion Have a group of students make a collage that represents the wavelengths heard by different animals. Have students use a variety of materials: yarn, cord, ribbon, paper, aluminum foil, etc. It should have enough texture so that students with visual impairments can feel the difference in ranges of frequencies.

Advanced

R Challenge Reading, pp. 116–117

INVESTIGATE Sound Frequency

PURPOSE To discover how frequency relates to pitch

TIP *20 min.* Students can use two identical rubber bands, pulling one tighter than the other, to show that the difference in pitch is from a difference in frequency, not a difference in rubber bands. Students may need to use tape to hold one or both of the rubber bands in place.

WHAT DO YOU THINK? *The tighter rubber band produces a higher pitch. The higher the frequency, the higher the pitch.*

CHALLENGE *To make a guitar string sound higher in pitch, tighten the string. A tighter string has a higher frequency and therefore higher pitch.*

 Datasheet, Sound Frequency, p. 93

Technology Resources

Customize this student lab as needed or look for an alternative. Print rubrics to assess student lab reports.

 Lab Generator CD-ROM

Ongoing Assessment

Discover how frequency relates to pitch.

Ask: Does the tighter rubber band vibrate at a higher or lower frequency? *higher*

CHECK YOUR READING *Answer: Resonance is the adding of a sound wave with an object's natural frequency of vibration.*

INVESTIGATE Frequency

How is frequency related to pitch?

PROCEDURE

1. Stretch the rubber bands around the open box.
2. Pull one of the rubber bands tightly across the open part of the box so that it vibrates with a higher frequency than the looser rubber band. Tape the rubber band in place.
3. Pluck each rubber band and listen to the sound it makes.

WHAT DO YOU THINK?

- Which rubber band produces a sound wave with a higher pitch?
- How is frequency related to pitch?

CHALLENGE Suppose you are tuning a guitar and want to make one of the strings sound higher in pitch. Do you tighten or loosen the string? Explain your answer.

SKILL FOCUS
Inferring

MATERIALS
- 2 rubber bands of different sizes
- small open box
- tape

TIME
20 minutes

Natural Frequencies

You have read that sound waves are produced by vibrating objects. Sound waves also cause particles in the air to vibrate as they travel through the air. These vibrations have a frequency, or a number of wavelengths per second. All objects have a frequency at which they vibrate called a natural frequency.

You may have seen a piano tuner tap a tuning fork against another object. The tuner does this to make the fork vibrate at its natural frequency. He or she then listens to the pitch produced by the tuning fork's vibrations and tunes the piano string to match it. Different tuning forks have different frequencies and can be used to tune instrument to different pitches.

When a sound wave with a particular frequency encounters an object that has the same natural frequency, constructive interference takes place. The amplitude of the vibrating object adds together with the amplitude of the sound wave. The strengthening of a sound wave in this way is called **resonance.** When a tuning fork is struck, a nearby tuning fork with the same natural frequency will also begin to vibrate because of resonance.

CHECK YOUR READING How is natural frequency related to resonance?

DIFFERENTIATE INSTRUCTION

? **More Reading Support**

D What is produced when a sound wave combines with a natural vibration? *resonance*

Additional Investigation To reinforce Section 2.2 learning goals, use the following full-period investigation:

 Additional INVESTIGATION, Exploring Resonance, A, B, & C, pp. 132–140, 284–285 (Advanced students should complete Levels B and C.)

Below Level Make sure students understand how frequency is represented in the visuals *Frequency and Pitch* (p. 46) and *Sound Frequencies Heard by Animals* (p. 47).

Sound Quality

Have you ever noticed that two singers can sing exactly the same note, or pitch, and yet sound very different? The singers produce sound waves with their vocal cords. They stretch their vocal cords in just the right way to produce sound waves with a certain frequency. That frequency produces the pitch that the note of music calls for. Why, then, don't the singers sound exactly the same?

Each musical instrument and each human voice has its own particular sound, or quality. Another word for sound quality is timbre (TAM-buhr). Timbre can be explained by the fact that most sounds are not single waves but are actually combinations of waves. The pitch that you hear is called the fundamental tone. Other, higher-frequency pitches are called overtones. The combination of pitches is the main factor affecting the quality of a sound.

Another factor in sound quality is the way in which a sound starts and stops. Think about a musician who is crashing cymbals. The cymbals' sound blasts out suddenly. A sound produced by the human voice, on the other hand, starts much more gently.

CHECK YOUR READING What are two factors that affect sound quality? Which sentences above tell you?

The illustration below shows oscilloscope (uh-SIHL-uh-SKOHP) screens. An oscilloscope is a scientific instrument that tracks an electrical signal. The energy of a sound wave is converted into a signal and displayed on an oscilloscope screen. The screens below show sound waves made by musicians playing a piano and a clarinet. Both of these musical instruments are producing the same note, or pitch. Notice that the sound waves look slightly different from each other. Each has a different combination of overtones, producing a unique sound quality.

Oscilloscope Images

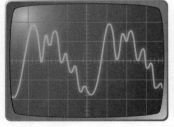

piano

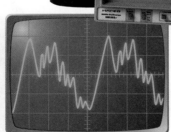

clarinet

Both oscilloscope images at left show sound waves of the same pitch produced on two different instruments. The waves, however, have different sound qualities.

DIFFERENTIATE INSTRUCTION

? More Reading Support

E What is another name for sound quality? *timbre*

F What instrument is used to track electrical signals? *an oscilloscope*

Advanced Have interested students research how electronic displays of sound are used in voice identification. Topics to investigate include security systems that require voice identification for admittance to a secure area and using voice patterns to identify suspects in criminal investigations.

R Challenge and Extension, p. 92

Real World Example

At a band or orchestra concert, a single player plays a note before the concert starts. All the other players match that note, tuning their instruments. Although the sound quality of the instruments may vary considerably, the pitch of each instrument must blend harmoniously with the pitches of the other instruments.

Develop Critical Thinking

Encourage students to study the images on the oscilloscopes. Have students apply their knowledge about waves to answer the following questions.

- What property of the wave appears on the horizontal axis? *frequency and wavelength*

- What property of the wave appears on the vertical axis? *the height of the wave, or amplitude*

- If a sound with a higher pitch appeared on the scope, how would the picture change? *More waves would be visible.*

EXPLORE (the BIG idea)

Revisit "What Gives a Sound Its Qualities?" on p. 35. Have students explain their results.

Ongoing Assessment

Contrast sound quality and pitch.

Ask: If two people sing at the same pitch, why does the quality of the sound differ? *It differs because there are qualities of sound other than pitch, such as timbre (tone color).*

CHECK YOUR READING *Answer: Sound quality is affected by the combination of pitches (last sentence of paragraph 2) and the way sound starts and stops (first sentence of paragraph 3).*

Integrate the Sciences

Because the Doppler effect applies to all waves, astronomers can use it to determine the speed and direction of a galaxy's movement. If the galaxy is getting nearer, the light it generates will shift toward blue (shorter) wavelengths. If it is moving away, its light will shift toward red (longer) wavelengths.

Teach Difficult Concepts

Some students may have a hard time understanding the pattern of waves in the Doppler effect. To clarify the concept, have them visualize or recall the pattern of waves that surround a moving boat. Waves to the front of the boat are close together. To the back of the boat, the waves are farther apart.

Ongoing Assessment

Explain how the Doppler effect works.

Ask: As a siren moves away from you, does the pitch become higher or lower? *lower*

CHECK YOUR READING *Answer: If a sound moves toward the listener, the pitch rises. If it moves away from the listener, the pitch lowers.*

The motion of the source of a sound affects its pitch.

Sometimes in traffic, a screeching siren announces that an ambulance must pass through traffic. Drivers slow down and pull over to the side, leaving room for the ambulance to speed by. Suppose you are a passenger in one of these cars. What do you hear?

When the ambulance whizzes past you the pitch suddenly seems to drop. The siren on the ambulance blasts the same pitches again and again. What has made the difference in what you hear is the rapid motion of the vehicle toward you and then away from you. The motion of the source of a sound affects its pitch.

The Doppler Effect

In the 1800s an Austrian scientist named Christian Doppler hypothesized about sound waves. He published a scientific paper about his work. In it, he described how pitch changes when a sound source moves rapidly toward and then away from a listener. Doppler described the scientific principle we notice when a siren speeds by. The **Doppler effect** is the change in perceived pitch that occurs when the source or the receiver of a sound is moving.

Before long, a Dutch scientist learned of Doppler's work. In 1845 he staged an experiment to test the hypothesis that Doppler described. In the experiment, a group of trumpet players were put on a train car. Other musicians were seated beside the railroad track. Those musicians had perfect pitch—that is, the ability to identify a pitch just by listening to it. The train passed by the musicians while the trumpeters on the train played their instruments. The musicians recorded the pitches they heard from one moment to the next. At the end of the demonstration, the musicians reported that they had heard the pitch of the trumpets fall as the train passed. Their experiment showed that the Doppler effect exists.

CHECK YOUR READING How does the motion of a sound affect its pitch?

DESCRIPTION WHEEL Make a description wheel in your notebook for the Doppler effect.

? **G**

To listeners outside the train, the noise made by this train sounds higher in pitch while it approaches them than while it speeds away.

DIFFERENTIATE INSTRUCTION

? **More Reading Support**

G When a siren approaches you, what happens to its pitch? *The pitch seems higher.*

English Learners English learners may have difficulty using the words *affect* and *effect* correctly. Explain that *affect* is a verb and *effect* is a noun. Compare the following two sentences from this page. *The motion of the source of a sound affects its pitch. The Doppler effect is the change in perceived pitch that occurs when the source or receiver of a sound is moving.* The first sentence uses the verb *affect*. The second sentence uses the noun *effect*.

The Doppler Effect

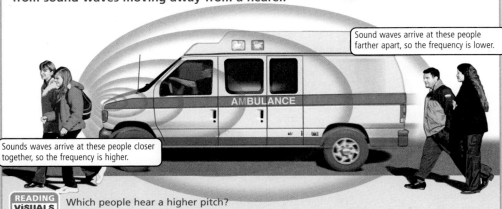

Sound waves moving toward a hearer have a different pitch from sound waves moving away from a hearer.

Sound waves arrive at these people farther apart, so the frequency is lower.

Sounds waves arrive at these people closer together, so the frequency is higher.

READING VISUALS Which people hear a higher pitch?

Frequency and Pitch

Again imagine sitting in a car as an ambulance approaches. The siren on the ambulance continually sends out sound waves. As the ambulance pulls closer to you, it catches up with the sound waves it is sending out. As a result, the sound waves that reach your ears are spaced closer together. The frequency, and therefore the pitch, is higher when it reaches you. As the ambulance continues, it gets farther and farther away from you, while the sound waves still move toward you. Now the waves arrive farther and farther apart. As the frequency decreases, you hear a lower pitch.

 **VISUALIZATION**
CLASSZONE.COM

Explore the Doppler effect.

2.2 Review

KEY CONCEPTS

1. Describe what is different about the sound waves produced by a low note and a high note on a musical instrument.

2. Explain why two people singing the same pitch do not sound exactly the same.

3. How does perceived pitch change as a sound source passes a listener?

CRITICAL THINKING

4. **Apply** How could you produce vibrations in a tuning fork without touching it? Explain your answer.

5. **Predict** Suppose you could view the waves produced by a high-pitched and a low-pitched voice. Which wave would display the greater number of crests and troughs? Why?

⚠ CHALLENGE

6. **Infer** Offer a possible explanation for why no one noticed the Doppler effect before the 1800s.

ANSWERS

1. The sound waves have different frequencies—high frequency for the high note and low frequency for the low note.

2. The quality, or timbre, of the sounds made by the voices is different, even though the pitch is the same.

3. The frequency changes as the source of the sound moves past a listener, changing the pitch.

4. by striking an identical tuning fork nearby; Resonance would cause the tuning fork to vibrate.

5. the high-pitched voice because the sound waves have higher frequency and thus more waves per unit of time

6. None of the modes of transportation went fast enough for the Doppler effect to happen.

Ongoing Assessment

READING VISUALS *Answer: the people on the left*

Reinforce (the **BIG** idea)

Have students relate the section to the Big Idea.

R Reinforcing Key Concepts, p. 94

2.2 ASSESS & RETEACH

Assess

A Section 2.2 Quiz, p. 23

Reteach

Have students try the following activity:

- Tightly stretch a thread or light string about 50 cm long between two supports, parallel to the floor. There should be at least 25 cm of space beneath the string.

- Tie a piece of string to each of seven identical metal washers. Three pieces of string should be 20 cm long, two should be 15 cm, one 10 cm, and one 5 cm. Tie the strings evenly spaced on the long thread.

- Start swinging one washer. Notice that other washers on the same length of string will also swing because they have the same natural frequency. The other washers will not move.

- Repeat with each different length of string.

Technology Resources

Have students visit **ClassZone.com** for reteaching of Key Concepts.

 CONTENT REVIEW

 CONTENT REVIEW CD-ROM

◆ Set Learning Goals

Students will

• Explain how the intensity of a wave affects its loudness.

• Describe how sound intensity can be controlled.

• Explain how loudness can affect hearing.

• Observe through an experiment how amplitude relates to loudness.

◯ 3-Minute Warm-Up

Display Transparency 13 or copy this exercise on the board:

Match each of the definitions in the first column to one of the terms in the second column.

Definitions

1. sound waves with frequency greater than 20,000 hertz *c*

2. the measure of the height of a wave's crest *b*

3. how high or low a sound is *a*

Terms

a. pitch

b. amplitude

c. ultrasound

 3-Minute Warm-Up, p. T13

2.3 MOTIVATE

THINK ABOUT

PURPOSE To understand how to produce different types of sound

DISCUSS Ask: Which part of the drum vibrates? *all of it (especially the drum head)*

Answers: drumstick moving up and down; louder: move drumsticks with more energy; softer: with less energy

Ongoing Assessment

CHECK YOUR READING *Answer: the more energy, the louder the sound*

2.3 Intensity determines loudness.

◀ **BEFORE, you learned**

• Sound waves are produced by vibrations
• Frequency determines the pitch of a sound
• Amplitude is a measure of the height of a wave crest

▶ **NOW, you will learn**

• How the intensity of a wave affects its loudness
• How sound intensity can be controlled
• How loudness can affect hearing

VOCABULARY

intensity p. 52
decibel p. 52
amplification p. 55
acoustics p. 55

THINK ABOUT

What makes a sound louder?

A drum player has to play softly at some times and loudly at others. Think about what the drummer must do to produce each type of sound. If you could watch the drummer in the photograph in action, what would you see? How would the drummer change the way he moves the drumsticks to make a loud, crashing sound? What might he do to make a very soft sound?

Intensity depends on the amplitude of a sound wave.

OUTLINE
Make an outline for this heading. Remember to include main ideas and details.

I. Main idea
   A. Supporting idea
      1. Detail
      2. Detail
   B. Supporting idea

Earlier you read that all waves carry energy. The more energy a sound wave carries, the more intense it is and the louder it will sound to listeners. The **intensity** of a sound is the amount of energy its sound wave has. A unit called the **decibel** (dB) is used to measure sound intensity. The faint rustling of tree leaves on a quiet summer day can hardly be heard. Some of the softest sounds measure less than 10 decibels. On the other hand, the noise from a jet taking off or the volume of a TV set turned all the way up can hurt your ears. Very loud sounds measure more than 100 decibels. Remember that amplitude is also a measure of wave energy. The greater the amplitude, the more intensity a sound wave has and the louder the sound will be.

 How is energy related to loudness?

RESOURCES FOR DIFFERENTIATED INSTRUCTION

Below Level

UNIT RESOURCE BOOK
• Reading Study Guide A, pp. 97–98
• Decoding Support, p. 120

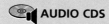 **AUDIO CDS**

Advanced

UNIT RESOURCE BOOK
Challenge and Extension, p. 103

English Learners

UNIT RESOURCE BOOK
Spanish Reading Study Guide, pp. 101–102

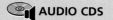 **AUDIO CDS**

• Audio Readings in Spanish
• Audio Readings (English)

INVESTIGATE Loudness

How is amplitude related to loudness?

PROCEDURE

1. Cut a notch in the middle of both ends of the cardboard. Stretch the rubber band around the cardboard so that it fits into the notches as shown.

2. Mark lines on the cardboard at one and four centimeters away from the rubber band.

3. Slide the pencils under the rubber band at each end.

4. Pull the rubber band to the one-centimeter line and let it go so that it vibrates with a low amplitude. Notice the sound it makes. Pull the rubber band to the four-centimeter line and let it go again. This time the amplitude is higher. Notice the sound it makes this time.

WHAT DO YOU THINK?
- How did the loudness of the sounds compare?
- How is amplitude related to loudness?

CHALLENGE Using what you learned from experimenting with the rubber band, explain why swinging a drumstick harder on a drum would make a louder sound than swinging a drumstick lightly.

SKILL FOCUS
Observing

MATERIALS
- piece of cardboard
- scissors
- large rubber band
- 2 pencils
- ruler

TIME
15 minutes

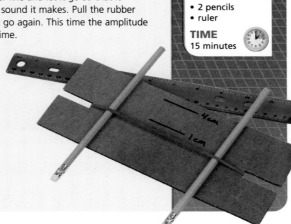

The drummer varies the loudness of a sound by varying the energy with which he hits the drum. Loudness is also affected by the distance between the source and the listener.

Have you ever wondered why sound gradually dies out over distance? Think about someone walking away from you with a radio. When the radio is close, the radio seems loud. As the person walks away, the sound grows fainter and fainter. Sound waves travel in all directions from their source. As the waves travel farther from the radio, their energy is spread out over a greater area. This means that their intensity is decreased. The sound waves with lower intensities are heard as quieter sounds.

Other forces can take energy away from sound waves, too. The force of friction can act on the medium of a sound wave to decrease the intensity of the waves. This effect of friction on sound is probably a good thing. Imagine what the world would be like if every sound wave continued forever!

INVESTIGATE Loudness

PURPOSE To observe how amplitude relates to loudness

TIPS *15 min.*

- The cardboard must be thick. It should remain flat when the rubber band is stretched on it.
- Emphasize safety. The notch should be narrow and deep enough that the rubber band stays in the notch when stretched.
- It might be easier to use a knife to notch the cardboard before the lab.
- For accurate observations, students should pull the rubber bands from the middle.

WHAT DO YOU THINK? *The sound was louder when the rubber band was pulled back 4 cm than when it was pulled back 1 cm. The greater the amplitude, the louder the sound.*

CHALLENGE *Swinging a drumstick harder would make a louder sound because the drum skin would vibrate with greater amplitude.*

 Datasheet, Loudness, p. 104

Technology Resources

Customize this student lab as needed or look for an alternative. Print rubrics to assess student lab reports.

 Lab Generator CD-ROM

Ongoing Assessment

Explain the effect of intensity of a wave on its loudness.

Ask: Which sound will have greater intensity, a whisper or a yell? *a yell*

Observe how amplitude relates to loudness.

Ask: How can you make a sound louder? *by increasing the amplitude of the sound wave*

DIFFERENTIATE INSTRUCTION

? More Reading Support

A How does distance affect intensity? *the less distance, the more intensity*

English Learners Often readers are asked to imagine a situation and consider its implications. Consider the example from this page: *Have you ever wondered why sound gradually dies out over distance? Think about someone walking away from you with a radio.* These sentences require the reader to imagine a hypothetical scenario. When this occurs explain to English learners that they should imagine the situation. Also, tell students that the phrasal verb *dies out* means "fades."

History of Science

Galileo, a famous physicist, studied resonance in the 1600s. Galileo noted that the swings of a pendulum increasingly grow with repeated, timed applications of a small force. When the frequency of an applied force matches the natural frequency of a system, large-amplitude vibrations result in what's called resonance. Resonance explains why a glass shatters at a pitch that matches its natural frequency.

Teaching with Technology

If you have a tape recorder, you might want to record various sounds and play them back for student analysis of pitch, intensity, and quality. Examples might include birdcalls or musical instruments.

Ongoing Assessment

Describe how to control sound intensity.

Ask: How does a muffler affect the intensity of sound? *It absorbs energy from the sound wave, decreasing the intensity.*

READING VISUALS *Answer: The sound of the jet taking off is the most intense; the rustling of leaves is the least intense.*

CHECK YOUR READING *Answer: Change its amplitude but not its frequency.*

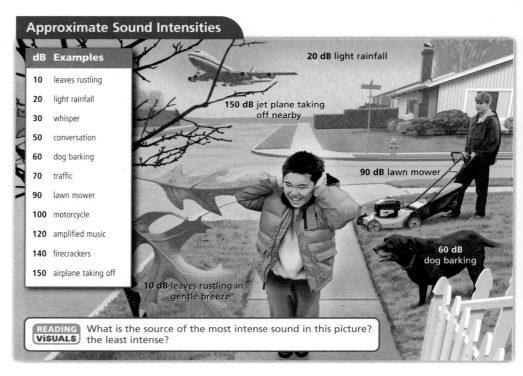

Approximate Sound Intensities

dB	Examples
10	leaves rustling
20	light rainfall
30	whisper
50	conversation
60	dog barking
70	traffic
90	lawn mower
100	motorcycle
120	amplified music
140	firecrackers
150	airplane taking off

20 dB light rainfall

150 dB jet plane taking off nearby

90 dB lawn mower

60 dB dog barking

10 dB leaves rustling in gentle breeze

READING VISUALS What is the source of the most intense sound in this picture? the least intense?

The intensity of sound can be controlled.

Over time and distance, a sound wave gets weaker and weaker until the sound becomes undetectable. The pitch, however, does not typically change as the sound grows weaker. In other words, even as the amplitude decreases, the frequency stays the same.

REMINDER
Remember, amplitude is a measure of wave energy.

Sometimes it is desirable to change sound intensity without changing the pitch and quality of a sound. We can do this by adding energy to or taking energy away from a sound wave. As you have already seen, intensity is the amount of energy in a sound wave. Changing the intensity of a sound wave changes its amplitude.

 B

 C

Sound intensity can be controlled in many ways. Mufflers on cars and trucks reduce engine noise. Have you ever heard a car with a broken muffler? You were probably surprised at how loud it was. Burning fuel in an engine produces hot gases that expand and make a very loud noise. A muffler is designed to absorb some of the energy of the sound waves and so decrease their amplitude. As a result, the intensity of the sound you hear is much lower than it would be without the muffler.

CHECK YOUR READING How could you change the intensity of a sound without changing the pitch?

DIFFERENTIATE INSTRUCTION

? More Reading Support

B Changing the intensity of sound also changes what else? *amplitude*

C What does a muffler in a car do? *reduces the sound of the engine*

Below Level Ask students what properties they think would be necessary for a device that could be worn over the ears to control sound intensity. *Sample answer: being made of dense material that would either absorb sound or keep vibrations from reaching the eardrums.* If possible, show students examples of such devices, such as earplugs or the devices used by people who work with loud industrial machinery or near airplanes at airports.

Amplification

In addition to being reduced, as they are in a muffler, sound waves can be amplified. The word *amplify* may remind you of *amplitude*, the measure of the height of a wave's crest. These words are related. To amplify something means to make it bigger. **Amplification** is the increasing of the strength of an electrical signal. It is often used to increase the intensity of a sound wave.

When you listen to a stereo, you experience the effects of amplification. Sound input to the stereo is in the form of weak electrical signals from a microphone. Transistors in an electronic circuit amplify the signals. The electrical signals are converted into vibrations in a coil in your stereo's speaker. The coil is attached to a cone, which also vibrates and sends out sound waves. You can control the intensity of the sound waves by adjusting your stereo's volume.

sound waves in

Amplifier

coil

cone

sound waves out

Acoustics

The scientific study of sound is called **acoustics** (uh-KOO-stihks). Acoustics involves both how sound is produced and how it is received and heard by humans and animals.

Acoustics also refers to the way sound waves behave inside a space. Experts called acoustical engineers help design buildings to reduce unwanted echoes. An echo is simply a reflected sound wave. To control sound intensity, engineers design walls and ceilings with acoustical tiles. The shapes and surfaces of acoustical tiles are designed to absorb or redirect some of the energy of sound waves.

The pointed tiles in this sound-testing room are designed to absorb sound waves and prevent any echoes.

The shapes and surfaces in this concert hall direct sound waves to the audience.

READING VISUALS COMPARE AND CONTRAST Imagine sound waves reflecting off the surfaces in the two photographs above. How do the reflections differ?

Here are a few branches of acoustics.

- Communications acoustics incorporates radio and other sound reproduction.
- Architectural acoustics is the control of sound waves inside buildings.
- Environmental acoustics applies sound theories to control noise pollution.

Integrate the Sciences

Physiological acoustics integrates the physics of sound with its biological application to human hearing. In the 1800s, German physicist Georg Ohm stated that the human ear is sensitive to the amplitude of sound waves. In 1863, Hermann von Helmholtz later expanded upon Ohm's theory by relating the physics of sound and music.

Address Misconceptions

IDENTIFY Ask: If sound damages your ears, how long does the damage last? If student answers indicate that any damage is not permanent, they hold the misconception that any hearing damage resulting from loud noises is temporary.

CORRECT Make two lists on the board. Label one "Temporary damage" and the other one "Permanent damage." Ask students to give instances when permanent physical damage occurs, such as some heart attacks, and instances of temporary damage, such as a cut finger.

REASSESS Ask: When delicate cells like hair cells in the ear are damaged, do you think the damage is permanent or temporary? *If the damage is slight, injured cells might be repaired, but such delicate cells could be irreversibly damaged.*

Technology Resources

Visit **ClassZone.com** for background on common student misconceptions.

 MISCONCEPTION DATABASE

Ongoing Assessment

READING VISUALS *Answer: There is more absorption than reflection of sound waves in the sound-testing room, but more reflection in the concert hall.*

DIFFERENTIATE INSTRUCTION

? More Reading Support

D When sound is amplified, what happens to it? *It becomes bigger, or louder.*

E What is the study of sound called? *acoustics*

Alternative Assessment Have students sketch a room interior. Then have them draw arrows showing how sound reflects off surfaces in the room. Ask: How would the addition of sound-absorbing materials affect your drawing? *It would reduce reflected sound.*

Intense sound can damage hearing.

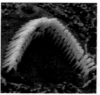
healthy hair cells

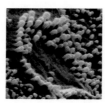
damaged hair cells

 RESOURCE CENTER
CLASSZONE.COM
Find out more about sound and protecting your hearing.

When a train screeches to a stop in a subway station, the sound of the squealing brakes echoes off the tunnel walls. Without thinking about it, you cover your ears with your hands. This response helps protect your ears from possible damage.

In the first section of this chapter, you read about the main parts of the human ear. The part of the inner ear called the cochlea is lined with special cells called hair cells. As you have seen, these cells are necessary for hearing.

The hair cells are extremely sensitive. This sensitivity makes hearing possible, but it also makes the cells easy to damage. Continual exposure to sounds of 90 dB or louder can damage or destroy the cells. This is one reason why being exposed to very loud noises, especially for more than a short time, is harmful to hearing.

 CHECK YOUR READING How do high-intensity sounds damage hearing?

Using earplugs can prevent damage from too much exposure to high-intensity sounds such as amplified music. The intensity at a rock concert is between 85 and 120 dB. Ear protection can also protect the hearing of employees in factories and other noisy work sites. In the United States, there are laws that require employers to reduce sounds at work sites to below 90 dB or to provide workers with ear protection.

Even a brief, one-time exposure to an extremely loud noise can destroy hair cells. Noises above 130 dB are especially dangerous. Noises above 140 dB are even painful. It is best to avoid such noises altogether. If you find yourself exposed suddenly to such a noise, covering your ears with your hands may be the best protection.

2.3 Review

KEY CONCEPTS

1. Explain how the terms *intensity, decibel,* and *amplitude* are related.
2. Describe one way in which sound intensity can be controlled.
3. How do loud sounds cause damage to hearing?

CRITICAL THINKING

4. **Synthesize** A wind chime produces both soft and loud sounds. If you could see the waves, how would they differ?
5. **Design an Experiment** How could you demonstrate that sound dies away over distance? Suppose you could use three volunteers, a boom box, and a tape recorder.

CHALLENGE

6. **Apply** Which of these acoustical designs would be best for a concert hall? Why?
a. bare room with hard walls, floor, and ceiling,
b. room padded with sound-absorbing materials such as acoustical tile,
c. room with some hard surfaces and some sound padding

D 56 Unit: Waves, Sound, and Light

SKILL: INTERPRETING GRAPHS

MATH TUTORIAL
CLASSZONE.COM

Click on Math Tutorial for more help with interpreting line graphs.

Measuring Hearing Loss

An audiogram is a graph that can be used to determine if a patient has hearing loss. The vertical axis shows the lowest intensity, in decibels, that the patient can hear for each frequency tested. Notice that intensity is numbered from top to bottom on an audiogram.

To determine the lowest intensity heard at a given frequency, find the frequency on the horizontal axis. Follow the line straight up until you see the data points, shown as ✳ for the right ear and ● for the left ear. Look to the left to find the intensity. For example, the lowest intensity heard in both ears at 250 Hz is 10 dB.

Audiogram for Patient A

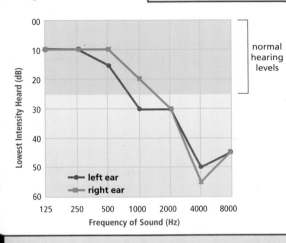

normal hearing levels

Lowest Intensity Heard (dB)

● left ear
✳ right ear

Frequency of Sound (Hz)
125 250 500 1000 2000 4000 8000

Use the graph to answer the following questions.

1. What is the lowest intensity heard in the patient's left ear at 1000 Hz? the right ear at the same frequency?

2. At which frequencies are the data points for both ears within normal hearing levels?

3. Data points outside the normal hearing levels indicate hearing loss. At which frequencies are the data points for both ears outside the normal levels?

CHALLENGE A dip in the graph at 3000 to 4000 Hz is a sign that the hearing loss was caused by exposure to loud noises. The patient is referred to a specialist for further testing. Should Patient A get further testing? Why or why not?

This air traffic ground controller wears ear protection to prevent hearing loss.

Chapter 2: **Sound** 57 **D**

ANSWERS

1. *30 dB, 20 dB*
2. *from 125 Hz to about 750 Hz*
3. *1500 (also accept 2000) to 8000 Hz*
CHALLENGE *Yes; patient A should get further testing because a dip in this audiogram occurs between 2000 and 4000 Hz.*

MATH IN SCIENCE
Math Skills Practice for Science

Set Learning Goal
To interpret a graph that relates intensity heard to frequency of sound

Present the Science
Few people with hearing difficulties lose the same amount of hearing at all frequencies. If the hearing loss is a consequence of exposure to loud noises, the frequencies that show greatest loss depend on the frequency of the noise that caused the damage. Hair cells that are permanently damaged by noise of certain frequencies might never again respond to sounds of that frequency.

Develop Graphing Skills
The vertical axis of the graph is labeled in an unconventional manner that should be explained. Ask students how many examples they can think of where lower is better. *Examples might include a game of Crazy 8s or miniature golf scores.*

In the audiogram, the greater the intensity necessary for hearing, the more hearing loss is present. The scale produces a graph that gives the correct impression—patient A has lost hearing at high frequencies in both ears.

Close
Ask: Why might an audiogram show quite different hearing results for the right and left ear of a patient? *Answers might include that the patient was consistently exposed to different intensities of noise in different ears.*

 • Math Support, p. 121
• Math Practice, p. 122

Technology Resources

Students can visit **ClassZone.com** for practice interpreting graphs.

MATH TUTORIAL

2.4 FOCUS

▶ Set Learning Goals

Students will

- Describe how ultrasound is used.
- Observe how musical instruments work.
- Explain how sound can be recorded and reproduced.
- Build a stringed instrument in an experiment.

◀ 3-Minute Warm-Up

Display Transparency 13 or copy this exercise on the board:

You own a movie theater. Explain how you could control the loudness of the sound of movies. How would you eliminate as much audience noise as possible?

The intensity of the sound from the movie should be measured so that it is loud enough to be heard but not loud enough to damage hearing. Because sound reflects, using sound-absorbing materials would eliminate echoing and minimize audience noise.

 3-Minute Warm-Up, p. T13

2.4 MOTIVATE

EXPLORE Echoes

PURPOSE To introduce the concept that sound reflects from an object

TIP *10 min.* Have students trade places so both partners have the opportunity to whisper and to listen.

WHAT DO YOU THINK? *The sound was louder with the book than without the book. An echo can be used to detect an object by reflecting off the object. If there is no object to reflect off, there would be no echo.*

Ongoing Assessment

CHECK YOUR READING *Acceptable answers include detecting objects, finding food, imaging the body.*

KEY CONCEPT

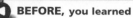

Sound has many uses.

◀ **BEFORE, you learned**

- Sound waves are produced by vibrations
- Sound waves have amplitude, frequency, and wavelength

▶ **NOW, you will learn**

- How ultrasound is used
- How musical instruments work
- How sound can be recorded and reproduced

VOCABULARY

echolocation p. 59
sonar p. 59

EXPLORE Echoes

How can you use sound to detect an object?

PROCEDURE

1. Tape the two cardboard tubes onto your desk at a right angle as shown.
2. Put your ear up to the end of one of the tubes. Cover your other ear with your hand.
3. Listen as your partner whispers into the outside end of the other tube.
4. Stand the book upright where the tubes meet. Repeat steps 2 and 3.

WHAT DO YOU THINK?

- How did the sound change when you added the book?
- How can an echo be used to detect an object?

MATERIALS

- 2 cardboard tubes
- tape
- book

Ultrasound waves are used to detect objects.

A ringing telephone, a honking horn, and the sound of a friend's voice are all reminders of how important sound is. But sound has uses that go beyond communication. For example, some animals and people use reflected ultrasound waves to detect objects. Some animals, such as bats, use the echoes of ultrasound waves to find food. People use ultrasound echoes to detect objects underwater or even to produce images of the inside of the body.

 Other than communication, what are three uses of sound?

RESOURCES FOR DIFFERENTIATED INSTRUCTION

Below Level

UNIT RESOURCE BOOK
- Reading Study Guide A, pp. 108–109
- Decoding Support, p. 120

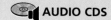 **AUDIO CDS**

Advanced

UNIT RESOURCE BOOK
Challenge and Extension, p. 114

English Learners

UNIT RESOURCE BOOK
Spanish Reading Study Guide, pp. 112–113

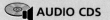 **AUDIO CDS**

- Audio Readings in Spanish
- Audio Readings (English)

Echolocation

Sending out ultrasound waves and interpreting the returning sound echoes is called **echolocation** (*echo* + *location*). Bats flying at night find their meals of flying insects by using echolocation. They send out as many as 200 ultrasound squeaks per second. By receiving the returning echoes, they can tell where prey is and how it is moving. They can also veer away from walls, trees, and other big objects.

VOCABULARY
Make description wheels for the terms *echolocation* and *sonar* to help you remember them later.

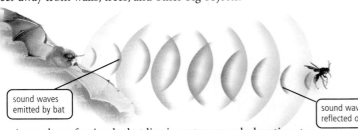

sound waves emitted by bat

sound waves reflected off prey

A number of animals that live in water use echolocation, too. Dolphins, toothed whales, and porpoises produce ultrasound squeaks or clicks. They listen to the returning echo patterns to find fish and other food in the water.

Sonar

People use the principles of echolocation to locate objects underwater. During World War I (1914–1918), scientists developed instruments that used sound waves to locate enemy submarines. Instruments that use echolocation to locate objects are known as **sonar.** Sonar stands for "sound navigation and ranging." The sonar machines could detect sounds coming from submarine propellers. Sonar devices could also send out ultrasound waves and then use the echoes to locate underwater objects. The information from the echoes could then be used to form an image on a screen.

Later, people found many other uses for sonar. Fishing boats use sonar to find schools of fish. Oceanographers—scientists who study the ocean—use it to map the sea floor. People have even used sonar to find ancient sunken ships in deep water.

This woman is using sonar to monitor for submarines.

Sonar is used to locate sunken ships. The image of the sunken ship above was produced on the basis of information from sonar.

DIFFERENTIATE INSTRUCTION

?) More Reading Support

A What happens to a sound wave in an echo? *It bounces back to its source.*

B What is the main use of sonar? *to locate objects underwater*

English Learners It can be beneficial for English learners and their classmates to draw upon other cultures when learning new concepts. For example, students studying how a guitar generates sound may also learn about stringed instruments from South America or Asia. Make an effort to incorporate cultural references into classroom discussion.

Teach from Visuals

To help students interpret the visual of sound waves:

- Tell students that a bat bounces sound waves off an insect. Ask: What does it tell the bat if the sound waves take longer to return to the bat than they did a few seconds ago? *The insect is moving away.*

- Ask: Why is *echolocation* a good term for how animals use sound to locate food and other objects? *"Echo" indicates the bouncing back of sound, and "location" shows that the bouncing back reveals where the object is.*

Real World Example

Both shape and distance must be considered in using sonar to locate and describe objects underwater. For example, a shape might appear to be a sunken boat, but if it is not as deep as the seafloor, it might be a school of fish with a similar shape swimming higher in the water.

Teacher Demo

Model how sound waves are used in echolocation and in sonar. Stand a student in the front of the class. Have another student act as a timekeeper. Have the first student walk at a steady rate from the front to the middle of the room and back to the same spot again. Record the time it took on the board. Have the same student walk at the same rate to the back of the room and back to the same spot again. Point out that sound will travel at the same speed in the same medium at the same temperature, so the time it takes to return to its source is an indication of how far it traveled.

Teach from Visuals

Have students examine the photograph of the ultrasound image and share what they see. Ask them to identify the three fetuses and any features, such as the heads, that they can identify. If possible, make a transparency of the photograph so various features can be pointed out on an overhead projector.

Integrate the Sciences

Show students a tessellation to explain the difference between noise and music. A tessellation is a pattern formed by regular-shaped pieces, such as equilateral triangles, with no gaps or overlaps. Examples can be found in many algebra or geometry books or on the Internet. Show students that tessellations produce a nonrandom pattern. Then show students that some shapes, such as pentagons, will not form a regular pattern and are analogous to noise.

Ongoing Assessment

Describe how ultrasound is used.

Ask: How might ultrasound be used to examine a human liver? *Ultrasound waves bounce off the liver, producing an image.*

 Answer: Both use the reflection of ultrasound waves to detect objects.

Medical Uses of Ultrasound

Ultrasound has many uses in medicine. Because ultrasound waves are not heard by humans, ultrasound can be used at very high intensities. For example, high-intensity vibrations from ultrasound waves are used to safely break up kidney stones in patients. The energy transferred by ultrasound waves is also used to clean medical equipment.

One of the most important medical uses of ultrasound is the ultrasound scanner. This device relies on the same scientific principle as sonar. It sends sound waves into a human body and then records the echoes that are reflected from inside the body. Information from the echoes forms a picture on a screen. The ultrasound scanner is used to examine internal organs such as the heart, pancreas, bladder, ovaries, and brain. Doppler ultrasound is a technology that can detect the movement of fluids through the body and is used to examine blood flow.

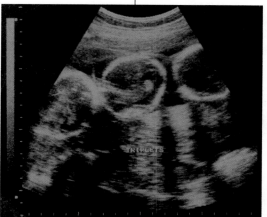

The image of these triplets was produced by reflected ultrasound waves.

 How is an ultrasound scanner similar to sonar?

One of the most well-known uses of ultrasound is to check on the health of a fetus during pregnancy. Problems that are discovered may possibly be treated early. The scan can also reveal the age and sex of the fetus and let the expecting parents know if they will be having twins or triplets. Ultrasound is safer than other imaging methods, such as the x-ray, which might harm the development of the fetus.

Sound waves can produce music.

Why are some sounds considered noise and other sounds considered music? Music is sound with clear pitches or rhythms. Noise is random sound; that means it has no intended pattern.

RESOURCE CENTER
CLASSZONE.COM
Explore instruments from around the world.

Musical instruments produce pitches and rhythms when made to vibrate at their natural frequencies. Some musical instruments have parts that vibrate at different frequencies to make different pitches. All of the pitches, together with the resonance of the instrument itself, produce its characteristic sound. The three main types of musical instruments are stringed, wind, and percussion. Some describe electronic instruments as a fourth type of musical instrument. Look at the illustration on the next page to learn more about how each type of musical instrument works.

DIFFERENTIATE INSTRUCTION

? More Reading Support

C What type of sound waves are used to check on the health of a fetus? *ultrasound*

D If sound is random, is it noise or music? *noise*

English Learners Discuss these words and their meanings with students: *scanner, fetus, expecting,* and *electronics.* Medical terms are often difficult for English learners because they usually are considerably different from one language to another. These terms also are not used frequently in everyday conversation. Use a bilingual dictionary to make a small medical dictionary of terms that might clarify the medical uses of ultrasound for English learners.

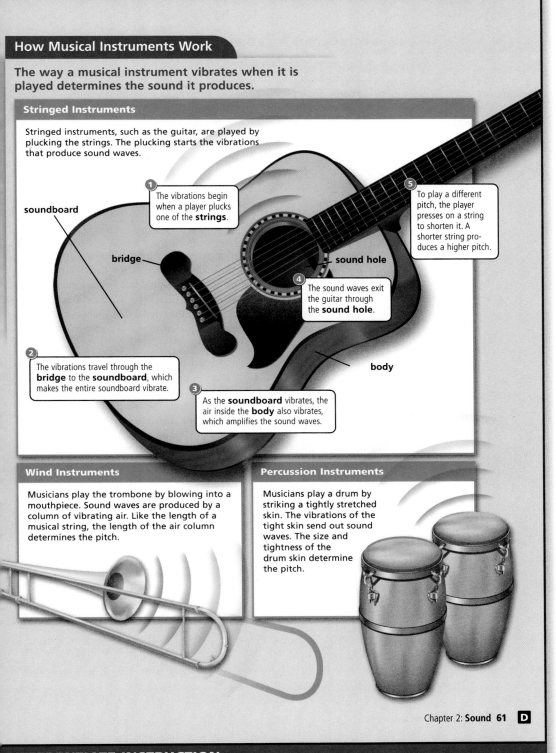

How Musical Instruments Work

The way a musical instrument vibrates when it is played determines the sound it produces.

Stringed Instruments

Stringed instruments, such as the guitar, are played by plucking the strings. The plucking starts the vibrations that produce sound waves.

soundboard

① The vibrations begin when a player plucks one of the strings.

⑤ To play a different pitch, the player presses on a string to shorten it. A shorter string produces a higher pitch.

bridge

sound hole

④ The sound waves exit the guitar through the sound hole.

② The vibrations travel through the bridge to the soundboard, which makes the entire soundboard vibrate.

body

③ As the soundboard vibrates, the air inside the body also vibrates, which amplifies the sound waves.

Wind Instruments

Musicians play the trombone by blowing into a mouthpiece. Sound waves are produced by a column of vibrating air. Like the length of a musical string, the length of the air column determines the pitch.

Percussion Instruments

Musicians play a drum by striking a tightly stretched skin. The vibrations of the tight skin send out sound waves. The size and tightness of the drum skin determine the pitch.

Chapter 2: Sound **61** **D**

Teach from Visuals

Have students examine the guitar in the visual. Explain that this acoustic guitar differs from an electric guitar. Show students an electric guitar or a picture of one.

- Ask them to predict the loudness of the electric guitar if the strings are plucked when it is not plugged in. *It will have a soft sound.*
- Ask students why an acoustic guitar is louder than the sound from an unamplified electric guitar. *It is designed to amplify the sound of the strings just the way it is.*

Teacher Demo

Use identical bottles, such as 1-liter or smaller transparent soft drink bottles, to make a wind instrument. Add different amounts of water to each bottle so that one bottle is nearly empty, one bottle is almost full, and the other bottles contain varying amounts of water. Blow across the mouth of each bottle, and ask students to explain their observations.

The pitch of the sound is lower as the amount of water decreases. The highest pitch comes from the bottle with the most water. Students should relate the pitch to the size of the column of air above the water.

EXPLORE (the BIG idea)

Revisit "How Does Size Affect Sound?" on p. 35. Have students explain their results.

Ongoing Assessment

Observe how musical instruments work.

Ask: What purpose does the bridge perform in a guitar? *It carries vibrations to the soundboard.*

DIFFERENTIATE INSTRUCTION

Alternative Assessment Have students label each of three pieces of paper with the name of a type of musical instrument. On half of the appropriate piece of paper, have them sketch an example of that type of instrument. Sketches should emphasize what part of the instrument is vibrating and producing sound. On the other half of the paper, have students list the characteristics and features of that type of instrument, and speculate as to how they determine pitch, quality, and intensity of sound.

Ask students to list as many examples as they can of sound being recorded and reproduced. Compile a class list so students can see how many ways sound reproduction is incorporated into daily life.

Teacher Demo

Have someone bring in an old phone that is no longer operating. Take the phone apart so that students can observe the different parts. Ask students to distinguish between the functions of the microphone and the diaphragm. *The microphone changes vibrations to electrical signals, and the diaphragm changes electrical signals into sound.*

If students have difficulty distinguishing between these roles, remind them that they speak into the microphone and hear sound from the diaphragm.

Ongoing Assessment

Explain how sound can be recorded and reproduced.

Ask: How are electrical signals related to sound waves? *Sound can be changed into electrical signals. These signals can be sent to other locations, where they are changed back into sound waves.*

 CHECK YOUR READING *Answer: the microphone (or mouthpiece)*

Sound can be recorded and reproduced.

For most of human history, people had no way to send their voices farther than they could shout. Nor could people before the 1800s record and play back sound. The voices of famous people were lost when they died. Imagine having a tape or a compact disk recording of George Washington giving a speech!

Then in the late 1800s, two inventions changed the world of sound. In 1876, the telephone was invented. And in 1877, Thomas Edison played the first recorded sound on a phonograph, or sound-recording machine.

READING TiP
The prefix *phono* means "sound," and the suffix *graph* means "writing."

? E

The Telephone

The telephone has made long-distance voice communication possible. Many people today use cell phones. But whether phone signals travel over wires or by microwaves, as in cell phones, the basic principles are similar. You will learn more about the signal that is used in cell phones when you read about microwaves in Chapter 3. In general, a telephone must do two things. It must record the sound that is spoken into it, and it must reproduce the sound that arrives as a signal from somewhere else.

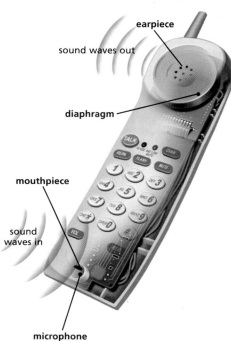

earpiece
sound waves out
diaphragm
mouthpiece
sound waves in
microphone

Suppose you are phoning your best friend to share some news. You speak into the mouthpiece. Sound waves from your voice cause a thin disk inside the mouthpiece to vibrate. A microphone turns these vibrations into electrical signals. Your handset sends these signals over wire to a switching station. Computers in the switching station connect phone callers and keep them connected until they finish their conversation.

Your friend receives the news by listening to the earpiece on his handset. There the process is more or less reversed. The electrical signals that arrive in the earpiece are turned into vibrations that shake another thin disk called a diaphragm. The vibrating diaphragm produces sound waves. The sound your friend hears is a copy of your voice, though it sounds like the real you.

 CHECK YOUR READING What part of a telephone detects sound waves?

DIFFERENTIATE INSTRUCTION

? More Reading Support

E What machine first recorded sound? *the phonograph*

F What in a telephone changes vibrations into electrical signals? *a microphone*

Advanced Have students make a Venn diagram that shows the similarities and differences between how sound travels through a telephone and how sound travels from one place to another in the human body. *In both cases, vibrations change into electrical signals. A telephone changes electrical signals back into sound. In the human body, however, the electrical signals travel to the brain, which interprets them.*

R Challenge and Extension, p. 114

Recorded Sound

Sound occurs in real time, which means it is here for a moment and then gone. That is why Thomas Edison's invention of the phonograph—a way to preserve sound—was so important.

Edison's phonograph had a needle connected to a diaphragm that could pick up sound waves. The vibrations transferred by the sound waves were sent to a needle that cut into a piece of foil. The sound waves were translated into bumps along the grooves cut into the foil. These grooves contained all the information that was needed to reproduce the sound waves. Look at the image on top at the right to view an enlargement of record grooves. To play back the sound, Edison used another needle to track along the grooves etched in the foil. Later, phonographs were developed that changed sound waves into electrical signals that could be amplified.

Most people today listen to music on audio tapes or CDs. Tape consists of thin strips of plastic coated with a material that can be magnetized. Sounds that have been turned into electrical signals are stored on the tape as magnetic information. A CD is a hard plastic disc that has millions of microscopic pits arranged in a spiral. The bottom photograph at the right shows an enlargement of pits on the surface of a CD. These pits contain the information that a CD player can change into electrical signals, which are then turned into sound waves.

needle

record grooves

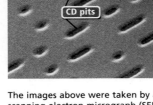

CD pits

The images above were taken by a scanning electron micrograph (SEM). Both the record grooves (top) and CD pits (bottom) store all of the information needed to reproduce sound.

CHECK YOUR READING Describe three devices on which sound is recorded.

2.4 Review

KEY CONCEPTS

1. Describe one medical use of ultrasound.
2. How are vibrations produced by each of the three main types of musical instruments?
3. How does a telephone record and reproduce sound?

CRITICAL THINKING

4. **Model** Draw a simple diagram to show how telephone communication works. Begin your diagram with the mouthpiece and end with the earpiece.
5. **Classify** The pitch of a musical instrument is changed by shortening the length of a vibrating column of air. What type of instrument is it?

CHALLENGE

6. **Synthesize** How is the earpiece of a telephone similar to the amplifier you read about in Section 3? Look again at the diagram of the amplifier on page 55 to help you find similarities.

Chapter 2: **Sound 63** **D**

ANSWERS

1. Sample answer: to detect the health of a fetus
2. Strings: by plucking or bowing the strings; Winds: by blowing air into a column; Percussion: by striking a part of the instrument
3. The microphone changes the vibrations to electrical signals that travel to another phone. The other phone changes the signals back into vibrations.
4. Diagrams should include sound waves entering the mouthpiece and being converted into electrical signals; electrical signals traveling to switching station; electrical signals traveling to listener's end; electrical signals being converted into sound waves in listener's earpiece.
5. a wind instrument
6. In both, electrical signals are converted into vibrations that produce sound waves.

CHAPTER INVESTIGATION

Focus

PURPOSE Students will make a stringed instrument and adjust it, so that changes in vibration produce two pitches.

OVERVIEW Students will observe how vibrations produce sound. They will build a stringed instrument and will find that

- the location of bridges on the instrument affects the pitch of its sound;
- cutting a hole in the instrument affects sound quality.

Lab Preparation

- Several days before performing the activity, have students bring appropriate boxes from home.
- Check rubber bands to be sure they are identical so that a difference in type or length of rubber band is not a variable.
- Prior to the investigation, have students read through the investigation, write their hypothesis, and prepare their data tables. Or you may wish to copy and distribute datasheets and rubrics.

 UNIT RESOURCE BOOK, pp. 123–131

 SCIENCE TOOLKIT, F14

Lab Management

SAFETY Emphasize the importance of using the rubber bands for their intended purpose only. Everyone in the classroom should wear safety goggles during the investigation.

INCLUSION Be sure any students who have hearing impairments are working with students who can describe the results to them. Have these students hold the sides of the box when the strings are plucked and note any differences in the vibrations they feel.

Teaching with Technology

Use a tape recorder to record the sounds of students' stringed instruments. Then play the tape back and ask students to recall what each pitch suggests about

---

Build a Stringed Instrument

OVERVIEW AND PURPOSE

People make music by plucking strings, blowing through tubes, and striking things. Part of each musical instrument vibrates to produce sounds that form the building blocks of music. In this lab, you will use what you have learned about sound to

- make a simple stringed instrument and see how the vibrating string produces sounds and
- change the design so that your stringed instrument produces more than one pitch.

Problem   Write It Up

How does the length of a string affect the pitch of the sound it produces when plucked?

Hypothesize   Write It Up

Write a hypothesis to explain how changing the length of the string affects the pitch of sound that is produced. Your hypothesis should take the form of an "If . . . , then . . . , because . . ." statement. Complete steps 1–3 before writing your hypothesis.

Procedure

MATERIALS
- book
- 3–5 rubber bands
- 2 pencils
- ruler
- shoebox
- scissors

1. Make a data table like the one shown. Try out the following idea for a simple stringed instrument. Stretch a rubber band around a textbook. Put two pencils under the rubber band to serve as bridges.

2. Put the bridges far apart at either end of the book. Find the string length by measuring the distance between the two bridges. Record this measurement in your **Science Notebook.** Pluck the rubber band to make it vibrate. Watch it vibrate and listen to the sound it makes.

3. Move the bridges closer together. What effect does this have on the length of the string? Measure and record the new length. How does this affect the tone that is produced? Record your observations.

INVESTIGATION RESOURCES

the instrument that made it.

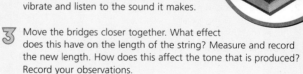

Technology Resources

Customize this student lab as needed or look for an alternative. Print rubrics to assess student lab reports.

 Lab Generator CD-ROM

 CHAPTER INVESTIGATION, Build a Stringed Instrument

• Level A, pp. 123–126
• Level B, pp. 127–130
• Level C, p. 131

Advanced students should complete Levels B & C.

Writing a Lab Report, D12–13

4 Make a musical instrument based on the principles you just identified. Begin by stretching rubber bands of the same weight or thickness over the box.

5 If necessary, reinforce the box with an extra layer of cardboard or braces so that it can withstand the tension of the rubber bands without collapsing.

6 Place pencils under the rubber bands at each end of the box. Arrange one pencil at an angle so that each string is a different length. Record the length of each string and your observations of the sounds produced. Experiment with the placement of the bridges.

7 You might also try putting one bridge at the center of the box and plucking on either side of it. How does this affect the range of pitches your instrument produces?

8 Experiment with the working model to see how you can vary the sounds. Try this variation: cut a hole in the center of the box lid. Put the lid back on the box. Replace the rubber bands and bridges. How does the hole change the sound quality?

▶ Observe and Analyze [Write It Up]

1. **RECORD OBSERVATIONS** Draw a picture of your completed instrument design. Be sure your data table is complete.

2. **ANALYZE** Explain what effect moving the bridges farther apart or closer together has on the vibrating string.

3. **SYNTHESIZE** Using what you have learned from this chapter, write a paragraph that explains how your instrument works. Be sure to describe how sound waves of different frequencies and different intensities can be produced on your instrument.

▶ Conclude [Write It Up]

1. **INTERPRET** Answer the question posed in the problem.

2. **ANALYZE** Compare your results with your hypothesis. Did your results support your hypothesis?

3. **EVALUATE** Describe any difficulties with or limitations of the materials that you encountered as you made your instrument.

4. **APPLY** Based on your experiences, how would you explain the difference between music and noise?

▶ INVESTIGATE Further

CHALLENGE Stringed instruments vary the pitch of musical sounds in several other ways. In addition to the length of the string, pitch depends on the tension, weight, and thickness of the string. Design an experiment to test one of these variables. How does it alter the range of sounds produced by your stringed instrument?

Build a Stringed Instrument

Problem How does the length of a string affect the pitch of the sound it produces when plucked?

Hypothesize

Observe and Analyze

Simple instrument: initial string length _____
Simple instrument: new string length _____

Table 1. Stringed Instrument Sound Observations

Stringed Instrument Designs	Length of Strings (cm)	Observations About Pitch and Sound Quality
Bridges at each end		
Bridge in middle		
After adding sound hole		

Conclude

BACK TO

the BIG idea

Have students describe a situation that demonstrates the transfer of energy by a sound wave. Students could use investigations in this chapter, the effect of vibrations on the human ear, or another example of how sound waves affect the medium they travel in.

◑ KEY CONCEPTS SUMMARY

SECTION 2.1

Ask: If the drum were in a vacuum, would sound be produced when it was struck? *No; sound needs a medium, and a vacuum has no medium.*

Ask: What part of the drum vibrates, producing sound? *the drum skin*

SECTION 2.2

Ask: If the frequency of the top wave is 200 hertz, what is the frequency of the bottom wave? *600 Hz*

SECTION 2.3

The waves shown were originally identical, but one of the waves has been amplified. Ask: Which wave was amplified? How do you know? *The bottom wave was amplified; its amplitude is greater.*

Ask: If energy is removed from the top wave, will it be more similar to the bottom wave or more different from it? *different since amplitude will be less*

SECTION 2.4

Ask: How is sound being used in the picture on the left? *Sound bounces off an object and is used to determine its location.*

Ask: What is the term for this practice? *echolocation*

Review Concepts

• Big Idea Flow Chart, p. T9
• Chapter Outline, p. T15–T16

---

2 Chapter Review

the BIG idea

Sound waves transfer energy through vibrations.

CONTENT REVIEW
CLASSZONE.COM

◑ KEY CONCEPTS SUMMARY

2.1 Sound is a wave.

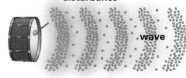

disturbance ⟶

wave

Sound is a longitudinal wave that travels through a material medium, such as air.

VOCABULARY
sound p. 37
vibration p. 37
vacuum p. 41

2.2 Frequency determines pitch.

A sound wave with a lower frequency and longer wavelength is perceived to have a lower pitch.

A sound wave with a higher frequency and shorter wavelength is perceived to have a higher pitch.

VOCABULARY
pitch p. 45
hertz p. 46
ultrasound p. 46
resonance p. 48
Doppler effect p. 50

2.3 Intensity determines loudness.

A sound wave with a lower amplitude and energy is perceived as a softer sound.

A sound wave with a higher amplitude and energy is perceived as a louder sound.

VOCABULARY
intensity p. 52
decibel p. 52
amplification p. 55
acoustics p. 55

2.4 Sound has many uses.

Human uses of sound:
sonar
ultrasound
music
telephone
recording

Bats use sound to locate objects.

VOCABULARY
echolocation p. 59
sonar p. 59

Technology Resources

Have students visit **ClassZone.com** or use the CD-ROM for a cumulative review of concepts.

 CONTENT REVIEW

 CONTENT REVIEW CD-ROM

Engage students in a whole-class interactive review of Key Concepts. Edit content as you wish.

 POWER PRESENTATIONS

Reviewing Vocabulary

Copy and complete the chart below by using vocabulary terms from this chapter.

Property of Wave	Unit of Measurement	Corresponding Quality of Sound
Frequency	1.	2.
3.	4.	loudness

Make a frame for each of the vocabulary words listed below. Write the word in the center. Decide what information to frame it with. Use definitions, examples, descriptions, parts, or pictures. An example is below.

5. resonance   **8.** acoustics

6. Doppler effect   **9.** echolocation

7. amplification   **10.** sonar

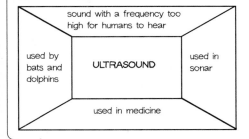

sound with a frequency too high for humans to hear

used by bats and dolphins   ULTRASOUND   used in sonar

used in medicine

Reviewing Key Concepts

Multiple Choice *Choose the letter of the best answer.*

11. Sound is a mechanical wave, so it always
  a. travels through a vacuum
  b. has the same amplitude
  c. is made by a machine
  d. travels through matter

12. Which unit is a measure of sound frequency?
  a. hertz
  b. decibel
  c. amp
  d. meters

13. In which of the following materials would sound waves move fastest?
  a. water
  b. cool air
  c. hot air
  d. steel

14. Which of the following effects is caused by amplification?
  a. wavelength increases
  b. amplitude increases
  c. frequency decreases
  d. decibel measure decreases

15. The frequency of a sound wave determines its
  a. pitch
  b. loudness
  c. amplitude
  d. intensity

16. As sound waves travel away from their source, their
  a. intensity increases
  b. energy increases
  c. intensity decreases
  d. frequency decreases

17. A telephone mouthpiece changes sound waves into
  a. electric signals
  b. vibrations
  c. CD pits
  d. grooves on a cylinder

Short Answer *Look at the diagrams of waves below. For the next two items, choose the wave diagram that best fits the description, and explain your choice.*

a.   **b.**   **c.**

18. the sound of a basketball coach blowing a whistle during practice

19. the sound of a cow mooing in a pasture

Reviewing Vocabulary

1. *hertz*
2. *pitch*
3. *intensity*
4. *decibel*

Sample answers:

5. *resonance: sound wave has the same frequency as an object's natural frequency; constructive interference; amplitudes combine with each other, sound wave is stronger*

6. *Doppler effect: lower pitch as source moves away; higher pitch as source approaches; pitch changes with movement of source; discovered by Christian Doppler*

7. *amplification: louder, greater intensity; greater height of sound wave; effect of an amplifier; used in TV, radio, stereo*

8. *acoustics: scientific study; study of sound; study of how sound is produced; study of how sound is detected*

9. *echolocation: use of sound; locates objects by reflection; used by bats; used by dolphins*

10. *sonar: used by humans to locate objects; used underwater; used in submarines, sound navigation, and mapping the sea floor*

Reviewing Key Concepts

11. *d*
12. *a*
13. *d*
14. *b*
15. *a*
16. *c*
17. *a*
18. *c; high frequency, pitch, and amplitude*
19. *a; low frequency, pitch, and amplitude*

Thinking Critically

20. By plucking the strings, which causes them to vibrate
21. by plucking the strings harder or more softly
22. The sound quality is different.
23. The order should be c, a, d, b.
24. Frequency and amplitude are both measures of a sound wave; frequency determines pitch, while amplitude determines loudness. Intensity and amplitude both relate to loudness of sound; intensity relates to amount of energy, while amplitude relates to the measure of wave energy. Pitch and quality concern an aspect of sound that we hear; pitch relates to the highness or lowness, while quality concerns other aspects of the sound. Fundamentals and overtones are pitches in a musical tone; the fundamental is the basic pitch we hear, while overtones are higher, fainter pitches that mix into the sound.

Using Math in Science

25. 7:45 A.M. to 8:00 A.M.
26. 103 dB; allow ±3 dB variance
27. yes, from about 7:30 A.M. to about 8:30 A.M.
28. You could use a bar to represent the noise-level "snapshot" at each quarter hour or half hour. The graph would not be as informative, because noise measures for times between would not be shown.

the BIG idea

29. by plucking the strings on the guitar; vibrations magnified by the amplifier; sound waves coming from the boom box; and sound caused by the shoes hitting the ground
30. Sample answer: Sound waves transfer energy through vibrations. Sound is a longitudinal wave that requires a medium. As frequency increases, pitch is higher. More intense sounds are louder.

UNIT PROJECTS

Collect schedules, materials lists, and questions. Be sure dates and materials are obtainable, and questions are focused.

Unit Projects, pp. 5-10

D 68 Unit: **Waves, Sound, and Light**

Thinking Critically

Look at the photograph of a lute above. Write a short answer to the next two questions.

20. **HYPOTHESIZE** How might sound waves be produced using the instrument in the photograph?

21. **APPLY** How might a person playing the instrument in the photograph vary the intensity?

22. **COMMUNICATE** Two people are singing at the same pitch, yet they sound different. Explain why.

23. **SEQUENCE** Copy the following sequence chart on your paper. Write the events in the correct sequence on the chart.

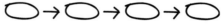

Events
a. Sound waves race out from the wind chime.
b. Air friction gradually weakens the chime sound.
c. A breeze makes a wind chime vibrate.
d. A person nearby hears the wind chime.

24. **COMPARE AND CONTRAST** Write a description of the similarities of and differences between each of the following pairs of terms: frequency—amplitude; intensity—amplitude; pitch—quality; fundamental tone—overtones.

Using Math in Science

Read the line graph below showing freeway noise levels at a toll collector's booth. Use the data in the graph to answer the next four questions.

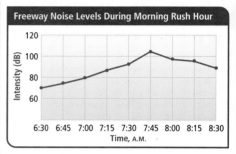

25. Which is the noisiest quarter-hour?

26. Estimate the loudest level of sound that the toll collector is exposed to.

27. If ear protection should be worn for a sound level above 90 dB, should the toll collector wear hearing protection? If so, during which times?

28. Describe how you could turn the line graph into a bar graph. Would the bar graph be as informative? Explain your answer.

the BIG idea

29. **ANALYZE** Look back at the picture at the start of the chapter on pages 34–35. How are sound waves being produced?

30. **SUMMARIZE** Write a paragraph summarizing this chapter. Use the Big Idea on page 34 as your topic sentence. Write examples of each key concept listed on page 34.

UNIT PROJECTS

Check your schedule for your unit project. How are you doing? Be sure that you've placed data or notes from your research in your project folder.

D 68 Unit: Waves, Sound, and Light

MONITOR AND RETEACH

If students have trouble applying the concepts in items 25-28, have them do the following activities:
• Make a bar graph as in problem 28 and compare the information.
• Trace the graph on a piece of paper. Use a ruler to carefully sketch grid lines on the graph, so that the numbers can be easily read.
• Make up additional questions about the graph. Trade questions, and answer someone else's questions.

Students may benefit from summarizing one or more sections of the chapter.

Summarizing the Chapter, pp. 141–142

Standardized Test Practice

For practice on your state test, go to . . .  **TEST PRACTICE** CLASSZONE.COM

Analyzing Experiments

Read the following description of the way scientists study animals' hearing. Then answer the questions below.

Scientists test the hearing ranges of a human by making a sound and asking the person to say whether it was heard. This cannot be done with animals. Scientists use different methods to find animals' hearing ranges. In some experiments, they train animals—by rewarding them with food or water—to make specific behaviors when they hear a sound. Another method is to study an animal's nervous system for electrical reactions to sounds.

Researchers have found that dogs and cats can hear a wide range of sounds. Both dogs and cats can hear much higher frequencies than humans can. Lizards and frogs can only hear sounds in a much narrower range than humans can. Elephants can hear a wider range than lizards and frogs but not as wide a range as dogs and cats. Elephants can hear the lowest frequency sounds of all these animals.

1. What type of behavior would be best for scientists to train animals to make as a signal they hear a sound?
   a. a typical motion that the animal makes frequently
   b. a motion that is difficult for the animal to make
   c. a motion the animal makes rarely but does make naturally
   d. a complicated motion of several steps

2. According to the passage, which animals can hear sounds with the highest frequencies?
   a. cats   c. frogs
   b. elephants   d. lizards

3. The high-pitched sounds of car brakes are sometimes more bothersome to pet dogs than they are to their owners. Based on the experimental findings, what is the best explanation for that observation?
   a. The dogs hear high-intensity sounds that their owners cannot hear.
   b. The dogs hear low-intensity sounds that their owners cannot hear.
   c. The dogs hear low-frequency sounds that their owners cannot hear.
   d. The dogs hear high-frequency sounds that their owners cannot hear.

4. Which animal will hear sounds with the longest wavelengths?
   a. cat   c. elephant
   b. dog   d. frog

Extended Response

Answer the two questions below in detail. Include some of the terms from the word box in your answer. Underline each term you use in your answer.

| amplitude | distance | Doppler effect |
| frequency | pitch | wavelength |

5. Suppose you are riding in a car down the street and pass a building where a fire alarm is sounding. Will the sound you hear change as you move up to, alongside, and past the building? Why or why not?

6. Marvin had six glass bottles that held different amounts of water. He blew air into each bottle, producing a sound. How would the sounds produced by each of the six bottles compare to the others? Why?

Chapter 2: **Sound** 69 **D**

Analyzing Experiments

1. c   3. d
2. a   4. c

Extended Response

5. RUBRIC
4 points for a response that correctly answers the question and uses the following terms accurately:
• pitch   • wavelength   • distance
Sample: The sound is higher in pitch as the car moves closer to the alarm and becomes lower in pitch as it moves farther away. This is because sound changes with distance. When the distance decreases, the wavelength is shorter and the pitch is higher. When the distance increases, the wavelength is longer and the pitch lowers. A change in pitch occurs when either the source or the receiver of the sound is moving. This is known as the Doppler effect.

3 points for a response that correctly answers the question and uses two terms accurately
2 points for a response that correctly answers the question and uses one term accurately
1 point for a response that correctly answers the question but does not use the terms

6. RUBRIC
4 points for a response that correctly answers the question and uses the following terms accurately:
• pitch   • frequency   • distance
Sample: The pitch, or frequency, increases as the amount of water increases. So the bottles with more water would have a higher pitch than the bottles with less water. The reason for this is that the sound is traveling through a smaller distance since there is a smaller amount of air.

3 points for a response that correctly answers the question and uses two terms accurately
2 points for a response that correctly answers the question and uses one term accurately
1 point for a response that correctly answers the question but does not use the terms

METACOGNITIVE ACTIVITY

Have students answer the following questions in their **Science Notebook:**

1. Was there a quality of sound that was new to you? If so, what?

2. What questions that you had about sound were answered in this chapter? What questions do you still have about sound?

3. Do the concepts in this chapter relate to your Unit Project? If so, how? If not, how might you revise your project to include sound?

CHAPTER
3 Electromagnetic Waves

Physical Science
UNIFYING PRINCIPLES

PRINCIPLE 1

Matter is made of particles too small to see.

PRINCIPLE 2

Matter changes form and moves from place to place.

PRINCIPLE 3

Energy changes from one form to another, but it cannot be created or destroyed.

PRINCIPLE 4

Physical forces affect the movement of all matter on Earth and throughout the universe.

Unit: Waves, Sound, and Light
BIG IDEAS

CHAPTER 1
Waves

Waves transfer energy and interact in predictable ways.

CHAPTER 2
Sound

Sound waves transfer energy through vibrations.

CHAPTER 3
Electromagnetic Waves
Electromagnetic waves transfer energy through radiation.

CHAPTER 4
Light and Optics

Optical tools depend on the wave behavior of light.

CHAPTER 3
KEY CONCEPTS

SECTION 3.1

Electromagnetic waves have unique traits.
1. An electromagnetic (EM) wave is a disturbance in a field.
2. EM waves can travel in a vacuum.
3. EM waves can interact with a material medium.

SECTION 3.2

Electromagnetic waves have many uses.
1. EM waves have different frequencies.
2. Long length, low frequency (radio waves, microwaves)
3. Mid-range length and frequency (infrared, visible and ultraviolet light)
4. Short length, high-frequency (x-rays, gamma rays)

SECTION 3.3

The Sun is the source of most visible light.
1. Light comes from the Sun and other natural sources.
2. Some living things produce visible light.
3. Human technologies produce visible light.

SECTION 3.4

Light waves interact with materials.
1. Light can be reflected, transmitted, or absorbed.
2. Wavelength determines color.

The Big Idea Flow Chart is available on p. T17 in the **UNIT TRANSPARENCY BOOK.**

Previewing Content

3.1 Electromagnetic waves have unique traits. pp. 73–78

1. An electromagnetic wave is a disturbance in a field.

An **electromagnetic wave** is a disturbance that transfers energy through a field.

A **field** is the area around an object where the object applies force on another object without touching it. The two fields of an EM wave—electric and magnetic—vibrate at right angles to each other and are perpendicular to the direction the wave is moving (as illustrated in the visual below).

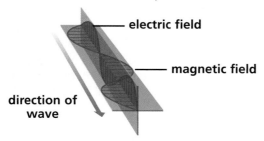

electric field

magnetic field

direction of wave

An EM wave is emitted whenever a charged atomic particle accelerates. The Sun and human technology are the two main sources of EM waves.

2. Electromagnetic waves can travel in a vacuum.

Once EM waves are produced, they travel on their own, independent of the source that emitted them. They don't need a medium and can travel in a vacuum at about 300,000 kilometers per second.

Radiation is the transfer of energy in the form of EM waves. EM waves do not lose energy as they travel in a vacuum. EM radiation from the Sun travels in a straight line through the vacuum of outer space.

3. Electromagnetic waves can interact with a material medium.

When EM waves encounter a material medium, they may transfer energy to it. Different mediums interact differently with EM waves, and can change the direction of the wave and affect energy transfer.

- In a vacuum, EM waves transfer energy by moving potential energy from one place to another.
- In a medium, an EM wave's potential energy can be converted into other forms, such as heat.

3.2 Electromagnetic waves have many uses. pp. 79–87

1. EM waves have different frequencies.

An EM wave's frequency determines the wave's characteristics. The higher the frequency, the more energy the wave carries. The **electromagnetic spectrum** is a continuum of waves from the lowest-frequency radio waves to the highest-frequency gamma waves.

2. Radio waves and microwaves have long wavelengths and low frequencies.

Radio waves are long, low-energy EM waves. They can be modified and converted into the sound and pictures of radios and TVs. **Microwaves** have more energy and shorter wavelengths than radio waves.

- In radar, microwaves are reflected off an object and returned to their source as a way of locating the object.
- Cell phone technology is like radio transmission but uses microwaves. A system of towers connects cell phones to each other and to the regular phone system.

3. Infrared, visible, and ultraviolet light have mid-range wavelengths and frequencies.

The range of frequencies that humans can see is just a tiny part of the EM spectrum.

- **Infrared waves** lie between visible light and microwaves. They are emitted by warm objects. Infrared technology is used to cook food, detect warm objects, and provide heat.
- **Ultraviolet light** carries more energy than visible light and can damage human tissue. It is used to sterilize medical equipment and kill bacteria in food.

4. X-rays and gamma rays have short wavelengths and high frequencies.

Because of their high frequencies, x-rays and gamma rays carry very high energies. They are naturally produced by stars.

- **X-rays** can penetrate soft tissues but not hard tissues of the body, making these waves useful for medical imaging.
- **Gamma rays** can penetrate all the tissues of the body, killing normal cells and causing cancer cells to grow.

Common Misconceptions

WAVE TRANSMISSION Students may think that EM waves must have a medium, or matter, to travel. EM waves can travel through matter, but do not need matter to travel.

 This misconception is addressed on p. 75.

MISCONCEPTION DATABASE
CLASSZONE.COM Background on student misconceptions

INFRARED LIGHT Students may think infrared light is part of the visible spectrum. Infrared light, however, is below the visible range and cannot be detected by the human eye.

 This misconception is addressed on p. 85.

Previewing Content

3.3 The Sun is the source of most visible light. pp. 88–92

1. Light comes from the Sun and other natural sources.
Almost all organisms depend on light for survival. Virtually all light on Earth initiates in sunlight. Green plants use sunlight to synthesize food that both plants and animals depend on for energy. The Sun's intense heat produces light through **incandescence.** This light is the ultimate source of almost all energy on Earth.

2. Some living things produce visible light.
Chemical reactions in some living organisms produce **bioluminescence.** Unlike incandescence, luminescence produces light without the high temperatures that could harm organisms.

3. Human technologies produce visible light.
The discovery of electricity has led to several artificial lighting technologies.

- Most incandescent light bulbs use tungsten filaments and produce light and heat.
- Halogen lighting also produces lots of heat, but the tungsten filament lasts longer than in ordinary incandescent light bulbs and produces more light.
- Fluorescent lighting is cool and efficient. The bulb is coated with a phosphor, which glows when it absorbs the UV waves generated within the bulb.
- LEDs are semiconductors that produce light when electricity passes through. They are cool, efficient, and long-lasting. Light produced by LEDs has many advantages over other forms of lighting.

The visual below shows the different parts of incandescent, halogen, and fluorescent bulbs.

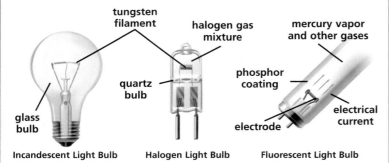

tungsten filament | halogen gas mixture | mercury vapor and other gases

quartz bulb | phosphor coating

glass bulb | electrode | electrical current

Incandescent Light Bulb   Halogen Light Bulb   Fluorescent Light Bulb

3.4 Light waves interact with materials. pp. 93–101

1. Light can be reflected, transmitted, or absorbed.
EM waves can interact with a material medium in the same ways that mechanical waves do. The medium can reflect, transmit, or absorb the waves.
Most objects are visible because they reflect light. **Transmission** and **absorption** affect how objects look.

- Objects that transmit most of the light that strikes them appear transparent.
- Objects that transmit some of the light that strikes them but cause it to scatter appear translucent.
- Opaque objects do not transmit light.

Scattering from fine particles in a material sends light in all directions and creates diffuse light. **Polarization** reduces glare. When all of the electric fields of a group of light waves vibrate in the same direction, the light is polarized.

2. Wavelength determines color.
Visible light is a spectrum that is usually divided into seven colors. Visible light reflected from an object gives it color; a green leaf reflects green wavelengths and absorbs all other visible wavelengths.
The three **primary colors** are light of different wavelengths that produce white light when mixed equally. They are red, green, and blue.
The **primary pigments** reflect wavelengths of cyan, yellow, and magenta. When you mix pigments, the mixture absorbs more colors, and reflects fewer wavelengths.
The visual below shows primary pigments.

Common Misconceptions

BIOLUMINESCENCE Students commonly think that bioluminescence is an electrical process that occurs within an organism. In truth, bioluminescence is the result of chemical reactions that produce energy in the form of light.

 This misconception is addressed on p. 89.

MISCONCEPTION DATABASE
CLASSZONE.COM Background on student misconceptions

THE SKY'S COLOR It is commonly thought that the sky is blue because of the reflection of blue light from Earth's oceans. In fact, particles in the atmosphere scatter the blue wavelengths of sunlight more than they scatter other wavelengths, making the sky appear blue.

TE This misconception is addressed on p. 95.

Previewing Labs

EXPLORE (the BIG idea)

What Melts the Ice Cubes? p. 71
Students will observe ice cubes melting to find out that black and white materials absorb different quantities of energy from light.

TIME 10 minutes
MATERIALS 2 ice cubes, 2 sandwich bags, sheet of white paper, sheet of black paper

What Is White Light Made Of? p. 71
Students will use a CD as a prism to find out that white light can separate into several colors.

TIME 10 minutes
MATERIALS compact disc, sheet of white paper, flashlight

Internet Activity: Electromagnetic Waves, p. 71
Students learn that the Sun emits different wavelengths.

TIME 20 minutes
MATERIALS computer with Internet access

SECTION 3.1

EXPLORE Electromagnetic Waves, p. 73
Students use a mirror to show that EM waves can be reflected.

TIME 10 minutes
MATERIALS TV with remote control unit, mirror with stand

INVESTIGATE Wave Behavior, p. 76
Students design an experiment to determine what makes the vanes of a radiometer move.

TIME 30 minutes
MATERIALS radiometer

SECTION 3.2

EXPLORE Radio Waves, p. 79
Students explore how radio waves form and how they are detected.

TIME 10 minutes
MATERIALS 25 cm copper wire (2 pieces), C or D battery, 5 cm electrical tape, metal fork, portable radio

INVESTIGATE The Electromagnetic Spectrum, p. 84
Students draw conclusions about the existence of invisible EM waves by measuring their temperature.

TIME 30 minutes
MATERIALS sheet of white paper, black marker, 3 thermometers, prism

---

SECTION 3.3

INVESTIGATE Artificial Lighting, p. 90
Students design an experiment to examine various kinds of bulbs.

TIME 30 minutes
MATERIALS a variety of bulb types and sizes

SECTION 3.4

EXPLORE Light and Matter, p. 93
Students observe the scattering of light in a translucent material.

TIME 10 minutes
MATERIALS clear plastic container with lid, 4–6 cups water, measuring spoons, 10 mL milk, flashlight

INVESTIGATE Mixing Colors, p. 98
Students discover the colors that make up black ink.

TIME 30 minutes
MATERIALS 3 coffee filters, scissors, 3 black felt-tip markers (different brands), 3 cups water

CHAPTER INVESTIGATION Wavelength and Color, pp. 100–101
Students use a light box to learn that an object's color is determined by the wavelengths of light it reflects.

TIME 40 minutes
MATERIALS sheets of acetate (red, blue, green), ruler, scissors, shoe box, masking tape, light source, solid-colored objects

---

 Additional INVESTIGATION, Seeing the Invisible, A, B, & C, pp. 203–211, Teacher Instructions, pp. 284–285

Previewing Chapter Resources

| | INTEGRATED TECHNOLOGY | LABS AND ACTIVITIES |

CHAPTER 3
Electro-magnetic Waves

 CLASSZONE.COM
- eEdition Plus
- EasyPlanner Plus
- Misconception Database
- Content Review
- Test Practice
- Visualization
- Resource Centers
- Internet Activity: EM Waves
- Math Tutorial

 SCILINKS.ORG
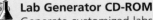 SCiLINKS

 CD-ROMS
- eEdition
- EasyPlanner
- Power Presentations
- Content Review
- Lab Generator
- Test Generator

 AUDIO CDS
- Audio Readings
- Audio Readings in Spanish

 EXPLORE the Big Idea, p. 71
- What Melts the Ice Cubes?
- What Is White Light Made Of?
- Internet Activity: Electromagnetic Waves

 UNIT RESOURCE BOOK
Unit Projects, pp. 5–10

 Lab Generator CD-ROM
Generate customized labs.

SECTION
3.1 Electromagnetic waves have unique traits.
pp. 73–78

Time: 2 periods (1 block)
 Lesson Plan, pp. 143–144

 • **VISUALIZATION,** EM Waves
- • **MATH TUTORIAL**

 UNIT TRANSPARENCY BOOK
- Big Idea Flow Chart, p. T17
- Daily Vocabulary Scaffolding, p. T18
- Note-Taking Model, p. T19
- 3-Minute Warm-Up, p. T20

 • EXPLORE Electromagnetic Waves, p. 73
- • INVESTIGATE Wave Behavior, p. 76
- • Math in Science, p. 78

 UNIT RESOURCE BOOK
- Datasheet, Wave Behavior, p. 152
- Math Support & Practice, pp. 192–193

SECTION
3.2 Electromagnetic waves have many uses.
pp. 79–87

Time: 2 periods (1 block)
 Lesson Plan, pp. 154–155

 RESOURCE CENTER, EM Spectrum

 UNIT TRANSPARENCY BOOK
- Daily Vocabulary Scaffolding, p. T18
- 3-Minute Warm-Up, p. T20
- "The Electromagnetic Spectrum" Visual, p. T22

 • EXPLORE Electromagnetic Waves, p. 79
- • INVESTIGATE The EM Spectrum, p. 84
- • Think Science, p. 87

 UNIT RESOURCE BOOK
- Datasheet, The EM Spectrum, p. 163
- Additional INVESTIGATION, Seeing the Invisible, A, B, & C, pp. 203–211

SECTION
3.3 The Sun is the source of most visible light.
pp. 88–92

Time: 2 periods (1 block)
 Lesson Plan, pp. 165–166

 RESOURCE CENTER, Visible Light

 UNIT TRANSPARENCY BOOK
- Daily Vocabulary Scaffolding, p. T18
- 3-Minute Warm-Up, p. T21

 INVESTIGATE Artificial Lighting, p. 90

 UNIT RESOURCE BOOK
Datasheet, Artificial Lighting, p. 174

SECTION
3.4 Light waves interact with materials.
pp. 93–101

Time: 4 periods (2 blocks)
 Lesson Plan, pp. 176–177

 UNIT TRANSPARENCY BOOK
- Big Idea Flow Chart, p. T17
- Daily Vocabulary Scaffolding, p. T18
- 3-Minute Warm-Up, p. T21
- Chapter Outline, pp. T23–T24

• EXPLORE Light and Matter, p. 93
- • INVESTIGATE Mixing Colors, p. 98
- • CHAPTER INVESTIGATION, Wavelength and Color, pp. 100–101

UNIT RESOURCE BOOK
- Datasheet, Mixing Colors, p. 185
- CHAPTER INVESTIGATION, A, B, & C, pp. 194–202

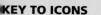
READING AND REINFORCEMENT

- Frame Game, B26–27
- Supporting Main Ideas, C42
- Daily Vocabulary Scaffolding, H1–8

 UNIT RESOURCE BOOK
- Vocabulary Practice, pp. 189–190
- Decoding Support, p. 191
- Summarizing the Chapter, pp. 212–213

 Audio Readings CD
Listen to Pupil Edition.

Audio Readings in Spanish CD
Listen to Pupil Edition in Spanish.

 **UNIT RESOURCE BOOK**
- Reading Study Guide, A & B, pp. 145–148
- Spanish Reading Study Guide, pp. 149–150
- Challenge and Extension, p. 151
- Reinforcing Key Concepts, p. 153
- Challenge Reading, pp. 187–188

UNIT RESOURCE BOOK
- Reading Study Guide, A & B, pp. 156–159
- Spanish Reading Study Guide, pp. 160–161
- Challenge and Extension, p. 162
- Reinforcing Key Concepts, p. 164

UNIT RESOURCE BOOK
- Reading Study Guide, A & B, pp. 167–170
- Spanish Reading Study Guide, pp. 171–172
- Challenge and Extension, p. 173
- Reinforcing Key Concepts, p. 175

UNIT RESOURCE BOOK
- Reading Study Guide, A & B, pp. 178–181
- Spanish Reading Study Guide, pp. 182–183
- Challenge and Extension, p. 184
- Reinforcing Key Concepts, p. 186

ASSESSMENT

- Chapter Review, pp. 103–104
- Standardized Test Practice, p. 105

 **UNIT ASSESSMENT BOOK**
- Diagnostic Test, pp. 40–41
- Chapter Test, A, B, & C, pp. 46–51
- Alternative Assessment, pp. 58–59

 Spanish Chapter Test, pp. 289–292

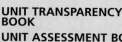 **Test Generator CD-ROM**
Generate customized tests.

 Lab Generator CD-ROM
Rubrics for Labs

 Ongoing Assessment, pp. 74–77

 Section 3.1 Review, p. 77

 UNIT ASSESSMENT BOOK
Section 3.1 Quiz, p. 42

 Ongoing Assessment, pp. 79, 81–82, 85

 Section 3.2 Review, p. 86

 UNIT ASSESSMENT BOOK
Section 3.2 Quiz, p. 43

 Ongoing Assessment, pp. 88–92

 Section 3.3 Review, p. 92

 UNIT ASSESSMENT BOOK
Section 3.3 Quiz, p. 44

 Ongoing Assessment, pp. 93–95, 97–99

 Section 3.4 Review, p. 99

 UNIT ASSESSMENT BOOK
Section 3.4 Quiz, p. 45

STANDARDS

National Standards
A.2–8, A.9.a–f, B.3.c, B.3.f, E.2–5, F.4.c, F.5.c

See p. 70 for the standards.

National Standards
A.2–8, A.9.a–f, E.2–5

National Standards
A.2–7, A.9.a–b, A.9.d–f, F.4.c, F.5.c

National Standards
A.2–7, A.9.a–b, A.9.d–f, B.3.f, E.2–5, F.5.c

National Standards
A.2–7, A.9.a–b, A.9.e–f, B.3.c

Previewing Resources for Differentiated Instruction

CHAPTER INVESTIGATION

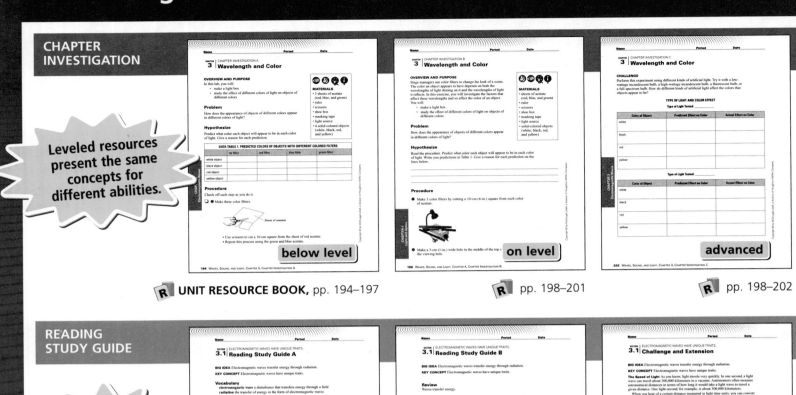

below level

on level

advanced

R **UNIT RESOURCE BOOK,** pp. 194–197   R pp. 198–201   R pp. 198–202

Leveled resources present the same concepts for different abilities.

READING STUDY GUIDE

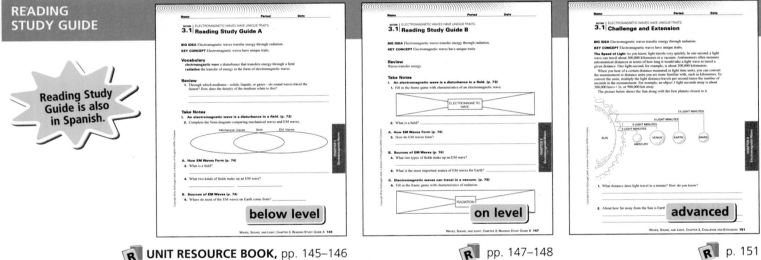

below level

on level

advanced

R **UNIT RESOURCE BOOK,** pp. 145–146   R pp. 147–148   R p. 151

Reading Study Guide is also in Spanish.

CHAPTER TEST

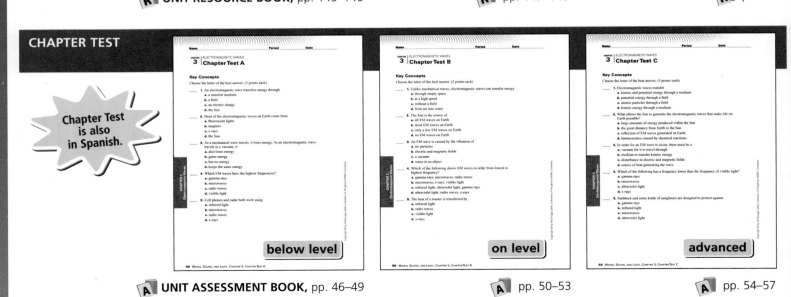

below level

on level

advanced

A **UNIT ASSESSMENT BOOK,** pp. 46–49   A pp. 50–53   A pp. 54–57

Chapter Test is also in Spanish.

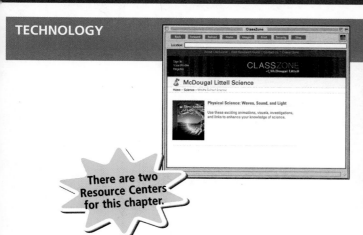

There are two Resource Centers for this chapter.

CLASSZONE.COM

CD/CD-ROMS

CLASSZONE.COM

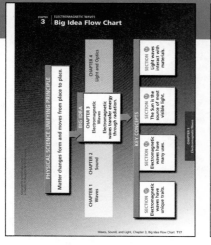

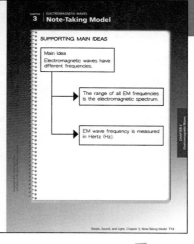

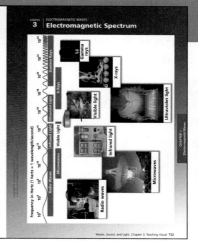

T UNIT TRANSPARENCY BOOK, p. T17

T p. T19

T p. T22

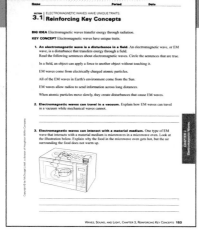

Reinforcing Key Concepts for each section

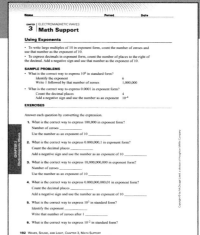

R UNIT RESOURCE BOOK, p. 153

R pp. 189–190

R p. 192

INTRODUCE

the BIG idea

Have students look at the photograph of the cell phone user and discuss how the question in the box links to the Big Idea:

- If waves are involved in cell phone technology, what kind of waves might they be?

- How can waves travel from one cell phone to another?

- What other devices use the same kind of waves as a cell phone?

National Science Education Standards

Content

B.3.c Light interacts with matter by transmission, absorption, and scattering.

B.3.f The Sun is a major source of energy for changes on Earth's surface. Energy from the Sun is transferred to Earth in the form of visible light, infrared, and ultraviolet radiation.

Process

A.2–8 Design and conduct an investigation; use tools to gather and interpret data; use evidence to describe, predict, explain, model; think critically to make relationships between evidence and explanation; recognize different explanations and predictions; communicate scientific procedures and explanations; use mathematics.

A.9.a–f Understand scientific inquiry by using different investigations, methods, mathematics, technology, and explanations based on logic, evidence, and skepticism.

E.2–5 Design, implement, and evaluate a solution or product; communicate technological design.

F.4.c–d Risks and benefits

F.5.c Science and technology in society

CHAPTER

3 Electromagnetic Waves

the BIG idea

Electromagnetic waves transfer energy through radiation.

Key Concepts

 SECTION 3.1 Electromagnetic waves have unique traits. Learn how electromagnetic waves differ from mechanical waves.

SECTION 3.2 Electromagnetic waves have many uses. Learn about the behaviors and uses of different types of electromagnetic waves.

 SECTION 3.3 The Sun is the source of most visible light. Learn about the natural and artificial production of light.

 SECTION 3.4 Light waves interact with materials. Learn how light waves behave in a material medium.

Internet Preview

CLASSZONE.COM

Chapter 3 online resources: Content Review, Simulation, Visualization, two Resource Centers, Math Tutorial, Test Practice.

How does this phone stay connected?

 INTERNET PREVIEW

CLASSZONE.COM For student use with the following pages:

Review and Practice
- Content Review, pp. 72, 102
- Math Tutorial: Positive and Negative Exponents, p. 78
- Test Practice, p. 105

Activities and Resources
- Internet Activity: EM Waves, p. 71
- Visualization: EM Waves, p. 74
- Resource Centers: EM Spectrum, p. 80; Visible Light, p. 88

 **NSTA** scilinks.org SC*L*INKS

Light and Color
Code: MDL029

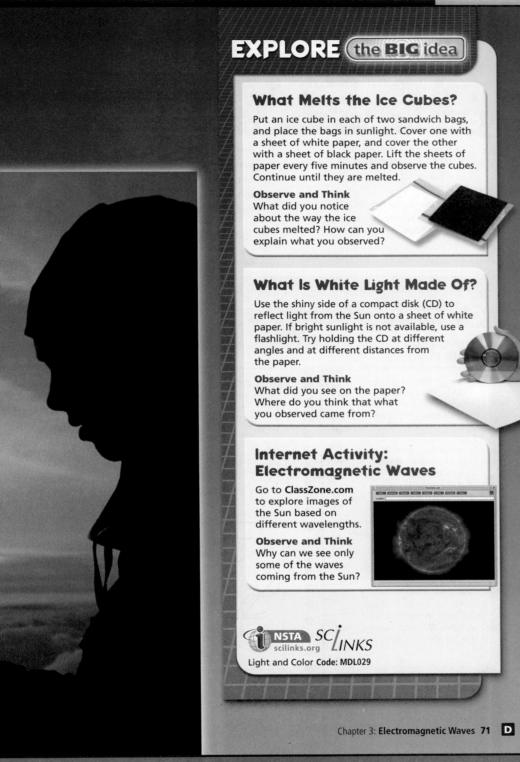

What Melts the Ice Cubes?

Put an ice cube in each of two sandwich bags, and place the bags in sunlight. Cover one with a sheet of white paper, and cover the other with a sheet of black paper. Lift the sheets of paper every five minutes and observe the cubes. Continue until they are melted.

Observe and Think
What did you notice about the way the ice cubes melted? How can you explain what you observed?

What Is White Light Made Of?

Use the shiny side of a compact disk (CD) to reflect light from the Sun onto a sheet of white paper. If bright sunlight is not available, use a flashlight. Try holding the CD at different angles and at different distances from the paper.

Observe and Think
What did you see on the paper? Where do you think that what you observed came from?

Internet Activity: Electromagnetic Waves

Go to **ClassZone.com** to explore images of the Sun based on different wavelengths.

Observe and Think
Why can we see only some of the waves coming from the Sun?

NSTA
scilinks.org
SCiLINKS

Light and Color **Code: MDL029**

TEACHING WITH TECHNOLOGY

BL and Probeware If probeware is available, students can use temperature-sensing probes instead of thermometers for "Investigate the Electromagnetic Spectrum" on p. 84.

Digital Camera Take photographs of each step during "Investigate Mixing Colors" on p. 98 to document the activity. Use the camera to show time-elapsed images of the ink samples, taking photographs every few minutes to show how the samples have changed. Compare the final images.

These inquiry-based activities are appropriate for use at home or as a supplement to classroom instruction.

What Melts the Ice Cubes?

PURPOSE To observe the difference between reflection and absorption of EM radiation. Students observe different amounts of light energy absorbed by black paper and white paper.

TIP *10 min.* The two ice cubes should be the same size when they are placed in the bags. Students can also try covering a bag with aluminum foil, shiny side out, and observe the rate of melting.

Answer: The cube under the black paper melts more quickly. The black paper gets hotter in the sun because it absorbs more light than the white paper.

REVISIT after p. 97.

What Is White Light Made Of?

PURPOSE To observe that white light is made up of different wavelengths and individual colors. Students use a CD as a prism to separate white light.

TIP *10 min.* Prompt students to think about why the silver CD reflects different colors and why the colors change as the CD moves.

Answer: A rainbow of colors from the light being reflected by the CD. The colors were part of the light.

REVISIT after p. 97.

Internet Activity: Electromagnetic Waves

PURPOSE To see that the Sun emits different wavelengths and that they must be detected and visualized in different ways.

TIP *20 min.* Before they use the simulation, have students predict how the different types of EM radiation coming from the Sun will look.

Answer: The human eye can detect waves only in the visible range.

REVISIT after p. 85.

◑ CONCEPT REVIEW

Activate Prior Knowledge

- Give students a list of phrases or objects that waves travel through. Be sure to include "vacuum" or "outer space."
- Ask if there is any place or object on the list that a mechanical wave cannot travel through. *yes; vacuum*
- Ask how speed of sound is affected by different mediums on the list. *Answers should recognize that sound travels faster in denser materials and not at all in a vacuum.*

◑ TAKING NOTES

Supporting Main Ideas

Making a chart of the main ideas will help students organize the material in the chapter. Using the section's heads is a good start; students can then add information, explanations, and examples that support these ideas.

Vocabulary Strategy

By surrounding each vocabulary term with examples and descriptions, students will develop a thorough understanding of the meaning of each term. Respellings should be included if appropriate.

Vocabulary and Note-Taking Resources

- Vocabulary Practice, pp. 189–190
- Decoding Support, p. 191

- Daily Vocabulary Scaffolding, p. T18
- Note-Taking Model, p. T19

- Frame Game, B26–27
- Supporting Main Ideas, C42
- Daily Vocabulary Scaffolding, H1–8

CHAPTER 3
Getting Ready to Learn

◑ CONCEPT REVIEW

- A wave is a disturbance that transfers energy.
- Mechanical waves have a medium.
- Waves can be measured.
- Waves react to a change in medium.

◑ VOCABULARY REVIEW

mechanical wave p. 11
wavelength p. 17
frequency p. 17
reflection p. 25
field *See Glossary.*

CONTENT REVIEW
CLASSZONE.COM
Review concepts and vocabulary.

◑ TAKING NOTES

SUPPORTING MAIN IDEAS

Make a chart to show main ideas and the information that supports them. Copy each blue heading. Below each heading, add supporting information, such as reasons, explanations, and examples.

VOCABULARY STRATEGY

Write each new vocabulary term in the center of a **frame game** diagram. Decide what information to frame it with. Use examples, descriptions, parts, sentences that use the term in context, or pictures. You can change the frame to fit each term.

See the Note-Taking Handbook on pages R45–R51.

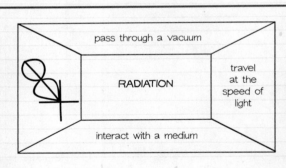

SCIENCE NOTEBOOK

MAIN IDEA
Electromagnetic waves have unique properties.

→ EM waves are disturbances in a field rather than in a medium.

→ EM waves can travel through a vacuum.

→ EM waves travel at the speed of light.

pass through a vacuum

RADIATION

travel at the speed of light

interact with a medium

CHECK READINESS

Administer the Diagnostic Test to determine students' readiness for new science content and their mastery of requisite math skills.

  Diagnostic Test, pp. 40–41

Technology Resources

Students needing content and math skills should visit **ClassZone.com**.

- **CONTENT REVIEW**
- **MATH TUTORIAL**

  CONTENT REVIEW CD-ROM

Electromagnetic waves have unique traits.

 BEFORE, you learned
- Waves transfer energy
- Mechanical waves need a medium to travel

▶ **NOW, you will learn**
- How electromagnetic waves differ from mechanical waves
- Where electromagnetic waves come from
- How electromagnetic waves transfer energy

VOCABULARY
electromagnetic wave p. 73
radiation p. 75

EXPLORE Electromagnetic Waves

How does the signal from a remote control travel?

PROCEDURE
1. Turn the TV on and off using the remote control.
2. Work with a partner to try to turn on the TV by aiming the remote control at the mirror.

MATERIALS
- TV with remote control unit
- mirror with stand

WHAT DO YOU THINK?
How did you have to position the remote control and the mirror in order to operate the TV? Why do you think this worked?

An electromagnetic wave is a disturbance in a field.

Did you know that you are surrounded by thousands of waves at this very moment? Waves fill every cubic centimeter of the space around you. They collide with or pass through your body all the time.

Most of these waves are invisible, but you can perceive many of them. Light is made up of these waves, and heat can result from them. Whenever you use your eyes to see, or feel the warmth of the Sun on your skin, you are detecting their presence. These waves also allow radios, TVs, and cell phones to send or receive information over long distances. These waves have the properties shared by all waves, yet they are different from mechanical waves in important ways. This second type of wave is called an electromagnetic wave. An **electromagnetic wave** (ih-LEHK-troh-mag-NEHT-ihk) is a disturbance that transfers energy through a field. Electromagnetic waves are also called EM (EE-EHM) waves.

VOCABULARY
Create a frame game diagram for the term *electromagnetic wave.*

Chapter 3: **Electromagnetic Waves** 73 **D**

RESOURCES FOR DIFFERENTIATED INSTRUCTION

Below Level
UNIT RESOURCE BOOK
- Reading Study Guide A, pp. 145–146
- Decoding Support, p. 191

AUDIO CDS

Advanced
UNIT RESOURCE BOOK
- Challenge and Extension, p. 162
- Challenge Reading, pp. 187–188

English Learners
UNIT RESOURCE BOOK
Spanish Reading Study Guide, pp. 149–150

AUDIO CDS
- Audio Readings in Spanish
- Audio Readings (English)

3.1 FOCUS

▶ Set Learning Goals
Students will
- Explain how EM waves differ from mechanical waves.
- Identify the sources of EM waves.
- Recognize how EM waves transfer energy.
- Observe through an experiment how EM waves interact with matter.

◐ 3-Minute Warm-Up
Display Transparency 20 or copy this exercise on the board:

Are these statements true? If not, correct them.

1. Mechanical waves transfer energy through a vacuum. *Mechanical waves transfer energy through a medium.*
2. A wave is a disturbance that transfers energy. *true*
3. Most EM waves are invisible but detectable. *true*

T 3-Minute Warm-Up, p. T20

3.1 MOTIVATE

EXPLORE Electromagnetic Waves

PURPOSE To show that EM waves can be reflected

TIP *10 min.* Stick a hand-held mirror into a blob of modeling clay if you don't have a mirror with a stand.

WHAT DO YOU THINK? *The remote control has to be aimed so that its beam reflects off the mirror in the same direction that light would, showing that EM waves can be reflected.*

Teach from Visuals

To help students interpret the visual of an electromagnetic wave, review the parts of a wave (such as trough and crest) and ask:

- How is the wavelength of the wave in the figure measured? *from one trough to another*
- What is the geometrical relationship between the magnetic and the electrical fields? *The fields are at right angles to each other.*
- How would a diagram of a wave with lower frequency differ from the wave in the figure? *The crests would be farther apart.*

Ongoing Assessment

Explain how EM waves differ from mechanical waves.

Ask: How do EM and mechanical waves differ in the way they form? *EM waves form when moving charged particles transfer energy through a field. Mechanical waves form when energy is transferred to particles of matter in a material medium.*

Identify the sources of EM waves.

Ask: Why do we receive so few EM waves from stars other than the Sun? *They are so far away from Earth.*

  Answer: electric and magnetic

 Answer: the Sun and human technology

A field is an area around an object where the object can apply a force—a push or a pull—to another object without touching it. You have seen force applied through a field if you have ever seen a magnet holding a card on the door of a refrigerator. The magnet exerts a pull on the door, even though it does not touch the door. The magnet exerts a force through the magnetic field that surrounds the magnet. When a disturbance occurs in a field rather than in a medium, the wave that results is an electromagnetic wave.

How EM Waves Form

 A

EM waves come from atomic particles that are electrically charged. Because of their charges, these particles can exert a force—a push or a pull—on one another through an electric field. These particles also create the magnetic fields that make magnets work.

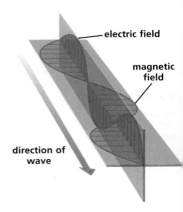
electric field
magnetic field
direction of wave

 B

When electrically charged particles move quickly, they can start a disturbance or vibration in their electric and magnetic fields. The fields vibrate at right angles to each other, as shown in the diagram above. The EM wave travels in the form of these vibrating fields. As you read in Chapter 1, all waves have the properties of amplitude, wavelength, and frequency. In an EM wave, as the diagram shows, both the electric and the magnetic fields have these three properties.

CHECK YOUR READING What are the two types of fields that make up an EM wave?

Sources of EM Waves

Many of the EM waves present in Earth's environment come from the Sun. The Sun's high energy allows it to give off countless EM waves. Other stars give off as many EM waves as the Sun, but because these bodies are so far away, fewer of their EM waves reach Earth. In addition to the Sun, technology is a source of EM waves that humans use for a wide variety of purposes.

When EM waves enter a material, the material often responds by giving off more EM waves. Many EM waves in the environment are given off by the surface of the Earth in response to EM waves from the Sun.

 CHECK YOUR READING What is the source of most EM waves in Earth's environment?

DIFFERENTIATE INSTRUCTION

? More Reading Support

A Where do EM waves come from? *electrically charged atomic particles*

B In what form does an EM wave travel? *as vibrating electric and magnetic fields*

English Learners English learners may have difficulty with nouns that are also used as adjectives. For example, "An EM wave can travel without any medium at all . . . " (p.75). Here, the word *medium* is used as a noun. Help English learners understand that for a word to be an adjective it must describe a noun.

Advanced Have students who are interested in the speed of light read the following article:

 Challenge Reading, pp. 187–188

Electromagnetic waves can travel in a vacuum.

The transfer of energy in the form of EM waves is called **radiation** (RAY-dee-AY-shuhn). Radiation is different from the transfer of energy through a medium by a mechanical wave. A mechanical wave must vibrate the medium as it moves, and this uses some of the wave's energy. Eventually, every mechanical wave will give up all of its energy to the medium and disappear. An EM wave can travel without any medium at all—that is, in a vacuum or space empty of matter—and does not lose energy as it moves. In theory, an EM wave can travel forever.

READING TiP

EM waves are also called rays. The words *radiation* and *radiate* come from the Latin word *radius,* which means "ray" or "spoke of a wheel."

How EM Waves Travel in a Vacuum

Because they do not need a medium, EM waves can pass through outer space, which is a near vacuum. Also, because they do not give up energy in traveling, EM waves can cross the great distances that separate stars and planets. For example, rays from the Sun travel about 150 million kilometers (93 million mi) to reach Earth. Rays from the most distant galaxies travel for billions of years before reaching Earth.

In a vacuum, EM waves spread outward in all directions from the source of the disturbance. The waves then travel in a straight line until something interferes with them. The farther the waves move from their source, the more they spread out. As they spread out, there are fewer waves in a given area and less energy is transferred. Only a very small part of the energy radiated from the Sun is transferred to Earth. But that energy is still a great amount—enough to sustain life on the planet.

The Speed of EM Waves in a Vacuum

In a vacuum, EM waves travel at a constant speed, and they travel very fast—about 300,000 kilometers (186,000 mi) per second. In 1 second, an EM wave can travel a distance greater than 7 times the distance around Earth. Even at this speed, rays from the Sun take about 8 minutes to reach Earth. This constant speed is called the speed of light. The vast distances of space are often measured in units of time traveled at this speed. For example, the Sun is about 8 light-minutes away from Earth. The galaxy shown in the photograph is 60 million light-years from Earth.

The light and other EM waves from this galaxy took approximately 60 million years to reach Earth.

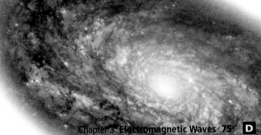

 CHECK YOUR READING How are EM waves used to measure distances in space?

Address Misconceptions

IDENTIFY Ask: What happens to an EM wave when it travels? If students respond that the EM wave must pass through matter, they may hold the misconception that a medium is necessary for an EM wave to continue through.

CORRECT Light waves from the Sun can travel through outer space to Earth, but sound waves, which need a medium, cannot. So we can see the Sun, but not hear it. (Actually, the Sun's sound waves are at frequencies too low for humans to hear.) Point out that light waves do not need matter to travel.

REASSESS Have students list examples of waves that need a medium to travel and waves that do not need a medium.

Technology Resources

Visit **ClassZone.com** for background on common student misconceptions.

 MISCONCEPTION DATABASE

History of Science

In the nineteenth century, scientists believed that light must travel through a hypothetical substance called "ether" that filled space. It was not until 1887 that Albert Michelson and Edward Morley demonstrated that ether does not exist.

Ongoing Assessment

CHECK YOUR READING *Answer: by the time it takes an EM wave to travel at the speed of light from one place to another*

DIFFERENTIATE INSTRUCTION

More Reading Support

C Why can EM waves travel in a vacuum? *They do not need a medium.*

D What do we call the time it takes for the Sun's rays to hit Earth? *the speed of light*

Advanced Challenge students to use a calculator to answer: If light from a galaxy takes 60 million years to reach Earth, about how far away is it? *about 6×10^{20} km*

Calculations: First, find seconds in a year: $60 \cdot 60 = 3,600 \cdot 24 = 86,400 \cdot 365 = 31,536,000$ Second, find distance for 1 year: $300,000 \cdot 31,536,000 = 9.4608 \cdot 10^{12}$ Third, find distance for 60 million years: $(9.4608 \cdot 10^{12}) \cdot (6 \cdot 10^7) = 5.67648 \cdot 10^{20} = 6 \cdot 10^{20}$ km

R Challenge and Extension, p. 151

INVESTIGATE Wave Behavior

PURPOSE To design an experiment to determine what makes a radiometer's vanes move

TIPS *30 min.* Allow students a few minutes to explore, then suggest the following:

- Aim a light at various parts of the radiometer.
- Remind students that the energy in light can be converted to heat.

WHAT DO YOU THINK? *The vanes turn with the white side forward and the black side on the back. Yes, light appears to push on the dark surfaces but not on the light surfaces. Answers will vary depending on student designs.*

CHALLENGE *The radiometer measures the relative brightness of the light falling on the radiometer. The brighter the light, the faster it spins.*

 Datasheet, Wave Behavior, p. 152

Metacognitive Strategy

Ask students if they are convinced that EM waves interact with matter. Have them write a paragraph that explains when they realized this interaction, or what doubts they still have and another experiment they would like to try.

Ongoing Assessment

Recognize how EM waves transfer energy.

Ask: What happens when EM waves encounter a material medium? *They transfer energy to the medium.*

CHECK YOUR READING *Answer: They transfer energy by moving potential energy from place to place.*

Electromagnetic waves can interact with a material medium.

When EM waves encounter a material medium, they can interact with it in much the same way that mechanical waves do. They can transfer energy to the medium itself. Also, EM waves can respond to a change of medium by reflecting, refracting, or diffracting, just as mechanical waves do. When an EM wave responds in one of these ways, its direction changes. When the direction of the wave changes, the direction in which the energy is transferred also changes.

Transferring Energy

REMINDER
Potential energy comes from position or form; kinetic energy comes from motion.

A mechanical wave transfers energy in two ways. As it travels, the wave moves potential energy from one place to another. It also converts potential energy into kinetic energy by moving the medium back and forth.

In a vacuum, EM waves transfer energy only by moving potential energy from one place to another. But when EM waves encounter matter, their energy can be converted into many different forms.

CHECK YOUR READING In what form do EM waves transfer energy in a vacuum?

INVESTIGATE Wave Behavior

How do EM waves interact with matter?

PROCEDURE

① Observe the radiometer on a table or desk.

② Write a hypothesis in the form of an "If . . . , then . . . , because . . ." statement to answer the question: What makes the radiometer vanes move?

③ Develop an experiment to test your hypothesis.

WHAT DO YOU THINK?

- How does light affect the vanes?
- Based on your observation of the vanes, does light affect the white and black surfaces differently? If so, how?
- How would you modify your design now that you have seen the results?

CHALLENGE Based on your observations, what does a radiometer measure? Explain your answer.

SKILL FOCUS
Designing experiments

MATERIALS
radiometer

TIME
30 minutes

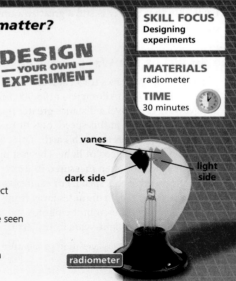

vanes
dark side
light side
radiometer

DIFFERENTIATE INSTRUCTION

 More Reading Support

E What does an EM wave transfer to the medium? *energy*

Below Level Students may have a hard time visualizing how a wave responds to a change of medium. Have them draw diagrams showing how a wave changes direction when it is reflected, refracted, and diffracted by a change of medium.

Converting Energy from One Form to Another

How EM waves interact with a medium depends on the type of the wave and the nature of the material. For example, a microwave oven uses a type of EM wave called microwaves. Microwaves pass through air with very little interaction. However, they reflect off the oven's fan and sides. But when microwaves encounter water, such as that inside a potato, their energy is converted into thermal energy. As a result, the potato gets cooked, but the oven remains cool.

 reflecting fan   **microwave source**

 microwaves

1 A device on the oven produces microwaves and sends them toward the reflecting fan.

2 Microwaves are reflected in many directions by the blades of the fan and then again by the sides of the oven.

3 Microwaves move through the air without transferring energy to the air.

4 Microwaves transfer energy to the water molecules inside the potato in the form of heat, cooking the potato.

EM waves usually become noticeable and useful when they transfer energy to a medium. You do not observe the microwaves in a microwave oven. All you observe is the potato cooking. In the rest of this chapter, you will learn about different types of EM waves, including microwaves, and about how people use them.

 CHECK YOUR READING   How does microwave cooking depend on reflection?

 **3.1 Review**

KEY CONCEPTS	CRITICAL THINKING	CHALLENGE
1. How are EM waves different from mechanical waves?	**4. Predict** What would happen to an EM wave that never came into contact with matter?	**6. Synthesize** EM waves can interact with a medium. How might this fact be used to make a device for detecting a particular type of EM radiation?
2. What are two sources of EM waves in Earth's environment?	**5. Infer** What might be one cause of uneven heating in a microwave oven?	
3. How can EM waves transfer energy differently in a material medium as compared to a vacuum?		

ANSWERS

1. EM waves can travel through a vacuum, while mechanical waves need a material medium; EM waves travel at speed of light, while speed of mechanical waves varies and is slower.

2. the Sun and human technology

3. In a vacuum, EM waves move potential energy from one place to another. In a material medium, they can both move potential energy and convert energy into other forms.

4. It would continue to travel indefinitely.

5. uneven distribution of microwaves by the fan, or uneven reflection by sides of the oven

6. Sample answer: Since infrared rays warm things, a device with a thermometer could be used to detect infrared radiation.

Ongoing Assessment

CHECK YOUR READING *Answer: Reflection causes waves to strike food from many directions so it cooks evenly.*

Teach from Visuals

To help students interpret the visual of a microwave, ask:

• What is the function of the fan? *It reflects the microwaves so they hit the food from all directions.*

• Why doesn't the inside of the microwave oven get hot? *The microwaves transfer their energy only to water molecules. There are no water molecules in the oven itself, so it does not absorb energy and does not get hot.*

Reinforce (the **BIG** idea)

All students can benefit from the following worksheet:

R Reinforcing Key Concepts, p. 153

3.1 ASSESS & RETEACH

Assess

A Section 3.1 Quiz, p. 42

Reteach

To help students visualize magnetic and electrical fields, have them do the following activities:

• Sprinkle iron filings on a glass or plastic plate; hold a magnet under the plate. Shake gently and observe.

• Rub a plastic comb vigorously with wool cloth. Hold the comb near a small continuous stream of water from a faucet. Observe the water.

• Ask: How do you describe these effects that you see? *Magnetic and electric fields affect objects near them.*

Technology Resources

Have students visit **ClassZone.com** for reteaching of Key Concepts.

CONTENT REVIEW

CONTENT REVIEW CD-ROM

Set Learning Goal

To gain skill in converting decimals to exponents and vice versa

Present the Science

The use of exponents makes working with very large and very small numbers easier. EM waves can have frequencies as great as a trillion trillion wavelengths per second and have wavelengths as small as trillionths of a centimeter. Working with so many zeros or decimal places is confusing and can lead to errors.

Develop Number Sense

Students should be proficient at using exponents before they begin the next section.

Emphasize that the final decimal place is counted when converting a decimal into exponent form and that the first integer of a large number is not counted when converting a large number.

DIFFERENTIATION TIP Demonstrate several conversion problems on the board before below-level students do the problems in the text.

Close

Ask: In what branches of science are numbers in exponent form used?
Sample answer: in astronomy to represent vast distances of space; in microbiology to count large numbers of microbes

• Math Support, p. 192
• Math Practice, p. 193

Technology Resources

Students can visit **ClassZone.com** for practice using exponents.

 MATH TUTORIAL

MATH in SCIENCE

MATH TUTORIAL
CLASSZONE.COM
Click on Math Tutorial for more help with positive and negative exponents.

The top photograph shows a visible-light image of the Crab Nebula. The bottom photograph shows the same nebula as it appears at higher x-ray frequencies.

SKILL: USING EXPONENTS

EM Frequencies

The Chandra X-Ray Observatory in the photograph is a space telescope that detects high-frequency EM waves called x-rays. A wave's frequency is the number of wavelengths that pass a given point in 1 second. EM frequencies usually run from about 100 wavelengths per second to about 1 trillion trillion wavelengths per second. If written in standard form (using zeros), 1 trillion trillion would look like this:

1,000,000,000,000,000,000,000,000

Because this number is hard to read, it would be helpful to write it more simply. Using exponents, 1 trillion trillion can be written as 10^{24}.

Exponents can also be used to simplify very small numbers. For example, the wavelength of a wave with a frequency of 10^{24} is about one ten-thousandth of one trillionth of a meter. That number can be written in standard form as **0.000,000,000,000,000,1 m.** Using exponents, the number can be written more simply as 10^{-16} **m.**

Examples

Large Numbers
To write a multiple of 10 in exponent form, just count the zeros. Then, use the total as the exponent.

(1) 10,000 has 4 zeros.

(2) 4 is the exponent.

ANSWER 10^4 is the way to write 10,000 using exponents.

Decimals
To convert a decimal into exponent form, count the number of places to the right of the decimal point. Then, use the total with a negative sign as the exponent.

(1) 0.000,001 has 6 places to the right of the decimal point.

(2) Add a negative sign to make the exponent –6.

ANSWER 10^{-6} is the way to write 0.000,001 using exponents.

Answer the following questions.

Write each number using an exponent.

1. 10,000,000	**3.** 100,000	**5.** 10,000,000,000
2. 0.000,01	**4.** 0.0001	**6.** 0.000,000,001

Write the number in standard form.

7. 10^8	**9.** 10^{11}	**11.** 10^{17}
8. 10^{-8}	**10.** 10^{-12}	**12.** 10^{-15}

CHALLENGE Using exponents, multiply 10^2 by 10^3. Explain how you got your result.

ANSWERS

1. 10^7	5. 10^{-4}	9. 100,000,000,000,000,000
2. 10^5	6. 10^{-9}	10. 0.000,000,01
3. 10^{10}	7. 100,000,000	11. 0.000,000,000,001
4. 10^{-5}	8. 100,000,000,000	12. 0.000,000,000,000,001

CHALLENGE 10^5; multiply powers of 10 by adding exponents.

KEY CONCEPT
3.2 Electromagnetic waves have many uses.

◀ **BEFORE, you learned**
- EM waves transfer energy through fields
- EM waves have measurable properties
- EM waves interact with matter

▶ **NOW, you will learn**
- How EM waves differ from one another
- How different types of EM waves are used

VOCABULARY

electromagnetic spectrum p. 80
radio waves p. 82
microwaves p. 83
visible light p. 84
infrared light p. 84
ultraviolet light p. 85
x-rays p. 86
gamma rays p. 86

EXPLORE Radio Waves

How can you make radio waves?

PROCEDURE

1. Tape one end of one length of wire to one end of the battery. Tape one end of the second wire to the other end of the battery.

2. Wrap the loose end of one of the wires tightly around the handle of the fork.

3. Turn on the radio to the AM band and move the selector past all stations until you reach static.

4. Hold the fork close to the radio. Gently pull the free end of wire across the fork's prongs.

MATERIALS
- two 25 cm lengths of copper wire
- C or D battery
- electrical tape
- metal fork
- portable radio

WHAT DO YOU THINK?
- What happens when you stroke the prongs with the wire?
- How does changing the position of the dial affect the results?

EM waves have different frequencies.

▼ **REMINDER**
Remember that frequency is the number of wavelengths that pass a given point per second. The shorter the wavelength, the higher the frequency.

It might seem hard to believe that the same form of energy browns your toast, brings you broadcast television, and makes the page you are now reading visible. Yet EM waves make each of these events possible. The various types of EM waves differ from each other in their wavelengths and frequencies.

The frequency of an EM wave also determines its characteristics and uses. Higher-frequency EM waves, with more electromagnetic vibrations per second, have more energy. Lower-frequency EM waves, with longer wavelengths, have less energy.

Chapter 3: **Electromagnetic Waves** 79 **D**

3.2 FOCUS

◉ Set Learning Goals
Students will
- Distinguish how EM waves differ from one another.
- Describe how different types of EM waves are used.
- Observe how to detect invisible light.

◉ 3-Minute Warm-Up
Display Transparency 20 or copy this exercise on the board:

Match each definition to the correct term.

Definitions
1. The transfer of energy by EM waves *b*
2. A disturbance that transfers energy through a field *a*

Terms
a. electromagnetic wave c. medium
b. radiation   d. vacuum

 3-Minute Warm-Up, p. T20

3.2 MOTIVATE

EXPLORE Radio Waves
PURPOSE To introduce radio waves, how they are formed, and how they are detected

TIP *10 min.* Have students listen to the static as they vary the speed at which the wire is drawn across the fork.

WHAT DO YOU THINK? *Static is produced. The sound of the static may change from one frequency to another.*

Ongoing Assessment
Distinguish how EM waves differ from one another.

Ask: How does a radio wave differ from a microwave? *in wavelength and frequency*

RESOURCES FOR DIFFERENTIATED INSTRUCTION

Below Level
UNIT RESOURCE BOOK
- Reading Study Guide A, pp. 156–157
- Decoding Support, p. 191

 AUDIO CDS

R **Additional INVESTIGATION,**
Seeing the Invisible, A, B, & C, pp. 203–211;
Teacher Instructions, pp. 284–285

Advanced
UNIT RESOURCE BOOK
Challenge and Extension, p. 162

English Learners
UNIT RESOURCE BOOK
Spanish Reading Study Guide, pp. 160–161

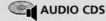 **AUDIO CDS**
- Audio Readings in Spanish
- Audio Readings (English)

Teach from Visuals

To help students interpret the diagram of the EM spectrum, point out the various parts. Ask:

- What is the smallest portion of the spectrum? *visible light*
- What is the frequency range of microwaves? *roughly 10^8 to 10^{12} hertz*
- Where do X-rays fall on the spectrum? *at the high-frequency end*

 The visual "The Electromagnetic Spectrum" is also available as T22 in the Unit Transparency Book.

 **RESOURCE CENTER**
CLASSZONE.COM

Learn more about the electromagnetic spectrum.

A

The Electromagnetic Spectrum

The range of all EM frequencies is known as the **electromagnetic spectrum** (SPEHK-truhm), or EM spectrum. The spectrum can be represented by a diagram like the one below. On the left are the waves with the longest wavelengths and the lowest frequencies and energies. Toward the right, the wavelengths become shorter, and the frequencies and energies become higher. The diagram also shows different parts of the spectrum: radio waves, microwaves, infrared light, visible light, ultraviolet light, x-rays, and gamma rays.

The EM spectrum is a smooth, gradual progression from the lowest frequencies to the highest. Divisions between the different parts of the spectrum are useful, but not exact. As you can see from the diagram below, some of the sections overlap.

The Electromagnetic Spectrum

Frequency in Hertz (1 hertz = 1 wavelength/second)

10^4	10^5	10^6	10^7	10^8	10^9	10^{10}	10^{11}	10^{12}	10^{13}

Radio Waves

Infrared L

Microwaves

This woman is speaking on the radio. **Radio waves** are used for radio and television broadcasts. They are also used for cordless phones, garage door openers, alarm systems, and baby monitors.

Not all astronomy involves looking at the sky. Telescopes like the one above pick up **microwaves** from space. Microwaves are also used for radar, cell phones, ovens, and satellite communications.

The amount of **infrared li**
object gives off depends
warmth. Above, different
indicate different amount
infrared light.

DIFFERENTIATE INSTRUCTION

 More Reading Support

A Describe the EM waves at the left end of the EM spectrum. *They have the longest wavelengths and the lowest frequencies and energies.*

Inclusion Make a large copy of the EM spectrum on shelf paper and display it in the classroom.

Divide the class into seven groups. Have each group make a poster, containing information and examples about one type of EM wave shown on the spectrum. Students who are visually impaired can contribute ideas for tactile or aural representations of waves. Those with hearing impairments can create or obtain visuals of examples for each type. Display posters in their proper position on the spectrum.

Measuring EM Waves

All EM waves move at the same speed in a vacuum. The frequency of an EM wave is determined from its wavelength. EM wavelengths run from about 30 kilometers for the lowest-frequency radio waves to trillionths of a centimeter for gamma rays. EM waves travel so quickly that even those with the largest wavelengths have very high frequencies. For example, a low-energy radio wave with a wavelength of 30 kilometers has a frequency of 10,000 wavelengths per second.

EM wave frequency is measured in hertz (Hz). One hertz equals one wavelength per second. The frequency of the 30-kilometer radio wave mentioned above would be 10,000 Hz. Gamma ray frequencies reach trillions of trillions of hertz.

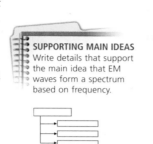

SUPPORTING MAIN IDEAS
Write details that support the main idea that EM waves form a spectrum based on frequency.

CHECK YOUR READING Why is wavelength all you need to know to calculate EM wave frequency in a vacuum?

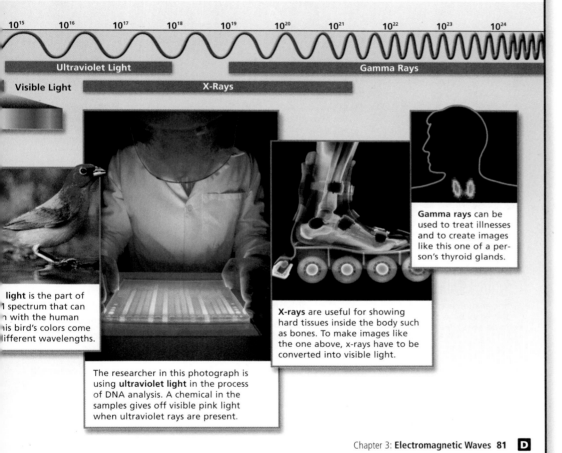

10^{15} 10^{16} 10^{17} 10^{18} 10^{19} 10^{20} 10^{21} 10^{22} 10^{23} 10^{24}

Ultraviolet Light

Gamma Rays

Visible Light

X-Rays

Gamma rays can be used to treat illnesses and to create images like this one of a person's thyroid glands.

X-rays are useful for showing hard tissues inside the body such as bones. To make images like the one above, x-rays have to be converted into visible light.

light is the part of spectrum that can with the human his bird's colors come ifferent wavelengths.

The researcher in this photograph is using **ultraviolet light** in the process of DNA analysis. A chemical in the samples gives off visible pink light when ultraviolet rays are present.

Chapter 3: Electromagnetic Waves **81** **D**

Integrate the Sciences

Many objects in space emit X-rays, radio waves, infrared light, or ultraviolet light as well as visible light. Newly developed devices can detect these EM waves and provide pictures of objects in the universe. You can see on the Internet some of the photographs that have been taken with these devices.

Develop Critical Thinking

APPLY Ask students to describe the mathematical relationship between frequency and wavelength. *As the frequency of a wave increases, its wavelength decreases. This is an inverse relationship. The frequency of any EM wave times its wavelength is always a constant—the speed of the wave in a vacuum (300,000 kilometers per second).*

Ongoing Assessment

CHECK YOUR READING *Answer: The speed of an EM wave in a vacuum is a constant. If you know the wavelength, you know both wavelength and speed, which are enough to calculate frequency.*

Real World Example

In the United States, the Federal Communications Commission (FCC) assigns frequencies in the EM spectrum for commercial radio and TV stations, ship-to-shore radio, citizen's band radio, and emergency police and fire channels. Because the number of frequencies is limited, they are highly prized.

Teach from Visuals

To help students interpret the diagram of radio broadcasting, have them describe the conversion of waves during radio broadcasting. *Sound waves are converted into electrical impulses, which are then converted into radio waves and sent out by the transmitter. These radio waves are converted back into sound by a radio receiver.*

Teach Difficult Concepts

Students often have trouble comprehending how fast EM waves travel compared to the speed of sound. You can demonstrate this on the football field with a balloon and two walkie-talkies.

- Have two students walk across the football field, each with a walkie-talkie turned on. One student should also have an inflated balloon.

- The student with the balloon should press and hold the talk button before popping the balloon.

- The student on the other end should simply watch and listen.

- Have the students record their observations. *They should hear the sound of the balloon popping first over the walkie-talkie and shortly afterward through the air.*

Ongoing Assessment

Radio waves and microwaves have long wavelengths and low frequencies.

Radio waves are EM waves that have the longest wavelengths, the lowest frequencies, and the lowest energies. Radio waves travel easily through the atmosphere and many materials. People have developed numerous technologies to take advantage of the properties of radio waves.

Radio Waves

Radio was the first technology to use EM waves for telecommunication, which is communication over long distances. A radio transmitter converts sound waves into radio waves and broadcasts them through the air in different directions. Radio receivers in many locations pick up the radio waves and convert them back into sound waves.

1. Sound waves enter the microphone and are converted into electrical impulses.

2. The electrical impulses are converted into radio waves and broadcast by the transmitter.

3. The radio waves reach a radio receiver and are converted back into sound.

Different radio stations broadcast radio waves at different frequencies. To pick up a particular station, you have to tune your radio to the frequency for that station. The numbers you see on the radio—such as 670 or 99.5—are frequencies.

Simply transmitting EM waves at a certain frequency is not enough to send music, words, or other meaningful sounds. To do that, the radio transmitter must attach information about the sounds to the radio signal. The transmitter attaches the information by modulating—that is, changing—the waves slightly. Two common ways of modulating radio waves are varying the amplitude of the waves and varying the frequency of the waves. Amplitude modulation is used for AM radio, and frequency modulation is used for FM radio.

You might be surprised to learn that broadcast television also uses radio waves. The picture part of a TV signal is transmitted using AM waves. The sound part is transmitted using FM waves.

CHECK YOUR READING
What two properties of EM waves are used to attach information to radio signals?

VOCABULARY
Make a frame game diagram for *radio waves* and the other types of EM waves.

AM Signal

Information is encoded in the signal by varying the radio wave's amplitude.

FM Signal

Information is encoded in the signal by varying the radio wave's frequency.

DIFFERENTIATE INSTRUCTION

 More Reading Support

D What does a radio transmitter do? *Converts sound waves to radio waves; broadcasts them.*

E What kind of waves broadcast TV? *radio waves*

Below Level Modeling will help students visualize how radio transmission occurs. Have them make paper representations of sound waves and radio waves and "walk through" the three-step process of radio transmission pictured on this page.

Microwaves

A type of EM waves called microwaves comes next on the EM spectrum. **Microwaves** are EM waves with shorter wavelengths, higher frequencies, and higher energy than other radio waves. Microwaves get their name from the fact that their wavelengths are shorter even than those of radio waves. Two important technologies that use microwaves are radar and cell phones.

READING **TiP**

As you read about the different categories of EM waves, refer to the diagram on pages 80 and 81.

Radar The term *radar* stands for "radio detection and ranging." Radar came into wide use during World War II (1939–1945) as a way of detecting aircraft and ships from a distance and estimating their locations. Radar works by transmitting microwaves, receiving reflections of the waves from objects the waves strike, and converting these reflections into visual images on a screen. Today, radar technology is used to control air traffic at airports, analyze weather conditions, and measure the speed of a moving vehicle.

Radar led to the invention of the microwave oven. The discovery that microwaves could be used to cook food was made by accident when radar waves melted a candy bar inside a researcher's pocket.

Cell Phones A cell phone is actually a radio transmitter and receiver that uses microwaves. Cell phones depend on an overlapping network of cells, or areas of land several kilometers in diameter. Each cell has at its center a tower that sends and receives microwave signals. The tower connects cell phones inside the cell to each other or to the regular wire-based telephone system. These two connecting paths are shown below.

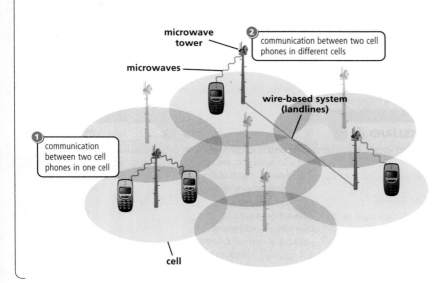

microwave tower

2 communication between two cell phones in different cells

microwaves

wire-based system (landlines)

1 communication between two cell phones in one cell

cell

History of Science

Radar, an acronym for radio detection and ranging, was first used during World War II to detect incoming enemy aircraft. A radar station sends out EM waves. Electronic equipment measures the time it takes for the waves to travel to the plane, reflect, and return to the station. The plane can then be located.

Teach from Visuals

To help students interpret the visual illustrating cellular communication, ask:

- What kind of waves are used in cell phones? *microwaves*

- How large is an overlapping network of cells? *several kilometers in diameter*

- What is the function of a tower in cell phone operation? *The tower receives microwave signals from the phone and sends the signal to another cell phone or to the wire-based telephone system.*

DIFFERENTIATE INSTRUCTION

More Reading Support

F What technologies use microwaves? *radar and cell phones*

G What connects cell phones to each other? *a tower that sends and receives signals*

Advanced Have interested students research and compare Doppler radar and conventional radar. Ask students to find out how Doppler radar is used and why it is used in weather prediction.

R Challenge and Extension, p. 162

INVESTIGATE The Electromagnetic Spectrum

PURPOSE To demonstrate the existence of invisible light waves

TIPS *30 min.* Suggest the following:

- Position the prism so that the spectrum is large enough to allow the thermometers not to touch each other.

- Use a glass or crystal optical-quality prism, the larger the better; don't use a plastic prism. The thermometers should show tenths of a degree.

- This activity will work best outdoors in bright sunlight. Window glass absorbs infrared radiation and will reduce results.

- The paper must be in the shade; students can put the paper in an open cardboard box in the shade of one side and mount the prism in a small rectangle cut into the upper edge.

WHAT DO YOU THINK? *The temperature in the shade was lower. The thermometers in the light, including the thermometer in the invisible infrared range, show higher temperatures because they absorb energy from EM radiation.*

CHALLENGE *Put three thermometers in different parts of the area outside the color spectrum next to the red area, and measure the temperatures.*

 Datasheet, The Electromagnetic Spectrum, p. 163

Technology Resources

Customize this student lab as needed or look for an alternative. Print rubrics to assess student lab reports.

 Lab Generator CD-ROM

Teaching with Technology

This investigate can be done using CBL probeware. Place temperature probes in different parts of the spectrum to measure the temperature of various wavelengths.

Infrared, visible, and ultraviolet light have mid-range wavelengths and frequencies.

Visible light is the part of the EM spectrum that human eyes can see. This part extends roughly from 10^{14} Hz, which we perceive as red, to 10^{15} Hz, which we perceive as violet. This narrow band is very small compared with the rest of the spectrum. In fact, visible light is only about 1/100,000 of the complete EM spectrum. Below visible light and above microwaves, is the infrared part of the EM spectrum. Above visible light is the ultraviolet part of the spectrum. You will read more about visible light in the next section.

READING TIP
Infrared means "below red." *Ultraviolet* means "beyond violet."

Infrared Light

The **infrared light** part of the spectrum consists of EM frequencies between microwaves and visible light. Infrared radiation is the type of EM wave most often associated with heat. Waves in this range are sometimes called heat rays. Although you cannot see infrared radiation, you can feel it as warmth coming from the Sun, a fire, or a radiator. Infrared lamps are used to provide warmth in bathrooms and to keep food warm after it is cooked. Infrared rays are also used to cook food—for example, in a toaster or over charcoal.

INVESTIGATE The Electromagnetic Spectrum

How can you detect invisible light?

PROCEDURE

1. Find a place that has both bright sunlight and shade, such as a window sill. Place the white paper in the shade.

2. Using the marker, paint the bulbs of the thermometers black and place one thermometer on the paper. After three minutes, record the temperature it shows.

3. Position the prism so that it shines a bright color spectrum on the white paper. Place the thermometers so that one bulb is in the blue area, one in the red, and one just outside the red.

4. After five minutes, record the three temperatures.

WHAT DO YOU THINK?

- How did the four temperature readings differ?
- How might you explain the temperature readings you noted?

CHALLENGE How could you modify the experiment to find the hottest part of the infrared range?

SKILL FOCUS
Drawing conclusions

MATERIALS
- white paper
- black marker
- 3 thermometers
- prism

TIME
30 minutes

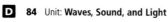

DIFFERENTIATE INSTRUCTION

Reading Support

H How large is the visible light part of the spectrum? *very small*

I What EM waves are sometimes called heat rays? *infrared waves*

Additional Investigation To reinforce Section 3.2 learning goals, use the following full-period investigation:

 Additional INVESTIGATION, Seeing the Invisible, A, B, & C, pp. 203–211, 284–285
(Advanced students should complete Levels B and C.)

Alternative Assessment Have students draw a diagram showing how they would modify the experiment as described in the Challenge question. *The diagram should show three thermometers placed on a spectrum in different places to the left of the visible red waves.*

Some animals, such as pit viper snakes, can actually see infrared light. Normally, human beings cannot see infrared light. However, infrared scopes and cameras convert infrared radiation into visible wavelengths. They do this by representing different levels of infrared radiation with different colors of visible light. This technology can create useful images of objects based on the objects' temperatures.

 CHECK YOUR READING How do human beings perceive infrared radiation?

Ultraviolet Light

The **ultraviolet light** part of the EM spectrum consists of frequencies above those of visible light and below those of x-rays. Because ultraviolet (UV) light has higher frequencies than visible light, it also carries more energy. The waves in this range, having higher energies, can damage tissue, burning your skin or hurting your eyes. Sunblock and UV-protection sunglasses are designed to filter out these frequencies.

Ultraviolet light has beneficial effects as well. Because it can damage cells, UV light can be used to sterilize medical instruments and food by killing harmful bacteria. In addition, UV light causes skin cells to produce vitamin D, which is essential to good health. Ultraviolet light can also be used to treat skin problems and other medical conditions.

Like infrared light, ultraviolet light is visible to some animals. Bees and other insects can see higher frequencies than people can. They see nectar guides—marks that show where nectar is located—that people cannot see in visible light. The photographs below show how one flower might look to a person and to a bee.

In this infrared photograph, warmer areas appear red and orange, while cooler ones appear blue, green, and purple.

This photograph shows the flower as it appears in visible light.

This photograph shows the flower as it might appear to a bee in ultraviolet light. Bees are able to see nectar guides in the UV range.

Chapter 3: Electromagnetic Waves **85** **D**

Chapter 3 **85** **D**

Revisit "Internet Activity: Electromagnetic Waves" on p. 71. Have students explain their results.

Develop Critical Thinking

INFER Ask students why gamma rays cannot be used to image bones and teeth. *Gamma rays have so much energy that they pass through hard tissue such as bones and teeth as well as soft tissue. If gamma rays were used to image teeth, no image would appear on the film.* Ask what exposed photographic film would look like if gamma rays were used to image teeth. *The film would be overexposed (white).*

Reinforce (the **BIG** idea)

All students can benefit from the following worksheet:

 Reinforcing Key Concepts, p. 164

3.2 ASSESS & RETEACH

Assess

 Section 3.2 Quiz, p. 43

Reteach

Create a puzzle for your students.

- Write a quiz consisting of ten to fifteen statements such as "These waves can pop popcorn" and "These waves have the shortest frequency."

- Include statements that are definitions of wavelength, frequency, and energy.

- Create a hidden-word puzzle in which the answers to the quiz statements are embedded.

Technology Resources

Have students visit **ClassZone.com** for reteaching of Key Concepts.

 CONTENT REVIEW

 CONTENT REVIEW CD-ROM

X-rays and gamma rays have short wavelengths and high frequencies.

At the opposite end of the EM spectrum from radio waves are x-rays and gamma rays. Both have very high frequencies and energies. **X-rays** have frequencies from about 10^{16} Hz to 10^{21} Hz. **Gamma rays** have frequencies from more than 10^{19} Hz to more than 10^{24} Hz. Like other EM waves, x-rays and gamma rays are produced by the Sun and by other stars. People have also developed technologies that use these EM frequencies.

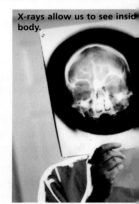
X-rays allow us to see insid[e] body.

X-rays pass easily through the soft tissues of the body, but many are filtered out by denser matter such as bone. If photographic film is placed behind the body and x-rays are aimed at the film, only the x-rays that pass through the body will expose the film. This makes x-ray images useful for diagnosing bone fractures and finding dense tumors. But too much exposure to x-rays can damage tissue. Even in small doses, repeated exposure to x-rays can cause cancer over time. When you have your teeth x-rayed, you usually wear a vest made out of lead to protect your organs. Lead blocks high-frequency radiation.

Gamma rays have the highest frequencies and energies of any EM waves. Gamma rays are produced by some radioactive substances as well as the Sun and other stars. Gamma rays can penetrate the soft and the hard tissues of the body, killing normal cells and causing cancer cells to develop. If carefully controlled, this destructive power can be beneficial. Doctors can also use gamma rays to kill cancer cells and fight tumors.

3.2 Review

KEY CONCEPTS

1. What two properties of EM waves change from one end of the EM spectrum to the other?

2. Describe two uses for microwave radiation.

3. How are EM waves used in dentistry and medicine?

CRITICAL THINKING

4. **Infer** Why do you think remote controls for TVs, VCRs, and stereos use infrared light rather than ultraviolet light?

5. **Apply** For a camera to make visible images of heat escaping from a building in winter, what type of EM wave would it need to photograph?

CHALLENGE

6. **Synthesize** When a person in a car is talking on a cell phone, and the car moves from one cell to another, the conversation continues without interruption. How might this be possible?

ANSWERS

1. wavelength and frequency

2. Answers might include ovens to cook and heat food; radar to measure distance, location, or speed; cell phones to communicate.

3. Sample answer: X-rays are used to make images of hard tissues such as bones and teeth.

4. Infrared waves have low energies. They do not harm human tissue.

5. infrared light

6. The connection to the cell phone is switched from the microwave tower in the first cell to the microwave tower in the second cell without any disruption of the signal.

SKILL: DETERMINING RELEVANCE

Are Cell Phones Harmful?

In 1993, a man appearing on a popular television talk show claimed that cell phone radiation had caused his wife's brain cancer. Since that time, concerned scientists have conducted more than a dozen studies. None of them have shown clear evidence of a connection between cell phones and cancer. However, researchers have made a number of experimental observations.

● Experimental Observations

Here are some results from scientists' investigations.

1. Substances that cause cancer work by breaking chemical bonds in DNA.
2. Only EM radiation at ultraviolet frequencies and above can break chemical bonds.
3. Microwave radiation may make it easier for molecules called free radicals to damage DNA bonds.
4. Other factors such as psychological stress may cause breaks in DNA bonds.
5. Performing multiple tasks like driving and talking on the phone reduces the brain's ability to perform either task.
6. Exposing the brain to microwave radiation may slow reaction times.

● Hypotheses

Here are some hypotheses that could be used for further research.

A. Microwaves from cell phones can break DNA bonds.
B. Cell phones may contribute to cancer.
C. Holding and talking into a cell phone while driving increases a person's risk of having an accident.
D. Worrying about cell phones may be a health risk.

Talking on a cell phone while driving may increase the risk of accidents.

● Determining Relevance

On Your Own On a piece of paper, write down each hypothesis. Next to the hypothesis write each observation that you think is relevant. Include your reasons.

As a Group Discuss how each observation on your list is or is not relevant to a particular hypothesis.

CHALLENGE Based on the observations listed above, write a question that you think would be a good basis for a further experiment. Then explain how the answer to this question would be helpful.

THINK SCIENCE
Scientific Methods of Thinking

Set Learning Goal
To evaluate hypotheses in terms of experimental observations

Present the Science
Scientists currently believe that a major cause of cancer is genetic, that is, breakage of the chemical bonds in the DNA molecule. Students will read about some of the facts related to the breaking of bonds in DNA and evaluate these facts to develop hypotheses about the safety of cell phones.

Guide the Activity
- Remind students that to evaluate means to judge a statement based on criteria.
- Ask students whether all four hypotheses listed are testable. *Hypothesis D would be difficult to test.*
- Ask students whether a hypothesis must explain all of the observations or just some of them. *A hypothesis must be supported by all of the observations. If some observations contradict a hypothesis, the hypothesis must be revised or discarded.*

COOPERATIVE LEARNING STRATEGY
Divide the class into small groups. In each group, assign a facilitator, a recorder, and a reporter. Assign each group one hypothesis to evaluate. The facilitator ensures that everyone has a chance to respond. The recorder writes the group's consensus. The reporter presents each observation in class.

Close
Ask: Why is it important to form hypotheses that are relevant to experimental observations? *You can build on information already acquired, instead of starting from scratch.*

ANSWERS

For hypothesis A, relevant observations are 1, 2, 3.

For hypothesis B, relevant observations are 1, 2, 3.

For hypothesis C, relevant observations are 5 and 6.

For hypothesis D, relevant observations are 1 and 4.

CHALLENGE Qestions will vary. Sample question: Can DNA damage by free radicals cause cancer? If the answer is yes, then observation 3 supports the view that microwaves can contribute to cancer. If the answer is no, then observation 3 does not support the view that microwaves contribute to cancer.

● Set Learning Goals

Students will

- Explain how visible light is produced.
- Describe how living organisms produce light.
- Describe how humans produce light artificially.
- Observe in an experiment the different types of artificial light.

◀ 3-Minute Warm-Up

Display Transparency 21 or copy this exercise on the board:

Predict what would happen if you kept a green plant in the dark for one month. Explain why. *It would probably die. Plants require light from the Sun to survive.*

T 3-Minute Warm-Up, p. T21

3.3 MOTIVATE

THINK ABOUT

PURPOSE To understand why light is important to living organisms

DISCUSS Have students brainstorm the following scenario: What would happen to living things if the Sun suddenly went dark? *Everything on Earth would die. There would be no energy to produce food or, in the long run, to keep from freezing.*

Ongoing Assessment

CHECK YOUR READING *Answer: Green plants use light to make food. Plant material provides food, directly or indirectly, for most animals.*

KEY CONCEPT

3.3 The Sun is the source of most visible light.

◀ **BEFORE, you learned**

- Visible light is part of the EM spectrum
- EM waves are produced both in nature and by technology

▶ **NOW, you will learn**

- How visible light is produced by materials at high temperatures
- How some living organisms produce light
- How humans produce light artificially

VOCABULARY

incandescence p. 89
luminescence p. 89
bioluminescence p. 89
fluorescence p. 91

THINK ABOUT

Why is light important?

This railroad worm has eleven pairs of green lights on its sides and a red light on its head. The animal probably uses these lights for illumination and to frighten away predators. Almost every living organism, including humans, depends on visible light. Think of as many different ways as you can that plants, animals, and people use light. Then, think of all the sources of visible light that you know of, both natural and artificial. Why is light important to living organisms?

Light comes from the Sun and other natural sources.

RESOURCE CENTER
CLASSZONE.COM

Learn more about visible light.

It is hard to imagine life without light. Human beings depend on vision in countless ways, and they depend on light for vision. Light is the only form of EM radiation for which human bodies have special-ized sensory organs. The human eye is extremely sensitive to light and color and the many kinds of information they convey.

Most animals depend on visible light to find food and to do other things necessary for their survival. Green plants need light to make their own food. Plants, in turn, supply food directly or indirectly for nearly all other living creatures. With very few exceptions, living creatures depend on light for their existence.

CHECK YOUR READING How is plants' use of light important to animals?

RESOURCES FOR DIFFERENTIATED INSTRUCTION

Below Level
UNIT RESOURCE BOOK
- Reading Study Guide A, pp. 167–168
- Decoding Support, p. 191

 AUDIO CDS

Advanced
UNIT RESOURCE BOOK
Challenge and Extension, p. 173

English Learners
UNIT RESOURCE BOOK
Spanish Reading Study Guide, pp. 170–171

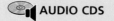 **AUDIO CDS**

- Audio Readings in Spanish
- Audio Readings (English)

Most of the visible light waves in the environment come from the Sun. The Sun's intense heat produces light of every wavelength. The production of light by materials at high temperatures is called **incandescence** (IHN-kuhn-DEHS-uhns). When a material gets hot enough, it gives off light by glowing or by bursting into flames.

Other than the Sun, few natural sources of incandescent light strongly affect life on Earth. Most stars give off as much light as the Sun, or even more, but little light from stars reaches Earth because they are so far away. Lightning produces bright, short-lived bursts of light. Fire, which can occur naturally, is a lower-level, longer-lasting source of visible light. The ability to make and use fire was the first light technology, making it possible for human beings to see on a dark night or inside a cave.

 CHECK YOUR READING Why does little light reach Earth from stars other than the Sun?

Some living things produce visible light.

Many organisms produce their own visible light, which they use in a variety of ways. They produce this light through luminescence. **Luminescence** is the production of light without the high temperatures needed for incandescence. The production of light by living organisms is called **bioluminescence.** Bioluminescent organisms produce light from chemical reactions rather than from intense heat. Bioluminescence enables organisms to produce light inside their tissues without being harmed.

Bioluminescent organisms include insects, worms, fish, squid, jellyfish, bacteria, and fungi. Some of these creatures have light-producing organs that are highly complex. These organs might include light-producing cells but also reflectors, lenses, and even color filters.

The firefly, a type of beetle, uses bioluminescence to attract mates. A chemical reaction in its abdomen allows the firefly to glow at specific intervals. The pattern of glowing helps fireflies of the same species identify each other at night when they cannot see each other. Most often, the male flashes a specific signal while flying around. Females, usually sitting on vegetation near the ground, wait a certain amount of time and then respond with a flash. After they have identified each

VOCABULARY
Don't forget to make word frames for the terms *luminescence* and *bioluminescence.*

Chapter 3: Electromagnetic Waves 89 **D**

Address Misconceptions

IDENTIFY Ask: Where does the energy for bioluminescence come from? If students respond that the light comes from heat or electricity produced by the organism, they may hold the misconception that bioluminescence is an electrical process rather than a chemical process.

CORRECT Have students experiment with a chemical glow stick (available at a hardware store). Show students that the stick is activated when a container inside the stick is broken. Have them observe that the stick remains cool to the touch. If possible, let them see a stick after it has stopped making light. Explain that the light in the stick comes from chemical reactions rather than heat or electricity.

REASSESS Ask students the following question: How does a glowstick model bioluminescence? *It glows because of a chemical reaction, not because of electricity.*

Technology Resources

Visit **ClassZone.com** for background on common student misconceptions.

MISCONCEPTION DATABASE

Ongoing Assessment

Explain how visible light is produced.

Ask: How does the Sun produce light? *It emits radiation.*

Describe how living organisms produce light.

How do living things produce light? *chemicals in body*

CHECK YOUR READING *Answer: Except for the Sun, stars are too far away for much of their light to reach Earth.*

DIFFERENTIATE INSTRUCTION

More Reading Support

A Where do most of the visible light waves in the environment come from? *the Sun*

English Learners English learners may be unfamiliar with some of the non-literal phrases used in the text. For example, the direction *Write up your experiment and carry it out* (p. 90) may be confusing to an English learner when taken literally. Be watchful for confusing language such as "carry it out," and explain unfamiliar phrases and idioms in clear terms.

INVESTIGATE Artificial Lighting

PURPOSE To examine properties of various types of artificial lighting

TIPS *30 min.* Allow students a few minutes to explore, then suggest the following:

- Use direct sunlight to test how different colored materials appear.
- Compare the properties of artificial lighting to those of natural sunlight.

WHAT DO YOU THINK? *Light qualities vary among different bulbs and lighting types. Light from any source can be broken up into a spectrum. One source's spectrum may be very different from another's. Answers will vary depending on student experiments.*

 Datasheet, Artificial Lighting, p. 174

Metacognitive Strategy

Ask students to make a diagram of the experimental steps they tried and discarded. Have them write briefly about the problems that arose.

Ongoing Assessment

Design an experiment to compare different types of artificial light.

Ask: How could you compare the amount of heat given off by a halogen bulb and a regular incandescent bulb? *Measure the temperature at the same distance from two bulbs of equal wattage.*

CHECK YOUR READING *Incandescence is the production of light from high temperatures, or intense heat. Bioluminescence is the production of light from chemical reactions rather than from intense heat.*

A female firefly responds to a male's signal.

other, the fireflies may continue to exchange flashes until the male has located the female.

The process of bioluminescence is very efficient. Almost all of the energy released by the chemical reactions of bioluminescence is converted into light. Very little heat is produced. Researchers in lighting technology have wanted for years to imitate this efficiency, but that has become possible only in recent years.

 CHECK YOUR READING How is bioluminescence different from incandescence?

Human technologies produce visible light.

B Human beings invented the first artificial lighting when they learned to make and control fire. For most of human history, people have made light with devices that use fire in some form, such as oil lamps, candles, and natural gas lamps. After the discovery of electricity, people began to make light through a means other than fire. However, the technique of using a very hot material as a light source stayed the same until the invention of fluorescent lighting. In recent years, "cool" lighting has become much more common.

INVESTIGATE Artificial Lighting

Is all artificial light the same?

Many types of artificial light sources are available. These sources differ in the amount of light they produce, the way the light beams are directed, and the quality of the light itself.

DESIGN —YOUR OWN— EXPERIMENT

SKILL FOCUS
Designing experiments

MATERIALS
Artificial lighting with a variety of bulb types and sizes

TIME
30 minutes

PROCEDURE

1. Design a procedure to discover and record differences among several different types of artificial lighting. Your procedure should test how different colored materials appear in different types of lighting. You should compare the results with how these materials appear in direct sunlight.

2. Write up your experiment and carry it out.

WHAT DO YOU THINK?

- What differences did you discover among bulbs of different types and sizes?
- How would you improve your design if you were to repeat your experiment?

DIFFERENTIATE INSTRUCTION

? More Reading Support

B What was the first artificial lighting used by humans? *fire*

Below Level Have students make a flow diagram showing how making light has changed throughout human history. *Diagrams should begin with fire, progress through oil lamps, candles, and natural gas, and end with electricity.*

Advanced

 Challenge and Extension, p. 173

Incandescent and Fluorescent Lighting

The development of the electric light bulb in the late 1800s made light available at the flip of a switch. An ordinary light bulb is a sealed glass tube with a thin tungsten wire running through it. This wire is called a filament. When electrical current passes through the filament, the tungsten gets hotter and begins to glow. Because these light bulbs need high temperatures to produce light, they are called incandescent bulbs.

Tungsten can become very hot—about 3500 degrees Celsius (6300°F)—without melting. At such high temperatures, tungsten can give off a bright light. However, the tungsten filament also produces much infrared radiation. In fact, the filament produces more infrared light than visible light. As a result, incandescent bulbs waste a lot of energy in the form of heat. At such high temperatures, tungsten also slowly evaporates and collects on the inside of the bulb. Eventually, the filament weakens and breaks, and the bulb burns out.

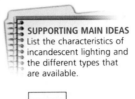

SUPPORTING MAIN IDEAS
List the characteristics of incandescent lighting and the different types that are available.

 CHECK YOUR READING What causes ordinary light bulbs to burn out?

Since the 1980s, halogen (HAL-uh-juhn) bulbs have come into wide use. Halogen bulbs have several advantages over ordinary incandescent bulbs. They contain a gas from the halogen group. This gas combines with evaporating tungsten atoms and deposits the tungsten back onto the filament. As a result, the filament lasts longer. The filament can also be raised to a higher temperature without damage, so it produces more light. Halogen bulbs, which are made of quartz, resist heat better than glass.

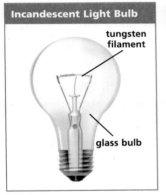

Incandescent Light Bulb
tungsten filament
glass bulb

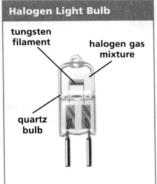

Halogen Light Bulb
tungsten filament
halogen gas mixture
quartz bulb

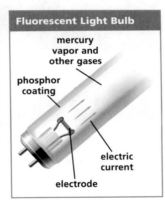

Fluorescent Light Bulb
mercury vapor and other gases
phosphor coating
electric current
electrode

Many electric lights in use today are fluorescent. **Fluorescence** (flu-REHS-uhns) occurs when a material absorbs EM radiation of one wavelength and gives off EM radiation of a different wavelength. Fluorescent bulbs are filled with a mixture of mercury vapor and other gases that give off ultraviolet light when an electric current passes

Chapter 3: **Electromagnetic Waves** 91 **D**

Integrate the Sciences

Tungsten has a melting point of about 3420°C, the highest of any metal. The name *tungsten* comes from the Swedish *tung sten,* meaning "heavy stone." Tungsten's chemical symbol is W, for wolfram. The name comes from wolframite, a tungsten mineral.

Teach from Visuals

To help students interpret the diagram of fluorescent lighting, point out the various parts of the bulb. Ask:

• What is the function of the phosphor? *It absorbs ultraviolet light and emits visible light.*

• What is the function of the mercury vapor? *It gives off ultraviolet light when a current passes through it.*

• What is the function of the current? *It causes the gases to give off ultraviolet light.*

Ongoing Assessment

Describe how humans produce light artificially.

Ask: What causes fluorescence? *Fluorescence occurs when a material absorbs EM radiation of one wavelength and gives off EM radiation of a different wavelength.*

CHECK YOUR READING *Answer: At high temperatures tungsten evaporates. This weakens the filament, which eventually breaks.*

Ongoing Assessment

 CHECK YOUR READING *Answer: Fluorescent light is cool and does not waste much energy as heat, unlike incandescent light.*

Reinforce (the **BIG** idea)

All students can benefit from the following worksheet:

 Reinforcing Key Concepts, p. 175

3.3 ASSESS & RETEACH

Assess

 Section 3.3 Quiz, p. 44

Reteach

Show students a regular incandescent light bulb and an LED. Have them list some visible differences between the two. Ask:

• What are the working elements of an incandescent bulb? *filament, vacuum or special gas enclosed in a bulb*

• Does the LED have these elements? What does it have instead? *no; semiconductor and wires*

• Compare heat, efficiency, and longevity of these two types of artificial lighting. *An LED lasts longer and is more efficient than an incandescent bulb; the incandescent bulb produces more heat.*

Technology Resources

Have students visit **ClassZone.com** for reteaching of Key Concepts.

 CONTENT REVIEW

CONTENT REVIEW CD-ROM

---

through them. The insides of the bulbs are coated with a powder called phosphor that fluoresces. Phosphor absorbs ultraviolet light and gives off visible light. Because fluorescent light is cool and does not waste much energy as heat, it is more efficient, more economical, and safer than incandescent lighting.

 CHECK YOUR READING Why are fluorescent lights more efficient than incandescent lights?

Other Types of Artificial Lighting

LEDs are being used more and more in place of incandescent bulbs.

Like fluorescent lights, many other artificial light sources use a gas in place of a filament. For example, neon lights use gas-filled tubes to produce light. However, instead of ultraviolet light, the gas gives off visible light directly. The colors of neon lights come from the particular mixtures of gases used. Vapor lights, which are commonly used for street lights, work in a similar way. In a vapor light, a material such as sodium is heated until it becomes a gas, or vapor. The vapor responds to electricity by glowing brightly.

One of fastest-growing types of artificial lighting is the light emitting diode, or LED. LEDs do not involve bulbs, filaments, or gases. Instead, they produce light electronically. A diode is a type of semiconductor—a device that regulates the flow of electricity. An LED is a semiconductor that converts electricity directly into visible light.

LEDs have many advantages over traditional forms of lighting. They produce a very bright light, do not break easily, use little energy, produce little heat, can save money, and can last for decades. Some technologists believe that LEDs will eventually replace most traditional forms of artificial lighting.

3.3 Review

KEY CONCEPTS

1. Describe natural, nonliving sources of incandescent light.

2. What advantages does bioluminescence have over incandescence as a way for living organisms to produce light?

3. What are some advantages and disadvantages of artificial incandescent lighting?

CRITICAL THINKING

4. **Classify** Make a chart summarizing the different types of artificial lighting discussed in this section.

5. **Infer** Why do you think moonlight does not warm you, even though the Moon reflects light from the hot Sun?

CHALLENGE

6. **Compare and Contrast** What does LED lighting have in common with bioluminescence? How are the two different?

ANSWERS

1. The Sun produces most of the natural light we use. Other sources are other stars, lightning, and fire.

2. Bioluminescence does not produce high temperatures, which could damage tissues.

3. advantages: give off bright light, no open flame; disadvantages: can get very hot, wastes energy

4. Charts should summarize the information from pp. 90–92.

5. The Moon reflects mostly visible light and not much infrared light. Only a small part of the radiation from the Sun strikes the Moon.

6. Both do not involve bulbs, filaments, or gases. They produce cool light. LEDs produce light electronically, bioluminescence chemically.

KEY CONCEPT

3.4 Light waves interact with materials.

◀ **BEFORE, you learned**

- Mechanical waves respond to a change in medium
- Visible light is made up of EM waves
- EM waves interact with a new medium in the same ways that mechanical waves do

▶ **NOW, you will learn**

- How the wave behavior of light affects what we see
- How light waves interact with materials
- Why objects have color
- How different colors are produced

VOCABULARY

transmission p. 93
absorption p. 93
scattering p. 95
polarization p. 96
prism p. 97
primary colors p. 98
primary pigments p. 99

EXPLORE Light and Matter

How can a change in medium affect light?

PROCEDURE

① Fill the container with water.

② Add 10 ml (2 tsp) of milk to the water. Put on the lid, and gently shake the container until the milk and water are mixed.

③ In a dark room, shine the light at the side of the container from about 5 cm (2 in.) away. Observe what happens to the beam of light.

WHAT DO YOU THINK?

- What happened to the beam of light from the flashlight?
- Why did the light behave this way?

MATERIALS

- clear plastic container with lid
- water
- measuring spoons
- milk
- flashlight

Light can be reflected, transmitted, or absorbed.

VOCABULARY
Don't forget to make word frames for *transmission* and *absorption*.

You have read that EM waves can interact with a material medium in the same ways that mechanical waves do. Three forms of interaction play an especially important role in how people see light. One form is reflection. Most things are visible because they reflect light. The two other forms of interaction are transmission and absorption.

Transmission (trans-MIHSH-uhn) is the passage of an EM wave through the medium. If the light reflected from objects did not pass through the air, windows, or most of the eye, we could not see the objects. **Absorption** (uhb-SAWRP-shun) is the disappearance of an EM wave into the medium. Absorption affects how things look, because it determines what light is available to be reflected or transmitted.

Chapter 3: Electromagnetic Waves **93** **D**

RESOURCES FOR DIFFERENTIATED INSTRUCTION

Below Level

UNIT RESOURCE BOOK
- Reading Study Guide A, pp. 178–179
- Decoding Support, p. 191

💿 **AUDIO CDS**

Advanced

UNIT RESOURCE BOOK
Challenge and Extension, p. 184

English Learners

UNIT RESOURCE BOOK
Spanish Reading Study Guide, pp. 182–183

💿 **AUDIO CDS**

- Audio Readings in Spanish
- Audio Readings (English)

◉ Set Learning Goals

Students will

- Describe how the wave behavior of light affects what we see.
- Recognize how light waves interact with materials.
- Recognize why objects have color.
- Explain how colors are produced.
- Observe through an experiment what makes up the color black.

◔ 3-Minute Warm-Up

Display Transparency 21 or copy this exercise on the board:

Match each definition to the correct term.

Definitions

1. the production of light without high temperatures *a*
2. the production of light by living organisms *b*
3. the production of light with high temperatures *c*

Terms

a. luminescence   c. incandescence
b. bioluminescence

 3-Minute Warm-Up, p. T21

3.4 MOTIVATE

EXPLORE Light and Matter

PURPOSE To introduce the concept of light scattering

TIP *10 min.* Tell students that milk is composed of large molecules, like fats and proteins.

WHAT DO YOU THINK? *light spreads out inside the container; the color and direction of the light beam changes when light strikes the particles of milk*

Ongoing Assessment

Describe how the wave behavior of light affects what we see.

Ask: How does absorption of light waves affect how things look? *by what light it reflects or transmits*

History of Science

Not all glass is transparent to visible light. Opaque glass, often called milk glass, is creamy white and feels silky. It was first developed by the Egyptians about 1500 B.C., when they added chemicals to glass. The Chinese made milk-glass snuff bottles around 140 B.C., and the Persians kept their spices and medicines in milk-glass jars beginning in the eighth century. Today, opaque glass is made in all colors and textures.

Ongoing Assessment

Recognize how light waves interact with materials.

Ask: Why do some materials appear opaque? *All the light waves that strike them are reflected, absorbed, or both.*

CHECK YOUR READING *Answer: Translucent materials let some light pass through, causing it to spread out in all directions. Objects can be seen indistinctly through a translucent material. Opaque materials do not allow any light to pass through because they reflect light, absorb light, or both.*

How Materials Transmit Light

Materials can be classified according to the amount and type of light they transmit.

1. Transparent (trans-PAIR-uhnt) materials allow most of the light that strikes them to pass through. It is possible to see objects through a transparent material. Air, water, and clear glass are transparent. Transparent materials are used for items such as windows, light bulbs, thermometers, sandwich bags, and clock faces.

2. Translucent (trans-LOO-suhnt) materials transmit some light, but they also cause it to spread out in all directions. You can see light through translucent materials, but you cannot see objects clearly through them. Some examples are lampshades, frosted light bulbs, frosted windows, fluorescent light coverings, sheer fabrics, and notepaper.

3. Opaque (oh-PAYK) materials do not allow any light to pass through them, because they reflect light, absorb light, or both. Heavy fabrics, construction paper, and ceramic mugs are opaque. Shiny materials may be opaque mainly because they reflect light. Other materials, such as wood and rock, are opaque mainly because they absorb light.

CHECK YOUR READING What is the difference between translucent and opaque materials?

This stained-glass window contains transparent, translucent, and opaque materials.

DIFFERENTIATE INSTRUCTION

? More Reading Support

A What two types of material transmit light? *transparent, translucent*

B What makes a material opaque? *no light passes through it*

English Learners Be aware that English learners do not always have the same background knowledge as the rest of the class. For example, the paragraph on scattering refers to what happens when students shine a flashlight into fog. Some students may have never had such an opportunity due to the climate in their native country or their lack of access to a flashlight.

A light filter is a material that is transparent to some kinds of light and opaque to others. For example, clear red glass transmits red light but absorbs other colors. Examples of light filters are the colored covers on taillights and traffic lights, infrared lamp bulbs, and UV-protected sunglasses. Filters that transmit only certain colors are called color filters.

Scattering

Sometimes fine particles in a transparent material interact with light passing through the material to cause scattering. **Scattering** is the spreading out of light rays in all directions, because particles reflect and absorb the light. Fog or dust in the air, mud in water, and scratches or smudges on glass can all cause scattering. Scattering creates glare and makes it hard to see through even a transparent material. Making the light brighter causes more scattering, as you might have noticed if you have ever tried to use a flashlight to see through fog.

Fine particles, such as those in fog, scatter light and reduce visibility.

Scattering is what makes the sky blue. During the middle of the day, when the Sun is high in the sky, molecules in Earth's atmosphere scatter the blue part of visible light more than they scatter the other wavelengths. This process gives the sky a blue tinge and makes it translucent. Light comes through the sky, but you cannot see through the sky. At dawn and dusk, light from the Sun must travel farther through the atmosphere before it reaches your eyes. By the time you see it, the greens and blues are scattered away and the light appears reddish. At night, because there is so little light, almost no scattering takes place, and you can see through the sky to the stars.

SUPPORTING MAIN IDEAS
Be sure to add to your chart the different ways light interacts with materials.

 CHECK YOUR READING How does scattering make the sky blue?

Address Misconceptions

IDENTIFY Ask: Why does the sky appear blue? If students respond that the oceans reflect blue light, they may hold the misconception that the oceans are responsible for the sky's color.

CORRECT Point out that, if the oceans were responsible for the sky's blue color, the sky would be less blue farther inland. This is obviously not true.

REASSESS Ask students to recall their results for "Explore Light and Matter" on p. 93. Ask:

• Why did the milky water appear blue? *The milk particles scattered the blue wavelengths in the light.*

• How is the sky similar to the milk and water? *Particles in the atmosphere scatter the blue wavelengths in sunlight.*

Technology Resources

Visit **ClassZone.com** for background on common student misconceptions.

MISCONCEPTION DATABASE

Ongoing Assessment

CHECK YOUR READING *Answer: Particles in Earth's atmosphere scatter the blue part of the visible spectrum more than they scatter the other wavelengths. This makes the sky appear blue.*

DIFFERENTIATE INSTRUCTION

More Reading Support

C What causes scattering of light? *fine particles in a transparent material*

Below Level Have students make a list of five materials that are transparent to visible light, five that are translucent, and five that are opaque. Ask students to write definitions of the three terms and to draw ray diagrams showing light rays being reflected, absorbed, or transmitted by each type of material.

To help students interpret the visual of light waves and filters, ask:

- In what direction are the light waves absorbed by the first filter? *horizontally*

- Why does no light pass through the second filter? *Only light waves vibrating vertically pass through the first filter. The second filter stops all waves except those vibrating horizontally.*

Teach Difficult Concepts

Students may have a hard time understanding polarization. It may help them visualize the process if they think of a polarizing filter as a picket fence. Only vertical objects can pass through an upright fence. If the fence were rotated to its side, only horizontal objects could pass through. To help students understand polarization, explain that light can become polarized when a nonmetallic surface reflects light. When the reflected light waves have a large concentration of horizontal vibrations, that is, in a plane parallel to the object's surface, they are called *glare*. Polarized sunglasses have microscopic vertical slits that block out the horizontally polarized light and therefore reduce glare. Have several pairs of polarized sunglasses available for students to take outdoors. Have students notice how the glare changes when they rotate the glasses.

Polarization

Polarization of light reduces glare and makes it easier to see objects. **Polarization** (POH-luhr-ih-ZAY-shuhn) is a way of filtering light so that all of its waves vibrate in the same direction. Remember that EM waves are made of electric and magnetic fields vibrating at right angles to each other. Polarization affects only the electric fields of a light wave. When all of the electric fields of a group of light waves vibrate in the same direction, the light is polarized.

Light can be polarized by a particular type of light filter called a polarizing filter. A polarizing filter acts on a light wave's electric field like the bars of a cage. The filter allows through only waves whose electric fields vibrate in one particular direction. Waves that pass through the filter are polarized. In the illustration below, these waves are shown in darker yellow.

Light reflecting off the surface of this pond causes glare.

A polarizing filter reduces glare, making it possible to see objects under the water.

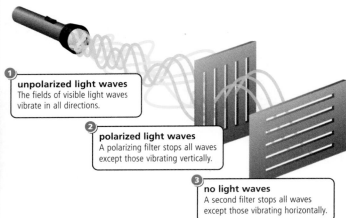

① unpolarized light waves
The fields of visible light waves vibrate in all directions.

② polarized light waves
A polarizing filter stops all waves except those vibrating vertically.

③ no light waves
A second filter stops all waves except those vibrating horizontally.

What do you think happens when polarized light passes into a second polarizing filter? If the direction of the bars in the second filter is the same as in the first, then all of the light will pass through the second filter. The light will still be polarized. If the direction of the bars in the second filter is at a right angle to the first, as in the illustration above, then no light at all will pass through the second filter.

Wavelength determines color.

The section of the EM spectrum called visible light is made up of many different frequencies and wavelengths. When all of these wavelengths are present together, as in light from the Sun or a light bulb, we see ordinary light, which is also called white light.

DIFFERENTIATE INSTRUCTION

 More Reading Support

D How do you make the electric fields of a wave vibrate in the same direction? *using polarization*

Advanced Unpolarized light (natural sunlight, for example) can become polarized to a certain degree when it is reflected from a flat surface such as a highway. This polarized light is termed glare and can be demonstrated by viewing the distant part of a highway on a sunny day. Have students make light diagrams showing how glare is formed. Have them investigate the conditions under which glare is formed, such as the angle at which the light approaches the surface, the time of day, and the material that is reflecting the light.

 Challenge and Extension, p. 184

Seen individually, the wavelengths appear as different colors of light. This fact can be demonstrated by using a prism. A **prism** is a tool that uses refraction to separate the different wavelengths that make up white light. The prism bends some of the wavelengths more than others. The lightwaves, bent at slightly different angles, form a color spectrum. The color spectrum could be divided into countless individual wavelengths, each with its own color. However, the color spectrum is usually divided into seven distinct color bands. In order of decreasing wavelength, the bands are red, orange, yellow, green, blue, indigo, and violet. You see a color spectrum whenever you see a rainbow.

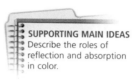

Prisms split light into colors by refracting wavelengths in different amounts.

Color Reflection and Absorption

The color of an object or material is determined by the wavelengths it absorbs and those it reflects. An object has the color of the wavelengths it reflects. A material that reflects all wavelengths of visible light appears white. A material that absorbs all wavelengths of visible light appears black. A green lime absorbs all colors except green and reflects green, so the lime looks green, as shown below.

SUPPORTING MAIN IDEAS
Describe the roles of reflection and absorption in color.

1 Light of all colors strikes the lime.

2 The lime absorbs all colors except green.

3 The lime reflects green, so it appears green.

The color that an object appears to the eye depends on another factor besides the wavelengths the object absorbs and reflects. An object can reflect only wavelengths that are in the light that shines on it. In white light, a white object reflects all the wavelengths of visible light and appears white. If you shine only red light on a white piece of paper, however, the paper will appear red, not white, because only red light is available to be reflected.

In summary, two factors determine the color of an object: first, the wavelengths that the object itself reflects or absorbs, and second, the wavelengths present in the light that shines on the object.

 CHECK YOUR READING What color wavelengths does a red apple absorb? What color wavelengths does a white flower absorb?

Chapter 3: Electromagnetic Waves **97** **D**

Teach from Visuals

To help students interpret the visual of the prism, ask:

- What color would the lime seem to be if you shined a red light on it? Why? *It would appear black. The green lime absorbs red, so no light would be reflected.*

- What would have to happen for the lime to appear white? *The lime would have to reflect all the wavelengths of visible light.*

Teach Difficult Concepts

Some students may have a hard time understanding that matter can absorb light. Remind them of the ice cube melting under black and white paper. Ask:

- Where did the ice melt faster? *under the black paper*

- Why was black paper warmer than white? *It absorbs more light waves.*

- A ginger (orange) cat and a blue (dark gray) cat are sitting on a window sill. Which cat's fur will absorb more light? How could you tell? *The blue cat; it will feel warmer to the touch.*

EXPLORE the **BIG** idea

Revisit "What Melts the Ice Cubes?" and "What Is White Light Made Of?" on p. 71. Have students explain the reasons for their results.

Ongoing Assessment
Recognize why objects have color.

Ask: Why does an object appear black? *The object absorbs all wavelengths of light that strike it.*

CHECK YOUR READING *Answer: all visible wavelengths except red; no visible wavelengths*

INVESTIGATE Mixing Colors

PURPOSE To observe the colors that make up black ink

TIPS *30 min.* To get the best results in this lab:

- Use cups with wide mouths so students can cut a long flap in the filter. The farther the ink moves, the more clearly the colors will separate.
- Handle the filters as little as possible. Oils from skin interfere with the movement of pigments.
- The felt-tip markers must be water soluble (washable). Try to get pens made by different manufacturers.

WHAT DO YOU THINK? *The colors in black ink separated. Each brand of black ink is made of a different combination of colors.*

CHALLENGE *If black ink is put on a filter, then it will spearate into colors because black is a combination of different colors of pigments.*

 Datasheet, Mixing Colors, p. 185

Technology Resources

Customize this student lab as needed or look for an alternative. Print rubrics to assess student lab reports.

 Lab Generator CD-ROM

Teaching with Technology

You may wish to photograph this lab if you have access to a digital camera. Students can see how the ink samples change over time.

Ongoing Assessment

Observe what makes up the color black.

Ask: Why does black ink appear black? *It absorbs all wavelengths of light.*

 Answer: produces all possible colors

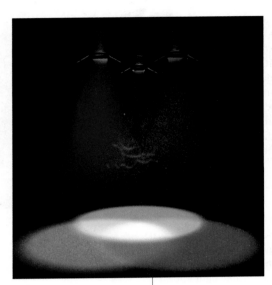

Primary colors of light combine to make the secondary colors yellow, cyan (light blue), and magenta (dark pink).

Primary Colors of Light

The human eye can detect only three colors: red, green, and blue. Your brain perceives these three colors and various mixtures of them as all the colors of the rainbow. These three colors of light, which can be mixed to produce all possible colors, are called **primary colors.** When all three colors are mixed together equally, they appear white, or colorless. Whenever colored light is added to a mixture, specific wavelengths are added. Mixing colors by adding wavelengths is called additive color mixing.

An example of the practical use of primary colors is a color television or computer monitor. The screen is divided into thousands of tiny bundles of red, green, and blue dots, or pixels. A television broadcast or DVD sends signals that tell the monitor which pixels to light up and when to do so. By causing only some pixels to give off light, the monitor can mix the three colors to create an amazing variety of colorful images.

 CHECK YOUR READING What does an equal mix of all three primary colors produce?

INVESTIGATE Mixing Colors

What is black ink made of?
PROCEDURE

1. Trim each of the filter papers to a disk about 10 cm (4 in.) in diameter. Make two parallel cuts about 1 cm (.5 in.) apart and 5 cm (2 in.) long from the edge of each disk toward the center. Fold the paper to make a flap at a right angle.

2. Use a different marker to make a dark spot in the middle of the flap on each disk.

3. Fill each of the cups with water. Set one of the disks on top of each cup so that the water covers the end of the flap but does not reach the ink spot.

4. After 15 minutes, examine each of the flaps.

WHAT DO YOU THINK?
- What did you observe about the effects of water on the ink spots?
- How do the three different samples compare?

CHALLENGE Write a hypothesis to explain what you observed about the colors in a black marker.

SKILL FOCUS
Observing

MATERIALS
- 3 coffee filters
- scissors
- 3 brands of black felt-tip marker
- 3 cups
- water

TIME
30 minutes

DIFFERENTIATE INSTRUCTION

 More Reading Support

G What are the three primary colors visible to human eyes? *red, green, and blue*

Alternative Assessment Have interested students videotape "Investigate Mixing Colors," then play back the tape at a slow speed. They can write and record a script to describe and explain what is happening as the separation of ink colors proceeds. The script should also explain why black ink has to be mixed this way.

Primary Pigments

Remember that two factors affect an object's color. One is the wavelengths present in the light that shines on the object. The other is the wavelengths that the object's material reflects or absorbs. Materials can be mixed to produce colors just as light can. Materials that are used to produce colors are called pigments. The **primary pigments** are cyan, yellow, and magenta. You can mix primary pigments just as you can mix primary colors to produce all the colors.

The primary pigment colors are the same as the secondary colors of light. The secondary pigment colors are red, blue, and green—the same as the primary colors of light.

The effect of mixing pigments is different from the effect of mixing light. Remember that a colored material absorbs all wavelengths except those of the color it reflects. Yellow paint absorbs all colors except yellow. Because pigments absorb wavelengths, whenever you mix pigments, you are subtracting wavelengths rather than adding them. Mixing colors by subtracting wavelengths is called subtractive color mixing. When all three primary pigments are mixed together in equal amounts, all wavelengths are subtracted. The result is black—the absence of color.

The inks used to make the circles on this page are primary pigments. They combine to make the secondary pigments red, blue, and green.

 CHECK YOUR READING How is mixing pigments different from mixing light?

3.4 Review

KEY CONCEPTS

1. What are some ways in which materials affect how light is transmitted?

2. How does a polarizing filter reduce glare?

3. In order for an object to appear white, which wavelengths must the light contain and the object reflect?

CRITICAL THINKING

4. **Apply** Imagine that you are a firefighter searching a smoke-filled apartment. Would using a stronger light help you see better? Explain your answer.

5. **Predict** Higher-energy EM waves penetrate farthest into a dense medium. What colors are more likely to penetrate to the bottom of a lake?

CHALLENGE

6. **Synthesize** If you focus a red light, a green light, and a blue light on the same part of a black curtain, what color will the curtain appear to be? Why?

Chapter 3: Electromagnetic Waves **99** D

CHAPTER INVESTIGATION

Focus

PURPOSE Students will learn that an object's color is determined by the wavelengths of light that shine upon it and those that it reflects.

OVERVIEW Students will make a light box and observe objects in blue, red, and green light. Students will find that

- a white object appears white in white light, red in red light, blue in blue light, and green in green light.
- a black object appears black in all colors of light.
- a red object appears red in white and red light, purple in blue light, and yellow in green light.
- a yellow object appears yellow in white light, orange in red light, green in blue light, and yellow-green in green light.

Lab Preparation

- Ask students to bring shoe boxes from home, or check with discount stores. Boxes without lids can be used upside down.
- Acetate sheets are available as report covers from school supply stores.
- Have students read the lab, write their hypotheses, and make data tables before class. Or copy and distribute datasheets and rubrics.

 UNIT RESOURCE BOOK, pp. 194–202

 SCIENCE TOOLKIT, F14

Lab Management

- This investigation can be set up with four viewing stations, each with a box testing one color of light. Students can circulate to each box.
- Remind students that they should examine each of the four objects in four colors of light.

INCLUSION If you have physically challenged students in your class, you might make their boxes before class. Team up color-blind students with those who can distinguish colors.

CHAPTER INVESTIGATION

Wavelength and Color

OVERVIEW AND PURPOSE Stage managers use color filters to change the look of a scene. The color an object appears to have depends on both the wavelengths of light shining on it and the wavelengths of light it reflects. In this exercise, you will investigate the factors that affect these wavelengths and so affect the color of an object. You will

- make a light box
- study the effect of different colors of light on objects of different colors

Problem

How does the appearance of objects of different colors change in different colors of light?

Hypothesize

Read the procedure below and look at the sample notebook page. Predict what color each object will appear to be in each color of light. Give a reason for each prediction.

Procedure

MATERIALS
- 3 sheets of acetate (red, blue, and green)
- ruler
- scissors
- shoe box
- masking tape
- light source
- 4 solid-colored objects (white, black, red, and yellow)

1. Draw a data table like the one in the sample **Science Notebook.**

2. Make 3 color filters by cutting a 10 cm (4 in.) square from each color of acetate.

3. Make a 3 cm (1 in.) wide hole in the middle of the top of the box. This will be the viewing hole.

4. Make an 8 cm (3 in.) hole in one end of the box. This will be the light hole.

5. You will observe each of the four colored objects four times—with no filter and with the red, blue, and green filters. Use masking tape to position the filters in the light hole, as shown.

step 5

INVESTIGATION RESOURCES

 CHAPTER INVESTIGATION, Wavelength and Color
- Level A, pp. 194–197
- Level B, pp. 198–201
- Level C, p. 202

Advanced students should complete Levels B & C.

 Writing a Lab Report, D12–13

Technology Resources

Customize this student lab as needed or look for an alternative. Print rubrics to assess student lab reports.

Lab Generator CD-ROM

6. Place the light box on a flat surface near a strong white light source such as sunlight or a bright lamp. Position the box with the uncovered light hole facing the light source. Place the white object inside the box, look through the eye hole, and observe the object's color. Record your observations.

step 7

7. Use the light box to test each of the combinations of object color and filter shown in the table on the sample notebook page. Record your results.

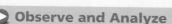
▶ Observe and Analyze | Write It Up

1. **RECORD OBSERVATIONS** Be sure your data table is complete.

2. **COMPARE** What color did the red object appear to be when viewed with a blue filter? a red filter?

▶ Conclude | Write It Up

1. **INTERPRET** Answer your problem question.

2. **ANALYZE** Compare your results to your hypotheses. How do the results support your hypotheses?

3. **IDENTIFY VARIABLES** What different variables affected the outcome of your experiment?

4. **INFER** Why do colors of objects appear to change in different types of light?

5. **IDENTIFY LIMITS** What possible limitations or sources of error could have affected your results?

6. **APPLY** If you were going to perform on a stage that was illuminated using several different color filters, what color clothing should you wear in order to look as bright and colorful as possible?

▶ INVESTIGATE Further

CHALLENGE Perform this experiment using different kinds of artificial light. Try it with a low-wattage incandescent bulb, a high-wattage incandescent bulb, a fluorescent bulb, or a full-spectrum bulb. How do different kinds of artificial light affect the colors that objects appear to be?

Wavelength and Color

Problem

How does the appearance of objects of different colors change in different colors of light?

Hypothesize

Observe and Analyze

Table 1. Predicted and Observed Colors of Objects with Different Colored Filters

Predicted	no filter	red filter	blue filter	green filter
white object				
black object				
red object				
yellow object				
Observed	no filter	red filter	blue filter	green filter
white object				

Chapter 3: **Electromagnetic Waves** 101 **D**

▶ Observe and Analyze | Write It Up

SAMPLE DATA White object: white in white light, red in red light, blue in blue light, green in green light; Black object: black in all colors of light; Red object: red in white and red light, purple in blue light, and yellow in green light; Yellow object: yellow in white light, orange in red light, green in blue light, and yellow-green in green light.

1. *See students' data tables.*

2. *purple, red*

▶ Conclude | Write It Up

1. *The color of an object depends both on the wavelengths of the light that shines on it and the wavelengths that it reflects.*

2. *Student answers will vary.*

3. *The variables are the color of the object and the color of the filter.*

4. *The color that an object reflects depends on the type of light shining on it.*

5. *Possible limitations or sources of error might include the shade of the colored acetate, the brightness of the light, leakage of white light into the box, and a person's sense of color.*

6. *You should wear white so your clothing will reflect any filter color.*

▶ INVESTIGATE Further

CHALLENGE Results will vary depending on the type of light bulb tested. Different kinds of light bulbs emit different wavelengths of light.

Post-Lab Discussion

- Copy the data table onto the board. Have students fill in the object colors they observed. Discuss why results might have varied. Ask if other variables were introduced in the experiment.

- Ask students why it is wise to examine a paint sample in the same kind of light that it will be used in. Discuss how the objects appeared under various kinds of artificial lighting.

BACK TO

the BIG idea

Have students compare the frequency of the following EM waves: a burning ember glowing red, a yellow candle flame, the blue-white flame of a blow-torch. *The colors represent different frequencies of light. The red glowing ember has low frequency, the yellow candle flame has a higher frequency, and the blue-white flame of the blow-torch has the highest frequency.*

◄ KEY CONCEPTS SUMMARY

SECTION 3.1
Ask: What is the spatial relationship between the electrical field and the magnetic field of an EM wave? *They are at right angles to each other.*

SECTION 3.2
Ask: What determines the characteristics of an EM wave? *frequency*

Ask: How would the wave pictured change if its frequency were increased? *The wavelength would be smaller.*

SECTION 3.3
Ask: What is the difference between the light produced from the two pictures? *The incandescent light bulb produces light from heat. The caterpillar produces light from chemical reactions.*

SECTION 3.4
Ask: What does a prism do? *separates white light into its component wavelengths*

Review Concepts

- Big Idea Flow Chart, p. T17
- Chapter Outline, p. T23–T24

 # Chapter Review

the BIG idea

Electromagnetic waves transfer energy through radiation.

CONTENT REVIEW
CLASSZONE.COM

◄ KEY CONCEPTS SUMMARY

3.1 Electromagnetic waves have unique traits.

- Electromagnetic (EM) waves are made of vibrating electric and magnetic fields.
- EM waves travel at the speed of light through a vacuum.
- EM waves transfer energy and can interact with matter.

VOCABULARY
electromagnetic wave p. 73
radiation p. 75

3.2 Electromagnetic waves have many uses.

- EM waves are grouped by frequency on the EM spectrum.
- The EM spectrum is divided into radio waves, microwaves, infrared light, visible light, ultraviolet light, x-rays, and gamma rays.

VOCABULARY
EM spectrum p. 80
radio waves p. 82
microwaves p. 83
visible light p. 84
infrared light p. 84
ultraviolet light p. 85
x-rays p. 86
gamma rays p. 86

3.3 The Sun is the source of most visible light.

- Most visible light comes from the Sun.
- Many living organisms produce visible light for their own use.
- Humans produce visible light artificially.

VOCABULARY
incandescence p. 89
luminescence p. 89
bioluminescence p. 89
fluorescence p. 91

3.4 Light waves interact with materials.

- Reflection, transmission, and absorption affect what light we see.
- Light can be scattered and polarized.
- Visible light is made up of individual color wavelengths.
- The primary colors are red, blue, and green.
- The primary pigments are yellow, cyan, and magenta.

VOCABULARY
transmission p. 93
absorption p. 93
scattering p. 95
polarization p. 96
prism p. 97
primary colors p. 98
primary pigments p. 99

D 102 Unit: Waves, Sound, and Light

Technology Resources

Have students visit **ClassZone.com** or use the CD-ROM for a cumulative review of concepts.

 CONTENT REVIEW

 CONTENT REVIEW CD-ROM

Engage students in a whole-class interactive review of Key Concepts. Edit content as you wish.

 POWER PRESENTATIONS

Reviewing Vocabulary

Make a four-square diagram for each of the listed terms. Write the term in the center. Define the term in one square. Write characteristics, examples, and nonexamples in other squares. A sample is shown below.

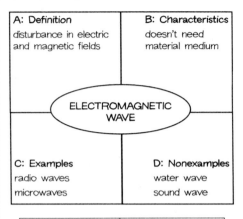

A: Definition	B: Characteristics
disturbance in electric and magnetic fields	doesn't need material medium

ELECTROMAGNETIC WAVE

C: Examples	D: Nonexamples
radio waves microwaves	water wave sound wave

1. gamma rays	**6.** radiation
2. infrared light	**7.** bioluminescence
3. transmission	**8.** EM spectrum
4. absorption	**9.** incandescence
5. pigment	**10.** polarization

Reviewing Key Concepts

Multiple Choice *Choose the letter of the best answer.*

11. An electromagnetic wave is a disturbance that transfers energy through a field. In this sense, a disturbance is the same as a
- **a.** confusion
- **b.** magnification
- **c.** vibration
- **d.** conflict

12. Unlike mechanical waves, EM waves can travel through
- **a.** a vacuum
- **b.** water
- **c.** the ground
- **d.** air

13. A light year is a measure of
- **a.** time
- **b.** distance
- **c.** speed
- **d.** wavelength

14. The Sun and a light bulb both produce light through
- **a.** bioluminescence
- **b.** incandescence
- **c.** luminescence
- **d.** polarization

15. Which of the following types of light bulb converts ultraviolet waves into visible light waves?
- **a.** incandescent
- **b.** fluorescent
- **c.** halogen
- **d.** tungsten

16. An object seen through a translucent material appears less clear than one seen through transparent material because the translucent material
- **a.** transmits none of the light coming from the object
- **b.** reflects all the light coming from the object
- **c.** transmits all the light coming from the object
- **d.** diffuses some light coming from the object

17. An object appears red because it
- **a.** reflects light waves of all colors
- **b.** reflects light waves of red
- **c.** absorbs light waves of red
- **d.** transmits light waves of all colors

18. Primary colors of light can combine to
- **a.** black light
- **b.** white light
- **c.** primary pigments
- **d.** ultraviolet light

Short Answer *Write a short answer to each question.*

19. What parts of an EM wave have amplitude?

20. How can EM waves be used to measure distance in space?

21. Describe how microwaves are used in communications.

22. What two properties of an EM wave change as you move from one part of the EM spectrum to another?

23. How does visible light differ from other EM waves? How is it similar?

24. Explain briefly how an incandescent light bulb works.

Reviewing Vocabulary

1. A: EM waves with highest frequencies and energies; B: 10^{19} Hz to 10^{24} Hz; C: the Sun; D: radio waves

2. A: EM frequencies between microwaves and visible light; B: invisible; C: lamps; D: x-rays

3. A: the passage of an EM wave through a medium; B: makes it possible to see through things; C: clear glass; D: wood

4. A: disappearance of an EM wave into the medium; B: causes materials to be opaque; C: wood; D: air

5. A: material used to mix colors; B: subtracts wavelengths rather than adding them; C: paint; D: colored light

6. A: transfer of energy in the form of EM waves; B: spreads outward in straight line from source; passes through vacuum; C: light from Sun; D: ocean waves

7. A: production of light by living organisms; B: light produced from chemical reactions; C: firefly; D: fire

8. A: range of all EM frequencies; B: smooth, gradual progression; C: microwaves, D: ultrasound

9. A: production of light from high temperatures; B: dangerous for living tissue; C: sunlight; D: firefly

10. A: filtering light so that all of its waves vibrate in the same direction; B: affects only electric fields of wave; C: sunglasses; D: glare

Reviewing Key Concepts

11. c	14. b	17. b
12. a	15. b	18. b
13. b	16. d	

19. electric and magnetic fields

20. by how far they travel in a certain amount of time

21. Cell phones and radar send and receive microwave signals.

22. wavelength and frequency

23. Visible light has different frequencies and wavelengths. Visible light, like all EM waves, is a disturbance in a field.

24. Electricity passes through the filament. Resistance causes the filament to heat up and glow.

Thinking Critically

25. absorption

26. Violet light should penetrate even deeper into the water than blue.

27. Objects near the ocean floor should appear blue.

28. when it interacts with a medium

29. The light source must contain red wavelengths and the object must reflect red light.

30. blue, all the other colors

31. Sample answer: Incandescent lighting wastes energy as heat, and lighting is more energy efficient.

32. The black numbers could be caused by polarizing filters at right angles that would screen out all light.

33. Like a sieve, a polarizing filter screens out certain kinds of light waves while letting others pass through. Unlike a sieve, a polarizing filter screens waves on the basis of orientation rather than size.

34. Fluorescent bulbs remain cooler than regular or halogen bulbs, so they waste less energy as heat.

35. blue, because only blue wavelengths pass through filter to your eye

the **BIG** idea

36. Sample answer: A cell phone uses microwaves. Cell phones depend on an overlapping network of cells. Each cell has a tower that sends and receives microwave signals. The tower connects cell phones to each other or the regular wired telephone system.

37. See students' summaries.

38. Infrared waves leave the radiator. They are transmitted by the air, which absorbs only a little of their energy. The kitten absorbs waves that travel toward it, and is warmed by energy transferred in the form of heat.

UNIT PROJECTS

Collect schedules, materials lists, and questions. Be sure dates and materials are obtainable, and questions are focused.

 Unit Projects, pp. 5–10

Thinking Critically

The diagram below shows how far different wavelengths of visible light penetrate into ocean water. Use information from this diagram to answer the next three questions.

25. OBSERVE An EM wave can interact with a material in different ways. Which type of interaction keeps some light waves from reaching the ocean floor?

26. PREDICT How would violet light behave in the same water? Think of where violet is on the color spectrum.

27. SYNTHESIZE How is the apparent color of objects near the ocean floor affected by the interactions shown in the diagram?

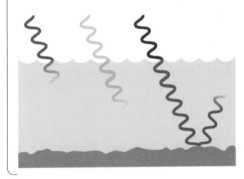

28. ANALYZE Under what circumstances can an EM wave begin to convert some of its electromagnetic energy into other forms of energy?

29. ANALYZE What two things must be true about the light source and the material of an object for you to see an object as red?

30. PREDICT If you shine a blue light on a white object, what color will the object appear to be? What color light would you need to add to make the white object appear white?

31. APPLY Why might incandescent lighting become less common in the future? Explain your reasoning.

---

32. CAUSE AND EFFECT Liquid crystal displays like the ones used in some calculators work by polarizing light. Describe how two polarizing filters could cause the numbers on the display panel to appear black.

33. COMPARE AND CONTRAST In what way would a sieve be a good model for a polarizing light filter? In what ways would it not be?

34. CONTRAST In what ways is a fluorescent bulb more efficient than incandescent and halogen bulbs?

35. PREDICT What color will a white object appear to be if you look at it through a blue filter?

the **BIG** idea

36. Return to the question on page 70. Answer the question again, using what you have learned in the chapter.

37. SUMMARIZE Write a summary of this chapter. Use the Big Idea statement from page 70 as the title for your summary. Use the Key Concepts listed on page 70 as the topic sentences for each paragraph. Provide an example for each key concept.

38. ANALYZE Describe all of the EM wave behaviors and interactions that occur when a radiator warms a kitten.

UNIT PROJECTS

Check your schedule for your unit project. How are you doing? Be sure that you've placed data or notes from your research in your project folder.

MONITOR AND RETEACH

If students have trouble applying the concepts in items 25–27, have them create a three-part visual aid showing how deeply various colors of light would penetrate the water.

Part 1 should show red, orange, yellow, green, blue, indigo, and violet light; **Part 2** should show relative frequencies of the various colors of light; **Part 3** should list the relative energies of the various colors of light. Students may benefit from summarizing one or more sections of the chapter.

 Summarizing the Chapter, pp. 212–213

Standardized Test Practice

For practice on your state test, go to . . .

TEST PRACTICE
CLASSZONE.COM

Interpreting Diagrams

The diagram below shows part of the electromagnetic (EM) spectrum. The lower band shows frequency in hertz. The upper band shows part of the spectrum used by different technologies.

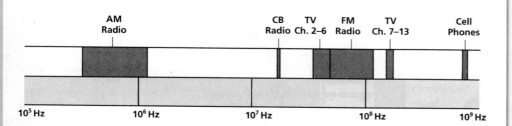

Use the diagram to answer the following questions.

1. Which of the technologies listed below uses the highest frequencies?
  a. AM radio
  b. CB radio
  c. FM radio
  d. TV channels 2–6

2. If you were receiving a signal at a frequency of nearly 10^9 Hz, what would you be using?
  a. a CB radio
  b. an AM radio
  c. an FM radio
  d. a cell phone

3. A television station broadcasts its video signal at 10^6 Hz and its audio signal at 10^8 Hz. To receive the broadcasts, your television would need to receive
  a. both AM and FM radio
  b. both CB and AM radio
  c. both CB and FM radio
  d. both CB radio and cell phone transmissions

4. Signals with similar frequencies sometimes interfere with each other. For this reason, you might expect interference in which of the following:
  a. lower television channels from cell phones
  b. upper television channels from FM radio
  c. lower television channels from FM radio
  d. upper television channels from cell phones

Extended Response

Answer the two questions below in detail. Include some of the terms from the word box. Underline each term you use in your answer.

frequency	energy	interaction
field	medium	vacuum

5. What are the similarities and differences between mechanical waves and electromagnetic waves?

6. What are some advantages and disadvantages of different types of artificial lighting?

METACOGNITIVE ACTIVITY

Have students answer the following questions in their **Science Notebook:**

1. What questions do you still have about electromagnetic waves?
2. What were you surprised to find out about visible light?
3. How do the concepts of this chapter relate to your Unit Project?

Interpreting Diagrams

1. c   3. a
2. d   4. c

Extended Response

5. RUBRIC
4 points for a response that correctly answers the question and that uses the following terms accurately:
  • wavelength   • medium
  • amplitude   • field
  • frequency

Sample: Mechanical and electromagnetic waves are disturbances that transfer energy. They have <u>wavelength</u>, <u>amplitude</u>, and <u>frequency</u>. Mechanical waves need a <u>medium</u>, while EM waves do not. EM waves are disturbances in electric and magnetic <u>fields</u>. Mechanical waves give up some energy when they travel, but EM waves can travel in a vacuum without losing energy.

3 points answers the question correctly and uses three of the listed terms accurately
2 points response is correct, but partial, and uses two terms correctly
1 point for a response that is partially correct, but contains some inaccuracies

6. RUBRIC
4 points for a response that correctly identifies advantages and disadvantages of the following types of artificial lighting:
  • incandescent   • halogen
  • fluorescent   • LED
Sample: <u>Incandescent</u> lighting is a good source of bright light, is relatively cheap, and is convenient to use. <u>Fluorescent</u> lights waste less energy than incandescent lights, because fluorescent bulbs produce less heat and they last longer. <u>Halogen</u> bulbs last longer than incandescent bulbs, but halogen bulbs are more of a fire and injury hazard because they get very hot. <u>LEDs</u> are very cheap, last a long time, are safe, and waste almost no energy.

3 points identifies advantages and disadvantages of three types of artificial lighting listed
2 points identifies advantages and disadvantages of two types of artificial lighting listed
1 point identifies advantages and disadvantages of one type of artificial lighting listed

● Set Learning Goals

Students will

- Observe how scientists historically studied light.
- Examine how observations about the properties of light led to explanations of its behavior.
- Make a camera obscura and write a news article about a discovery regarding light.

National Science Education Standards

A.9.a–g Understandings About Scientific Inquiry

E.6.a–c Understandings About Science and Technology

F.5.a–e, F.5.g Science and Technology in Society

G.1.a–b Science as a Human Endeavor

G.2.a Nature of Science

G.3.a–c History of Science

History Connection

Point out to students that the top half of the timeline shows some major events in the scientific study of light that were historically recorded. The bottom half of the timeline illustrates advances in the technology that enables study of light and practical applications of the results of this study. The two gaps in the timeline represent periods of time in which no advances in the study of light are highlighted.

Technology

REFLECTING TELESCOPES The first telescope was a refractor, built by Galileo, and many current telescopes use lenses to refract light. When light travels through lenses, it bends. The amount it is bent depends on the shape, composition, and thickness of the lens. Many telescopes are both reflecting and refracting because they use both mirrors and lenses.

TIMELINES in Science

THE STORY OF LIGHT

Light has fascinated people since ancient times. The earliest ideas about light were closely associated with beliefs and observations about vision. Over the centuries, philosophers and scientists developed an increasingly better understanding of light as a physical reality that obeyed the laws of physics.

With increased understanding of the nature and behavior of light has come the ability to use light as a tool. Many applications of light technology have led to improvements in human visual abilities. People can now make images of a wide range of objects that were invisible to earlier generations. The study of light has also led to technologies that do not involve sight at all.

This timeline shows just a few of the many steps on the road to understanding light. The boxes below the timeline show how these discoveries have been applied and developed into new technologies.

400 B.C.
Light Travels in a Straight Line
Observing the behavior of shadows, Chinese philosopher Mo-Ti finds that light travels in a straight line. His discovery helps explain why light passing through a small opening forms an upside-down image.

300 B.C.
Reflection Obeys Law
Greek mathematician Euclid discovers that light striking mirrors obeys the law of reflection. The angle at which light reflects off a mirror is equal to the angle at which it strikes the mirror.

EVENTS

| 450 B.C. | 425 B.C. | 400 B.C. | 375 B.C. | 350 B.C. | 325 B.C. | 300 B.C. |

APPLICATIONS AND TECHNOLOGY

APPLICATION

Camera Obscura

The principle described by Mo-Ti in 400 B.C. led to the development of the camera obscura. When light from an object shines through a small hole into a dark room, an image of the object appears on the far wall. The darkened room is called, in Latin, *camera obscura*. Because light travels in a straight line, the highest points on the object appear at the lowest points on the image; thus, the image appears upside down. Room-sized versions of the camera obscura like the one shown here were a popular attraction in the late 1800s.

DIFFERENTIATE INSTRUCTION

Advanced Tell students that a line that is perpendicular to a flat surface is called the *normal*. Have students state generalizations about the following angles: the angle formed by the normal and a beam of light hitting a surface and the angle formed by the normal and the beam of light reflecting from the surface. *Sample answer: The angle formed by the normal and a beam of light hitting a surface is equal to the angle formed by the normal and the beam of light reflecting from the surface.*

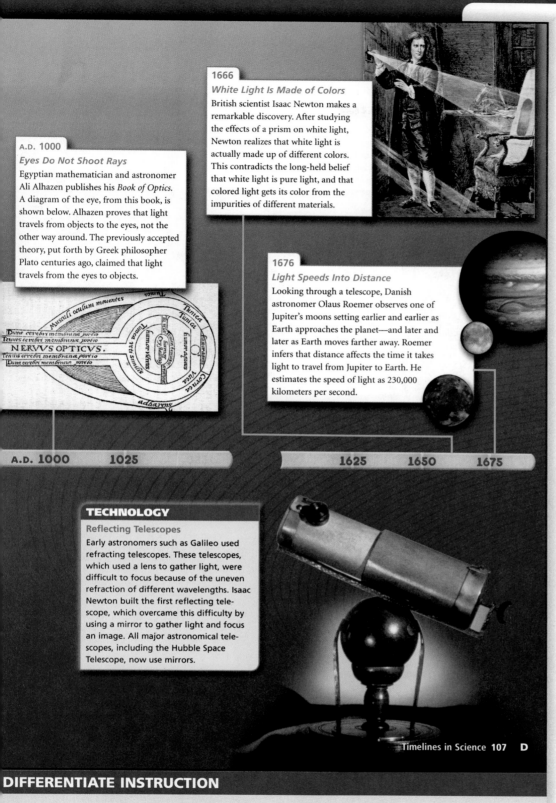

1666

White Light Is Made of Colors

British scientist Isaac Newton makes a remarkable discovery. After studying the effects of a prism on white light, Newton realizes that white light is actually made up of different colors. This contradicts the long-held belief that white light is pure light, and that colored light gets its color from the impurities of different materials.

A.D. 1000

Eyes Do Not Shoot Rays

Egyptian mathematician and astronomer Ali Alhazen publishes his *Book of Optics*. A diagram of the eye, from this book, is shown below. Alhazen proves that light travels from objects to the eyes, not the other way around. The previously accepted theory, put forth by Greek philosopher Plato centuries ago, claimed that light travels from the eyes to objects.

1676

Light Speeds Into Distance

Looking through a telescope, Danish astronomer Olaus Roemer observes one of Jupiter's moons setting earlier and earlier as Earth approaches the planet—and later and later as Earth moves farther away. Roemer infers that distance affects the time it takes light to travel from Jupiter to Earth. He estimates the speed of light as 230,000 kilometers per second.

A.D. 1000   1025   1625   1650   1675

TECHNOLOGY

Reflecting Telescopes

Early astronomers such as Galileo used refracting telescopes. These telescopes, which used a lens to gather light, were difficult to focus because of the uneven refraction of different wavelengths. Isaac Newton built the first reflecting telescope, which overcame this difficulty by using a mirror to gather light and focus an image. All major astronomical telescopes, including the Hubble Space Telescope, now use mirrors.

Timelines in Science **107** **D**

Mathematics Connection

1676 Working with the knowledge available in 1676, Roemer's estimated speed of light is relatively close to the currently accepted value of 3.00×10^8 m/s. The relationship among speed, frequency, and wavelength of light (or any other wave) can be expressed by the equation $S = f \lambda$, where S is the speed, f is the frequency, and λ is the wavelength.

Integrate the Sciences

Students are familiar with the spectrum of colors seen in a rainbow. This splitting of visible light into its various colors occurs when sunlight strikes droplets of water and light is refracted into different colors. Some of the light reflects off the back surface of the drop, and it refracts further when it leaves the water drop and reenters the air. Violet light is at one end of the visible spectrum, and it is seen coming from droplets lower in the atmosphere. Violet light emerges from the droplets at an angle of approximately 40°. Red light is at the opposite end of the visible spectrum, and it is seen coming from droplets higher in the air. Red light leaves the droplets at an angle of about 42°. The other colors of light are between these two extremes.

Language Arts Connection

All types of light make up the electromagnetic spectrum, of which visible light is only a small portion. Ask: Ultrasound is sound that is in a higher range than can be heard by humans. From this meaning of *ultra-*, what conclusion might you draw about ultraviolet light? *It has a frequency higher than that of violet light.* The prefix *infra-* has the opposite meaning of the prefix *ultra-*. Based on this information, where would infrared light be located in the electromagnetic spectrum? *It would have a frequency less than that of red light.*

DIFFERENTIATE INSTRUCTION

Advanced Have students use the mathematical relationship and the speed of light shown in the Mathematics Connection on this page to make the following calculations:

1. What is the wavelength of visible light that has a frequency of 5.0×10^{14} s? $(3.00 \times 10^8$ m/s$)(5.0 \times 10^{14}$/s$) = 6.0 \times 10^{-7}$ m

2. What is the frequency of x-rays that have a wavelength of 2.0×10^{-8} m? $(3.00 \times 10^8$ m/s$)(2.0 \times 10^{-8}$ m$) = 1.5 \times 10^{16}$ s

Scientific Process

Observations of light led to hypotheses that light is a wave. Observations also support light's being a particle. Experiments that test these hypotheses support both of them. As a result, light is considered to be both a wave and a particle made up of discrete packets of energy.

Application

HOLOGRAMS Because holograms look different from various angles, holograms are difficult to reproduce. As a result, they are used on certain credit cards and other documents to help prevent forgery.

Art Connection

Art forgery is a common problem for artists, art dealers, and art buyers. Some art forgeries are so well done that special technology is necessary to tell the real item from a forgery. What is done to tell forgeries from original art? There is a difference in modern paint and paints that were used years ago, but no one wants to chip off paint from an original painting to test it. One way to test a painting without damaging it is to use ultraviolet and infrared light and x-rays.

The human eye cannot see the varnish that is painted on the top of oil paintings, but it glows under ultraviolet light. Irregularities in the varnish and places that have had paint added show up as dark spots when this light is shined on it. Infrared light penetrates layers of oil paints, and if there is an image hidden under the surface oil paints, it shows up when infrared light is shined on the painting. X-rays show dense materials, such as metals and paints that contain metallic pigments. If metallic objects or materials that are inconsistent with when the painting was supposed to have been painted are seen, forgeries can be detected.

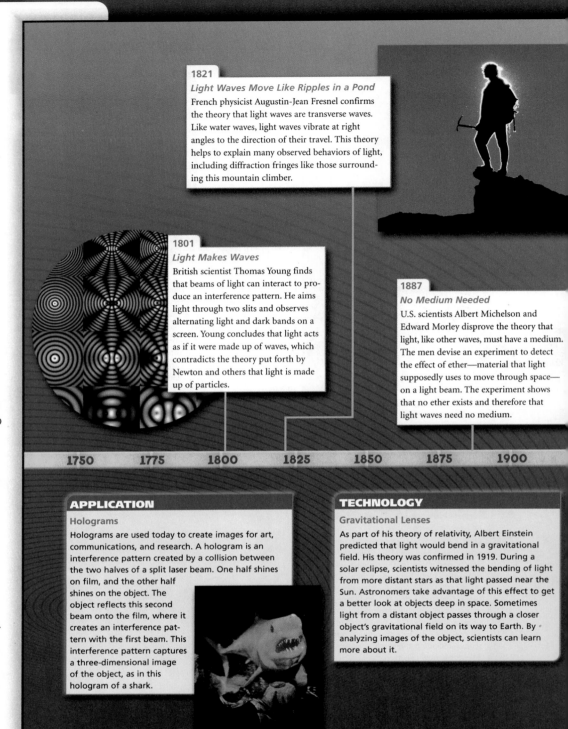

1821
Light Waves Move Like Ripples in a Pond
French physicist Augustin-Jean Fresnel confirms the theory that light waves are transverse waves. Like water waves, light waves vibrate at right angles to the direction of their travel. This theory helps to explain many observed behaviors of light, including diffraction fringes like those surrounding this mountain climber.

1801
Light Makes Waves
British scientist Thomas Young finds that beams of light can interact to produce an interference pattern. He aims light through two slits and observes alternating light and dark bands on a screen. Young concludes that light acts as if it were made up of waves, which contradicts the theory put forth by Newton and others that light is made up of particles.

1887
No Medium Needed
U.S. scientists Albert Michelson and Edward Morley disprove the theory that light, like other waves, must have a medium. The men devise an experiment to detect the effect of ether—material that light supposedly uses to move through space—on a light beam. The experiment shows that no ether exists and therefore that light waves need no medium.

| 1750 | 1775 | 1800 | 1825 | 1850 | 1875 | 1900 |

APPLICATION
Holograms
Holograms are used today to create images for art, communications, and research. A hologram is an interference pattern created by a collision between the two halves of a split laser beam. One half shines on film, and the other half shines on the object. The object reflects this second beam onto the film, where it creates an interference pattern with the first beam. This interference pattern captures a three-dimensional image of the object, as in this hologram of a shark.

TECHNOLOGY
Gravitational Lenses
As part of his theory of relativity, Albert Einstein predicted that light would bend in a gravitational field. His theory was confirmed in 1919. During a solar eclipse, scientists witnessed the bending of light from more distant stars as that light passed near the Sun. Astronomers take advantage of this effect to get a better look at objects deep in space. Sometimes light from a distant object passes through a closer object's gravitational field on its way to Earth. By analyzing images of the object, scientists can learn more about it.

D 108 Unit: Waves, Sound, and Light

DIFFERENTIATE INSTRUCTION

Below Level Have students show how light is separated into different colors by shining a light on bubbles made by using liquid dishwashing detergent. The colors can be seen better if this activity is done in a darkened room.

1960

Light Beams Line Up

U.S. inventor Theodore Harold Maiman builds a working laser by stimulating emission of light in a cylinder of ruby crystal. Laser light waves all have the same wavelength and their crests and troughs are lined up.

2001

Light Is Completely Stopped

After slowing light to the speed of a bicycle, Danish physicist Lene Vestergaard Hau brings it to a complete halt in a super-cold medium. Controlling the speed of light could revolutionize computers, communications, and other electronic technology.

RESOURCE CENTER
CLASSZONE.COM

Learn more about current research involving light.

1925   1950   1975   2000

APPLICATION

Lasers in Eye Surgery

For centuries, people have used corrective lenses to help their eyes focus images more clearly. Today, with the help of lasers, doctors can correct the eye itself. Using an ultraviolet laser, doctors remove microscopic amounts of a patient's cornea to change the way it refracts light. As a result, the eye focuses images exactly on the retina. For many nearsighted people, the surgery results in 20/20 vision or better.

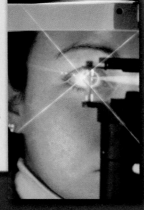

INTO THE FUTURE

Much of our current knowledge in science, from the workings of our bodies to the universe as a whole, is founded upon experiments that used light. Evidence from new light applications will continue to shape our knowledge. In the future, the nature of light, itself, may again come into question as new experiments are performed.

As new light microscopes are developed, scientists will gain more detailed information about how systems within our bodies work, such as how our brain cells interact with each other to perform a complex task. With powerful telescopes, scientists will gain a better understanding of the universe at its beginnings and how galaxies are formed.

Finally, as we continue to study the behavior of light, we may continue to modify its very definition. Sometimes considered a stream of particles, and other times considered waves, light is now understood to have qualities of both particles and waves.

ACTIVITIES

Make a Camera Obscura

Take a small box and paint the interior black. On one side, make a pinhole. On a side next to that one, make a hole about 5 cm in diameter.

On a bright, sunny day, hold the box so that sunlight enters the box through the pinhole. Fit your eye snugly against the larger hole and look inside.

Writing About Science

Lasers are currently used in entertainment, medicine, communication, supermarkets, and so on. Write a prediction about a specific use of lasers in the future. You might describe a new invention.

Timelines in Science **109** **D**

INTO THE FUTURE

Have students divide into two groups. Have one group list the ways they think that the use of light can provide information about the formation of the universe. Have the other group make a similar list that refers to using light to study the cells of the human body. Then have students in each group research actual scientific investigations in the area they have been assigned. Have each group prepare a presentation of their results. Presentations might consist of bulletin boards, videos done as news segments, or oral reports with visual aids.

ACTIVITIES

Make a Camera Obscura

Be sure the box does not allow any light to enter, other than the light that enters through the holes made in the box. A full-spectrum artificial light, such as a sunlamp, can be used instead of natural sunlight. Ask students to explain why a camera obscura might be a safe way to view a solar eclipse. *The gradual darkening of the Sun could be viewed on the side of the camera. Looking directly at the Sun can damage eyes, but the intensity of the light in the camera obscura is much less than direct sunlight.*

Writing About Science

Students can use Internet sources for information, as well as reference books. While writing their predictions, students should be aware of current uses of lasers.

Technology Resources

Students can visit **ClassZone.com** for current news about new advances in technology involving light.

DIFFERENTIATE INSTRUCTION

Inclusion Pair any visually impaired student with another student who can explain the timeline content to them. When making the camera, outline with yarn the area on the side of the camera where the image is shown. Have the visually impaired student feel the box, the two holes, and the area where the image is shown.

CHAPTER 4
Light and Optics

Physical Science
UNIFYING PRINCIPLES

PRINCIPLE 1

Matter is made of particles too small to see.

PRINCIPLE 2

Matter changes form and moves from place to place.

PRINCIPLE 3

Energy changes from one form to another, but it cannot be created or destroyed.

PRINCIPLE 4

Physical forces affect the movement of all matter on Earth and throughout the universe.

Unit: Waves, Sound, and Light
BIG IDEAS

CHAPTER 1
Waves

Waves transfer energy and interact in predictable ways.

CHAPTER 2
Sound

Sound waves transfer energy through vibrations.

CHAPTER 3
Electromagnetic Waves

Electromagnetic waves transfer energy through radiation.

CHAPTER 4
Light and Optics

Optical tools depend on the wave behavior of light.

CHAPTER 4
KEY CONCEPTS

SECTION 4.1	SECTION 4.2	SECTION 4.3	SECTION 4.4
Mirrors form images by reflecting light.	**Lenses form images by refracting light.**	**The eye is a natural optical tool.**	**Optical technology makes use of light waves.**
1. Optics is the science of light and vision.	**1.** A medium can refract light.	**1.** The eye gathers and focuses light.	**1.** Mirrors and lenses can be combined to make more powerful optical tools.
2. Mirrors use regular reflection.	**2.** Shape determines how lenses form images.	**2.** Corrective lenses can improve vision.	**2.** Lasers use light in new ways.
3. Shape determines how mirrors form images.			

 The Big Idea Flow Chart is available on p. T25 in the **UNIT TRANSPARENCY BOOK.**

Previewing Content

SECTION

4.1 Mirrors form images by reflecting light. pp. 113–118

1. Optics is the science of light and vision.
Optics is the study and application of visible light and its inter-action with the eye to produce vision. Optical tools can improve vision or use light to do work.

2. Mirrors use regular reflection.
A mirror forms an image by reflecting light from an object. Light reflecting off a mirror follows the **law of reflection,** which states that the angle formed by a line perpendicular to the mirror and the light hitting the mirror (the incident ray) is equal to the angle formed by the perpendicular and the light leaving the mirror (the reflected ray).

3. Shape determines how mirrors form images.
Different mirror shapes produce different kinds of images.
- A flat mirror produces a virtual image. The image seems to come from behind the mirror.
- A **convex** mirror reflects light waves so that they spread out, never meeting at a focal point.
- A **concave** mirror reflects light waves so that they converge at a **focal point.**

Convex Mirror **Concave Mirror, Far Away** **Concave Mirror, Up Close**

The images formed in mirrors depend on the curve of the mirror's surface and the distance of the object from the mirror.

SECTION

4.2 Lenses form images by refracting light. pp. 119–125

1. A medium can refract light.
When a light wave moves into a new medium, it may change speed. If it hits the medium at an angle, one side of the wave changes speed before the other side, bending the wave. This is refraction.
- When light waves refract, they can bend toward or away from the normal.
- Refraction can occur when light moves from cool air in the sky to warmer air near the ground. Mirages occur in this way.

2. Shape determines how lenses form images.
Lenses produce predictable images when waves of light pass through the lens and refract. The shape of the lens determines the way objects look.
- A convex lens refracts parallel light waves so they meet at a focal point. As with a concave mirror, the type of image formed depends on the distance between the object and the lens.
- A concave lens spreads out light waves, which do not meet at a focal point. As with a convex mirror, the image formed is upright and reduced in size.

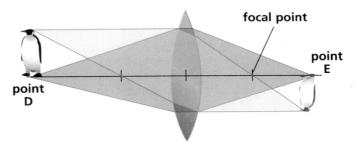

As the diagram shows, the image is inverted and reduced if the object is more than two focal lengths from the lens. Between one and two focal lengths from the lens, the image is inverted and enlarged. If the object is within one focal length from the lens, the image is virtual, upright, and enlarged.

Common Misconceptions

FULL-LENGTH REFLECTION Students might think that a flat mirror turns the image of a person's body around. Actually, in a mirror image a person's left hand still appears on the left. The image is reversed front to back, but not left to right.

 This misconception is addressed on p. 116.

MISCONCEPTION DATABASE
CLASSZONE.COM Background on student misconceptions

MAGNIFYING GLASSES Students may think that a magnifying glass always magnifies images of objects. However, a magnifying glass is simply a convex lens. It magnifies only when an object is the right distance from the lens.

 This misconception is addressed on p. 122.

Previewing Content

SECTION

 4.3 **The eye is a natural optical tool.**
pp. 126–130

1. The eye gathers and focuses light.
Light travels through the eye from the **cornea,** through the
pupil, through the lens, and hits the **retina.**
- The cornea and the lens are convex lenses that refract light in
the eye.
- Pupil size, controlled by the iris, determines how much light
enters.
- Light strikes the retina, forming a reduced, inverted image.
The retina has cells called rods that distinguish brightness
and cells called cones that detect color.

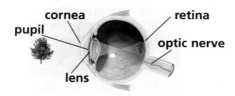

cornea   retina
pupil   optic nerve
lens

- The brain receives signals from the retina and interprets the
image as an object that is right side up.

2. Corrective lenses can improve vision.
Vision is blurry if an image does not fall exactly on the retina.
This occurs because the shape of the eye is imperfect or because
the eye's lens does not work correctly. Corrective lenses, contact
lenses, or surgery can correct these problems.
- Nearsightedness occurs when the image focuses in front of the
retina. A corrective concave lens spreads out the light rays
before they enter the eye and moves the image back toward
the retina.
- Farsightedness occurs when the image focuses behind the
retina. A corrective convex lens refracts the light rays inward
before they enter the eye and moves the image forward
toward the retina.
- Cornea shape can be changed by surgery so that the image
will focus on the retina.
- Contact lenses correct vision by changing the way the cornea
refracts light.

SECTION

 4.4 **Optical technology makes use of
light waves.** pp. 131–139

1. Mirrors and lenses can be combined to make more
powerful optical tools.
Microscopes enlarge tiny objects by combining convex lenses.
The objective lens produces an enlarged image. The eyepiece
lens forms an enlarged virtual image of the first image.

A **refracting telescope** combines convex lenses. Because the
object is more than two focal lengths from the objective lens,
the first image is reduced. The eyepiece lens forms an enlarged
virtual image of the first image.

A **reflecting telescope** has a concave mirror that focuses an
image of the object. The eyepiece lens then forms a virtual
enlarged image of the first image. A small flat mirror redirects
the image to the telescope's side.

A **film camera** uses a convex lens to focus an image on light-
sensitive film. A **digital camera** focuses the image on a sensor,
which converts light waves into electrical charges and sends the
information to a small computer. The computer reconstructs the
image and displays it.

2. Lasers use light in new ways.
A **laser** is intense and concentrated light that carries a lot of
energy. A laser beam has light waves with a single wavelength
and a pure color. Lasers are made in tubes containing a stimulus
that gives off light. The light is concentrated into a beam as it
passes back and forth between two mirrors.

Fiber optics use total internal reflection to send signals
through thin transparent fibers. Light reflects off the internal
surface of a fiber. Fiber optic technology is important in communications and in medical imaging.

Common Misconceptions

REFRACTION IN THE EYE Students commonly think that refraction takes place only in the lens of the eye. Actually, most of the
refraction takes place in the cornea, which is a membrane that
covers the eye. The lens makes additional focusing adjustments so
the image falls exactly on the retina.

[T E] This misconception is addressed on p. 127.

MISCONCEPTION DATABASE
CLASSZONE.COM Background on student misconceptions

REFRACTING TELESCOPES Some students may think that each
lens of a refracting telescope enlarges the image of the object
being observed and that the main function of a telescope is to
produce an enlarged image. The main function of the objective
lens is not to magnify the object, but to collect as much light from
the faraway object as possible to clarify the object's details. The
objective lens actually produces a slightly reduced image.

[T E] This misconception is addressed on p. 133.

Previewing Labs

Lab Generator CD-ROM
Edit these Pupil Edition labs and generate alternative labs.

EXPLORE the BIG idea

How Does a Spoon Reflect Your Face? p. 111
Students observe their reflections in the concave and convex surfaces of a spoon.

TIME 10 minutes
MATERIALS shiny metal spoon

Why Do Things Look Different Through Water? p. 111
Students examine objects while looking though a glass of water.

TIME 10 minutes
MATERIALS glass jar filled with water

Internet Activity: Optics, p. 111
Students are introduced to the science of light and vision.

TIME 20 minutes
MATERIALS computer with Internet access

SECTION 4.1

EXPLORE Reflection, p. 113
Students observe how different surface textures of aluminum foil affect reflection.

TIME 10 minutes
MATERIALS one square sheet of new aluminum foil

INVESTIGATE The Law of Reflection, p. 115
Students make a periscope and analyze how light travels through it.

TIME 30 minutes
MATERIALS quart-sized milk or juice carton, scissors, 0.5 m masking tape, 2 mirrors slightly smaller than the bottom of the carton, protractor

---

SECTION 4.2

EXPLORE Refraction, p. 119
Students explore refraction of light in water and in mineral oil.

TIME 10 minutes
MATERIALS clear plastic cup, pencil, 0.25 L water, 0.25 L mineral oil

---

CHAPTER INVESTIGATION
Looking at Lenses, pp. 124–125
Students use a convex lens to focus different types of images.

TIME 40 minutes
MATERIALS index card, marker, 0.25 lbs. modeling clay, convex lens, meter stick, flashlight, 6 cm masking tape, white poster board

---

SECTION 4.3

EXPLORE Focusing Vision, p. 126
Students explore the way the eye focuses close and distant images.

TIME 10 minutes
MATERIALS object to view

INVESTIGATE Vision, p. 128
Students use a magnifying glass to observe an image, as an analogy to the way an image forms in the eye.

TIME 10 minutes
MATERIALS convex lens, index card, white paper plate, 0.25 kg modeling clay, lamp

---

SECTION 4.4

EXPLORE Combining Lenses, p. 131
Students determine how two convex lenses work together to focus different types of images.

TIME 10 minutes
MATERIALS 2 convex lenses, 0.25 kg modeling clay, 2 index cards, object to view

---

INVESTIGATE Optical Tools, p. 134
Students make a model of a refracting telescope and determine how to position the lenses to produce a clear image.

TIME 30 minutes
MATERIALS 2 convex lenses, 2 cardboard tubes, 0.5 m duct tape

---

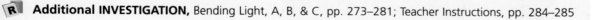

R **Additional INVESTIGATION,** Bending Light, A, B, & C, pp. 273–281; Teacher Instructions, pp. 284–285

Previewing Chapter Resources

	INTEGRATED TECHNOLOGY	**LABS AND ACTIVITIES**

CHAPTER 4
Light and Optics

 CLASSZONE.COM
- eEdition Plus
- EasyPlanner Plus
- Misconception Database
- Content Review
- Test Practice
- Visualization
- Simulation
- Resource Centers
- Internet Activity: Optics
- Math Tutorial

 SCILINKS.ORG

 CD-ROMS
- eEdition
- EasyPlanner
- Power Presentations
- Content Review
- Lab Generator
- Test Generator

 AUDIO CDS
- Audio Readings
- Audio Readings in Spanish

 EXPLORE the Big Idea, p. 111
- How Does a Spoon Reflect Your Face?
- Why Do Things Look Different Through Water?
- Internet Activity: Optics

UNIT RESOURCE BOOK
Unit Projects, pp. 5–10

 Lab Generator CD-ROM
Generate customized labs.

SECTION
4.1 Mirrors form images by reflecting light.
pp. 113–118

Time: 2 periods (1 block)
 Lesson Plan, pp. 214–215

 • **VISUALIZATION,** Reflection
• **MATH TUTORIAL**

 UNIT TRANSPARENCY BOOK
- Big Idea Flow Chart, p. T25
- Daily Vocabulary Scaffolding, p. T26
- Note-Taking Model, p. T27
- 3-Minute Warm-Up, p. T28

 • EXPLORE Reflection, p. 113
• INVESTIGATE The Law of Reflection, p. 115
• Math in Science, p. 118

 UNIT RESOURCE BOOK
- Datasheet, The Law of Reflection, p. 223
- Math Support & Practice, pp. 262–263

SECTION
4.2 Lenses form images by refracting light.
pp. 119–125

Time: 3 periods (1.5 block)
 Lesson Plan, pp. 225–226

 SIMULATION, Using Lenses to Form Images

 UNIT TRANSPARENCY BOOK
- Daily Vocabulary Scaffolding, p. T26
- 3-Minute Warm-Up, p. T28
- "How a Convex Lens Forms an Image" Visual, p. T30

 • EXPLORE Refraction, p. 119
• CHAPTER INVESTIGATION, Looking at Lenses, pp. 124–125

 UNIT RESOURCE BOOK
- CHAPTER INVESTIGATION, Looking at Lenses, Levels A, B, & C, pp. 264–272
- Additional INVESTIGATION, Bending Light, A, B, & C, pp. 273–281

SECTION
4.3 The eye is a natural optical tool.
pp. 126–130

Time: 2 periods (1 block)
 Lesson Plan, pp. 235–236

 UNIT TRANSPARENCY BOOK
- Daily Vocabulary Scaffolding, p. T26
- 3-Minute Warm-Up, p. T29

 • EXPLORE Focusing Vision, p. 126
• INVESTIGATE Vision, p. 128

 UNIT RESOURCE BOOK
Datasheet, Vision, p. 244

SECTION
4.4 Optical technology makes use of light waves.
pp. 131–139

Time: 3 periods (1.5 block)
 Lesson Plan, pp. 246–247

 RESOURCE CENTERS, Microscopes and Telescopes, Lasers

 UNIT TRANSPARENCY BOOK
- Big Idea Flow Chart, p. T25
- Daily Vocabulary Scaffolding, p. T26
- 3-Minute Warm-Up, p. T29
- Chapter Outline, pp. T31–T32

 • EXPLORE Combining Lenses, p. 131
• INVESTIGATE Optical Tools, p. 134
• Science on the Job, p. 139

 UNIT RESOURCE BOOK
Datasheet, Optical Tools, p. 255

KEY TO ICONS

 CD/CD-ROM

 INTERNET **Pupil Edition**

TE **Teacher Edition**

R **UNIT RESOURCE BOOK**

T **UNIT TRANSPARENCY BOOK**

A **UNIT ASSESSMENT BOOK**

SP A **SPANISH ASSESSMENT BOOK**

SCIENCE TOOLKIT

READING AND REINFORCEMENT

• Choose Your Own Strategy, B20–27
• Combination Notes, C36
• Daily Vocabulary Scaffolding, H1–8

 UNIT RESOURCE BOOK
• Vocabulary Practice, pp. 259–260
• Decoding Support, p. 261
• Summarizing the Chapter, pp. 282–283

 **Audio Readings CD**
Listen to Pupil Edition

 Audio Readings in Spanish CD
Listen to Pupil Edition in Spanish

 UNIT RESOURCE BOOK
• Reading Study Guide, A & B, pp. 216–219
• Spanish Reading Study Guide, pp. 220–221
• Challenge and Extension, p. 222
• Reinforcing Key Concepts, p. 224

 UNIT RESOURCE BOOK
• Reading Study Guide, A & B, pp. 227–230
• Spanish Reading Study Guide, pp. 231–232
• Challenge and Extension, p. 233
• Reinforcing Key Concepts, p. 234

UNIT RESOURCE BOOK
• Reading Study Guide, A & B, pp. 237–240
• Spanish Reading Study Guide, pp. 241–242
• Challenge and Extension, p. 243
• Reinforcing Key Concepts, p. 245
• Challenge Reading, pp. 257–258

UNIT RESOURCE BOOK
• Reading Study Guide, A & B, pp. 248–251
• Spanish Reading Study Guide, pp. 252–253
• Challenge and Extension, p. 254
• Reinforcing Key Concepts, p. 256

ASSESSMENT

• Chapter Review, pp. 141–142
• Standardized Test Practice, p. 143

 UNIT ASSESSMENT BOOK
• Diagnostic Test, pp. 60–61
• Chapter Test, Levels A, B, & C, pp. 66–77
• Alternative Assessment, pp. 78–79
• Unit Test, A, B, & C, pp. 80–91

• Spanish Chapter Test, pp. 241–244
• Spanish Unit Test, pp. 297–300

 Test Generator CD-ROM
Generate customized tests.

Lab Generator CD-ROM
Rubrics for Labs

 Ongoing Assessment, pp. 113, 114, 116

 Section 4.1 Review, p. 117

 UNIT ASSESSMENT BOOK
Section 4.1 Quiz, p. 62

 Ongoing Assessment, pp. 120, 122–123

 Section 4.2 Review, p. 123

UNIT ASSESSMENT BOOK
Section 4.2 Quiz, p. 63

 Ongoing Assessment, pp. 127–130

 Section 4.3 Review, p. 130

UNIT ASSESSMENT BOOK
Section 4.3 Quiz, p. 64

 Ongoing Assessment, pp. 132–137

 Section 4.4 Review, p. 138

 UNIT ASSESSMENT BOOK
Section 4.4 Quiz, p. 65

STANDARDS

National Standards
A.2–8, A.9.a–f, B.3.c, E.2–5,
E.6.a–f, F.5.a–c

See p. 110 for the standards.

National Standards
A.2–8, A.9.a–c, A.9.e–f, B.3.c

National Standards
A.2–7, A.9.a–b, A.9.e–f, B.3.c

National Standards
A.2–7, A.9.a–b, A.9.d–f, B.3.c

National Standards
A.2–7, A.9.a–b, A.9.d–f, B.3.c,
E.2–5, E.6.a–f, F.5.a–c

Previewing Resources for Differentiated Instruction

CHAPTER INVESTIGATION

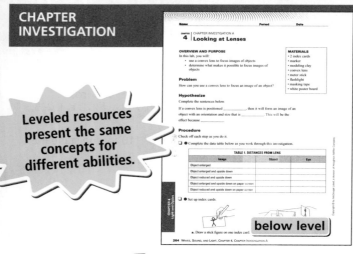

below level

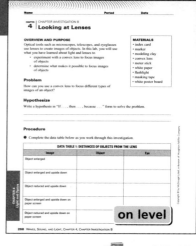

on level

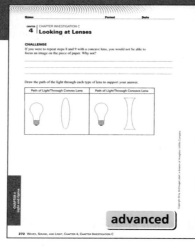

advanced

R **UNIT RESOURCE BOOK,** pp. 264–267

R pp. 268–271

R pp. 268–272

> Leveled resources present the same concepts for different abilities.

READING STUDY GUIDE

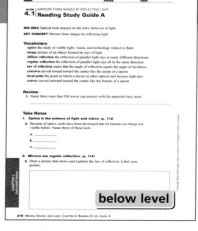

below level

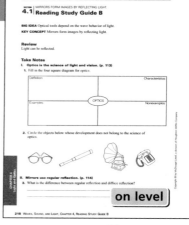

on level

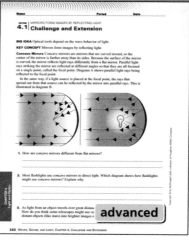

advanced

R **UNIT RESOURCE BOOK,** pp. 216–217

R pp. 218–219

R p. 222

> Reading Study Guide is also in Spanish.

CHAPTER TEST

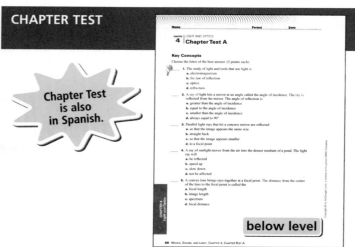

below level

on level

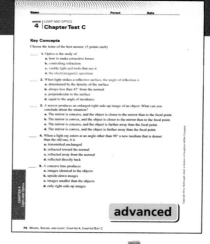

advanced

A **UNIT ASSESSMENT BOOK,** pp. 66–69

A pp. 70–73

A pp. 74–77

> Chapter Test is also in Spanish.

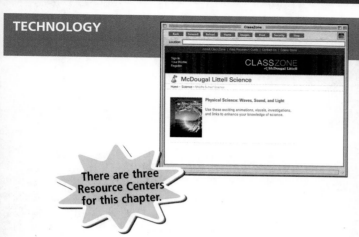

There are three Resource Centers for this chapter.

AUDIO READINGS

McDOUGAL LITTELL
LAB GENERATOR

Customize and edit labs with this easy-to-use CD-ROM
- Searchable database of all labs from the program
- Additional lab options
- Template for creating your own labs
- Rubrics and other resources

 **CLASSZONE.COM**

 **CD/CD-Roms**

 **CLASSZONE.COM**

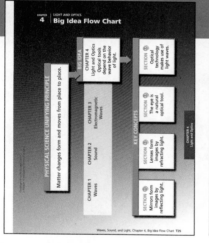

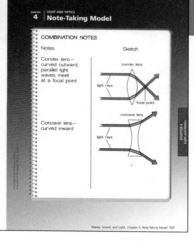

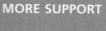 **UNIT TRANSPARENCY BOOK,** p. T25

 p. T27

 p. T30

Reinforcing Key Concepts for each section

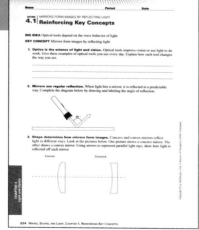

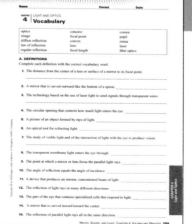

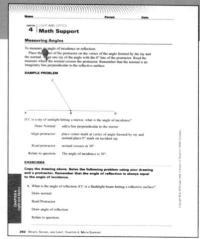

 **UNIT RESOURCE BOOK,** p. 224

 pp. 259–260

 p. 262

INTRODUCE

the **BIG** idea

Have students look at the photograph of the optical refractor and discuss how the question in the box links to the Big Idea:

* How is the device in the photograph used?
* Why is the device called a refractor?

National Science Education Standards

Content

B.3.c Light interacts with matter by transmission (including refraction), absorption, or scattering (including reflection). To see an object, light from that object—emitted or scattered from it—must enter the eye.

Process

A.2–8 Design and conduct an investigation; use tools to gather and interpret data; use evidence to describe, predict, explain, model; think critically to make relationships between evidence and explanation; recognize different explanations and predictions; communicate scientific procedures and explanations; use mathematics.

A.9.a–f Understand scientific inquiry by using different investigations, methods, mathematics, technology, and explanations based on logic, evidence, and skepticism.

E.2–5 Design, implement, and evaluate a solution or product; communicate technological design.

E.6.a–f Understandings about science and technology

F.5.a–c Science influences society; societal challenges inspire scientific research; technology influences society through its products and processes.

---

the **BIG** idea

Optical tools depend on the wave behavior of light.

How can this device help a person to see better?

Key Concepts

SECTION
4.1 **Mirrors form images by reflecting light.**
Learn how mirrors use reflection to create images.

SECTION
4.2 **Lenses form images by refracting light.**
Learn how lenses use refraction to create images.

SECTION
4.3 **The eye is a natural optical tool.**
Learn about how eyes work as optical tools.

SECTION
4.4 **Optical technology makes use of light waves.**
Learn about complex optical tools.

Internet Preview

CLASSZONE.COM

Chapter 4 online resources: Content Review, Simulation, Visualization, three Resource Centers, Math Tutorial, Test Practice.

INTERNET PREVIEW

CLASSZONE.COM For student use with the following pages:

Review and Practice
* Content Review, pp. 112, 140
* Math Tutorial: Measuring Angles, p. 118
* Test Practice, p. 143

Activities and Resources
* Internet Activity: Optics, p. 111
* Visualization: Reflection, p. 115; Simulation, p. 123
* Resource Centers: Microscopes and Telescopes, p. 132; Lasers, p. 136

NSTA
scilinks.org
SC*i*LINKS
Lenses **Code: MDL030**

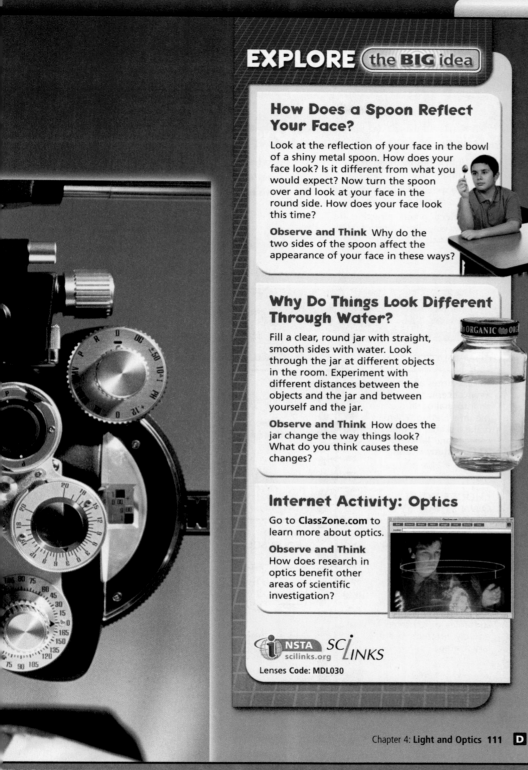

EXPLORE (the BIG idea)

How Does a Spoon Reflect Your Face?

Look at the reflection of your face in the bowl of a shiny metal spoon. How does your face look? Is it different from what you would expect? Now turn the spoon over and look at your face in the round side. How does your face look this time?

Observe and Think Why do the two sides of the spoon affect the appearance of your face in these ways?

Why Do Things Look Different Through Water?

Fill a clear, round jar with straight, smooth sides with water. Look through the jar at different objects in the room. Experiment with different distances between the objects and the jar and between yourself and the jar.

Observe and Think How does the jar change the way things look? What do you think causes these changes?

Internet Activity: Optics

Go to **ClassZone.com** to learn more about optics.

Observe and Think How does research in optics benefit other areas of scientific investigation?

NSTA
scilinks.org
SCiLINKS

Lenses Code: MDL030

TEACHING WITH TECHNOLOGY

CBL and Probeware If students have probeware, encourage them to use a light sensor with some of the activities in this chapter.

Telescope If you have access to a telescope, you may want to let students compare the images it shows to the images their telescope model from "Investigate Optical Tools" on p. 134 shows.

EXPLORE (the BIG idea)

These inquiry-based activities are appropriate for use at home or as a supplement to classroom instruction.

How Does a Spoon Reflect Your Face?

PURPOSE To demonstrate that a mirror's shape affects the image it produces. Students observe the images that concave and convex mirrors produce.

TIP *10 min.* Students should think about why the image is inverted inside the spoon but not on the back.

Answer: The inside reflects an inverted image of a face because the reflected light rays cross each other. The back of the spoon reflects a right-side-up image because the reflected rays do not meet.

REVISIT after p. 117.

Why Do Things Look Different Through Water?

PURPOSE To see how refraction of light affects images. Students look at objects through a jar of water.

TIP *10 min.* Students can also try moving the jar as they look at an object to observe how the image of the object moves in relation to the object.

Answer: The jar appears to change the position and shape of objects. The water in the jar bends the light rays that make up images of objects.

REVISIT after p. 123.

Internet Activity: Optics

PURPOSE To examine how advances in optics affect other sciences.

TIP *20 min.* Ask students how research in optics can benefit other areas of scientific research.

Answer: Better microscopes or telescopes allow microbiology or astronomy researchers to see small or faraway objects more clearly. The invention of the laser benefits many areas of science.

REVISIT after p. 138.

○ CONCEPT REVIEW

Activate Prior Knowledge

- Cut three equal circles out of cardboard and punch a pinhole in each.
- Cover a flashlight with one disk.
- Align the holes in the other two disks.
- Ask students why you can see the light only if the three pinholes and your eye are aligned.

▶ TAKING NOTES

Combination Notes

Making an outline of the main ideas of a concept will help students organize new material. Students who are visual learners will benefit by making a labeled sketch of a new concept.

Choose Your Own Strategy

Students can choose the strategies that best fit their individual learning styles. By surrounding a vocabulary term with information in a four square, description wheel, or frame game, students will develop a thorough understanding of the meaning of the term. Respellings should be included if appropriate.

Vocabulary and Note-Taking Resources

- Vocabulary Practice, p. 259–260
- Decoding Support, p. 261

- Daily Vocabulary Scaffolding, p. T26
- Note-Taking Model, p. T27

- Choose Your Own Strategy, B20–27
- Combination Notes, C36
- Daily Vocabulary Scaffolding, H1–8

CHAPTER 4
Getting Ready to Learn

◀ CONCEPT REVIEW

- Light tends to travel in a straight line.
- The speed of light is affected by a material medium.
- Reflection and refraction are two ways light interacts with materials.

◀ VOCABULARY REVIEW

reflection p. 25
refraction p. 25
visible light p. 84

CONTENT REVIEW
CLASSZONE.COM
Review concepts and vocabulary.

▶ TAKING NOTES

COMBINATION NOTES

To take notes about a new concept, first make an informal outline of the information. Then make a sketch of the concept and label it so you can study it later.

CHOOSE YOUR OWN STRATEGY

Take notes about new vocabulary terms, using one or more of the strategies from earlier chapters—**four square, description wheel,** or **frame game.** Feel free to mix and match the strategies, or to use an entirely different vocabulary strategy.

See the Note-Taking Handbook on pages R45–R51.

D 112 Unit: Waves, Sound, and Light

SCIENCE NOTEBOOK

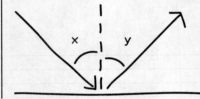

NOTES
The angle of incidence (x) equals the angle of reflection (y).

FOUR SQUARE

Definition	Characteristics

TERM

Examples	Nonexamples

DESCRIPTION WHEEL

feature feature
feature TERM feature
feature feature

FRAME GAME

example
sketch TERM
description

CHECK READINESS

Administer the Diagnostic Test to determine students' readiness for new science content and their mastery of requisite math skills.

 Diagnostic Test, pp. 60–61

Technology Resources

Students needing content and math skills should visit **ClassZone.com**.

- **CONTENT REVIEW**
- **MATH TUTORIAL**

  CONTENT REVIEW CD-ROM

KEY CONCEPT

Mirrors form images by reflecting light.

◀ **BEFORE, you learned**
- EM waves interact with materials
- Light can be reflected

▶ **NOW, you will learn**
- About the science of optics
- How light is reflected
- How mirrors form images

VOCABULARY

optics p. 113
law of reflection p. 114
regular reflection p. 114
diffuse reflection p. 114
image p. 115
convex p. 116
concave p. 116
focal point p. 117

EXPLORE Reflection

How does surface affect reflection?

PROCEDURE

1. Tear off a square sheet of aluminum foil. Look at your reflection in the shiny side of the foil.

2. Turn the foil over and look at your reflection in the dull side.

3. Crumple up the piece of foil, then smooth it out again, shiny side up. Again, look at your reflection in the foil.

WHAT DO YOU THINK?
- How did the three reflections differ from one another?
- What might explain these differences?

MATERIALS
aluminum foil

COMBINATION NOTES
Don't forget to include sketches of important concepts in your notebook.

Optics is the science of light and vision.

Optics (AHP-tihks) is the study of visible light and the ways in which visible light interacts with the eye to produce vision. Optics is also the application of knowledge about visible light to develop tools—such as eyeglasses, mirrors, magnifying lenses, cameras, and lasers—that extend vision or that use light in other ways.

Mirrors, lenses, and other optical inventions are called optical tools. By combining optical tools, inventors have developed powerful instruments to extend human vision. For example, the microscope uses a combination of mirrors and lenses to make very small structures visible. Telescopes combine optical tools to extend vision far into space. As you will see, some of the latest optical technology—lasers—use visible light in ways that do not involve human vision at all.

Chapter 4: **Light and Optics** 113 **D**

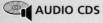
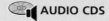

Chapter 4 **113** **D**

Teach from Visuals

To help students interpret the mirror visual, ask:

- If the angle of incidence is 1 degree, what is the angle of reflection? *1 degree*

- How would a diagram of diffuse reflection look different from the diagram of regular reflection? *The angle of reflection of the light rays would not equal the angle of incidence. The reflected rays would go in different directions.*

Teacher Demo

To illustrate the law of reflection, set a protractor at right angles to the tabletop in a blob of clay. Place a mirror on the table in front of the protractor. Use a flashlight to make a beam of light visible to the class. Shine the flashlight at an angle to the surface of the mirror so that it strikes the mirror and is reflected onto a piece of paper or a wall. Point out the normal. Show that the angle of incidence equals the angle of reflection.

Ongoing Assessment

Describe how mirrors control reflection.

Ask: How do mirrors reflect light? *Parallel light rays reflected off the surface of mirrors remain parallel to each other.*

Mirrors use regular reflection.

You have read that when light waves strike an object, they either pass through it or they bounce off its surface. Objects are made visible by light waves, or rays, bouncing off their surfaces. In section 3 you will see how the light waves create images inside the human eye.

Light rays bounce off objects in a very predictable way. For example, look at the diagram on the left below. Light rays from a flashlight strike a mirror at an angle of 60° as measured from the normal, an imaginary line perpendicular to the surface of the mirror. This angle is called the angle of incidence. The angle at which the rays reflect off the mirror, called the angle of reflection, is also 60° as measured from the normal. The example illustrates the **law of reflection,** which states that the angle of reflection equals the angle of incidence. As you can see in the second diagram, holding the flashlight at a different angle changes both the angle of incidence and the angle of reflection. However, the two angles remain equal.

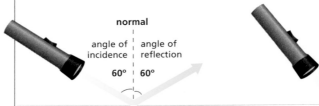

normal

angle of incidence | angle of reflection

60° | 60°

40° | 40°

The angle of reflection equals the angle of incidence.

The light rays striking the mirror bounce back by regular reflection. Rays striking everything else bounce back by diffuse reflection.

If the surface of an object is very smooth, like a mirror, light rays that come from the same direction will bounce off in the same new direction. The reflection of parallel light rays all in the same direction is called **regular reflection.**

If the surface is not very smooth—even if it feels smooth to the touch, like a piece of paper—light rays striking it from the same direction bounce off in many new directions. Each light ray follows the law of reflection, but rays coming from the same direction bounce off different bumps and hollows of the irregular surface. The reflection of parallel light rays in many different directions is called **diffuse reflection.**

DIFFERENTIATE INSTRUCTION

 More Reading Support

A What is regular reflection? *the reflection of parallel light rays all in the same direction*

English Learners Help students with the two "if-then" situations on this page by setting them up as cause and effect. (Smooth surfaces cause regular reflection. Unsmooth surfaces cause diffuse reflection.) Help them use the same type of reasoning for the different shaped mirrors on pp. 115–117.

INVESTIGATE The Law of Reflection

How can you use mirrors to see around a corner?

PROCEDURE

(1) To make a periscope, cut two flaps on opposite sides of the carton, one from the top and one from the bottom, as shown in the illustration.

(2) Fold each flap inward until it is at a 45-degree angle to the side cuts and tape it into place.

(3) Attach a mirror to the outside surface of each of the flaps.

(4) Holding the periscope straight up, look through one of the openings. Observe what you can see through the other opening.

MATERIALS
- paper milk or juice carton
- scissors
- tape
- 2 mirrors slightly smaller than the bottom of the carton
- protractor

TIME
30 minutes

WHAT DO YOU THINK?

- Where are the objects you see when you look through the periscope?
- How does the angle of the mirrors affect the path of light through the periscope?

CHALLENGE How would it affect what you see through the periscope if you changed the angle of the mirrors from 45 degrees to 30 degrees? Try it.

mirror

tape

flap 1
flap 2
fold
cut
45°
step 1

Shape determines how mirrors form images.

When you look in a mirror, you see an image of yourself. An **image** is a picture of an object formed by waves of light. The image of yourself is formed by light waves reflecting off you, onto the mirror, and back toward your eyes. Mirrors of different shapes can produce images that are distorted in certain ways.

Flat Mirrors

Your image in a flat mirror looks exactly like you. It appears to be the same size as you, and it's wearing the same clothes. However, if you raise your right hand, the image of yourself in the mirror will appear to raise its left hand. That is because you see the image as a person standing facing you. In fact, your right hand is reflected on the right side of the image, and your left on the left side.

CHECK YOUR READING If you wink your left eye while looking in the mirror, which eye in the image of you will wink?

DIFFERENTIATE INSTRUCTION

More Reading Support

B What is an image? *a picture of an object formed by waves of light*

C How does your reflection look in a flat mirror? *The image looks backward.*

Alternative Assessment Have students make diagrams to answer the questions in "Investigate The Law of Reflection." Diagrams should show the shifted position of the object and the path of light through the periscope.

Advanced

Challenge and Extension, p. 222

INVESTIGATE The Law of Reflection

PURPOSE To analyze the path of light through a periscope in order to learn about reflection

TIPS *30 min.* Suggest the following to students:

- Aluminum foil or squares of foil duct tape can be used in place of mirrors.
- Do not use the periscope to look at very bright light, including the Sun.

WHAT DO YOU THINK? *The objects are actually in front of and somewhat above the periscope. Having the mirrors aligned at 45 degrees means that the line of sight out of the periscope is parallel to the line of sight into the periscope.*

CHALLENGE *If the mirrors were positioned at an angle of 30 degrees, the line of sight out of the periscope would still be parallel to the line of sight into the periscope.*

Datasheet, The Law of Reflection, p. 223

Technology Resources

Customize this student lab as needed or look for an alternative. Print rubrics to assess student lab reports.

Lab Generator CD-ROM

Metacognitive Strategy

Ask students to write a paragraph describing a scenario in which a periscope would be useful.

Ongoing Assessment

CHECK YOUR READING *Answer: The eye to the left in the mirror will blink; if you think of the image as a person facing you, it would appear that the image's right eye was blinking.*

Address Misconceptions

IDENTIFY Ask: Is a person's body turned around in a mirror image? If students answer yes, they may hold the misconception that what you see when you look at yourself in a mirror is the same as what you would see if your body were turned around to face you.

CORRECT Have students experiment by standing in front of a mirror, raising a hand on one side, and observing on which side the movement is reflected. The movement will be reflected on the same side as the hand that moves. Point out that if the mirror image were the same as one's body turned around, the movement would be reflected on the opposite side.

REASSESS Students should understand that the appearance that one's body is turned around in a mirror image is an illusion. Ask: If you wear a watch on your right wrist, which side of a mirror image will the watch appear on? *the right*

Teach from Visuals

To help students interpret the mirror visuals on pp. 116 and 117, ask:

• Where does the image appear to be in the flat mirror? *behind the mirror*

• How do images appear in a convex mirror? *always right-side up and smaller*

• Does the image in a concave mirror always appear the same? *No, the image may be inverted, smaller, right-side up, or larger. This depends on the position of the object relative to the focal point and the mirror.*

Ongoing Assessment

Describe how mirrors produce images.

Ask: In a concave mirror, where do parallel light rays meet after reflecting off the mirror? *at the focal point*

The solid line shows the actual path of light. The broken line shows where the light appears to be coming from.

If you look closely at your image in a mirror, you will notice that it actually appears to be on the far side of the mirror, exactly as far from the mirror as you are. This is a trick of light. The solid yellow arrows in the photograph above show the path of the light rays from the boy's elbow to the mirror and back to his eyes. The light rays reflect off the mirror. The broken yellow line shows the apparent path of the light rays. They appear to his eyes to be coming through the mirror from a spot behind it.

Concave and Convex Mirrors

Unlike light rays hitting a flat mirror, parallel light rays reflecting off a curved mirror do not move in the same direction. A **convex** mirror is curved outward, like the bottom of a spoon. In a convex mirror, parallel light rays move away from each other, as you can see in the diagram below on the left. A **concave** mirror is curved inward toward the center, like the inside of a spoon. Parallel light rays reflecting off a concave mirror move toward each other, as shown on the right.

VOCABULARY
Try making sketches to help you remember the new terms on this page.

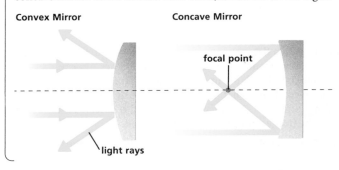

Convex Mirror Concave Mirror

focal point

light rays

DIFFERENTIATE INSTRUCTION

 More Reading Support

D How is a convex mirror curved? *A convex mirror is curved outward, like the underside of a spoon.*

Below Level If some students have trouble distinguishing concave and convex mirrors, tell them that a con*cave* mirror is shaped like a cave.

The rays striking a concave mirror cross and then move apart again. The point at which the rays meet is called the **focal point** of the mirror. The distance between the mirror and its focal point depends on the shape of the curve.

The images formed in these mirrors depend on the curve of the mirror's surface and the distance of the object from the mirror. Your image in a curved mirror may appear larger or smaller than you are, and it may even be upside down.

Convex Mirror

Your image in a convex mirror appears smaller than you.

Concave Mirror, Far Away

If you are standing far away, your image in a concave mirror appears upside down and smaller than you.

Concave Mirror, Up Close

If you are standing inside the focal point, your image in a concave mirror appears right-side up and larger.

All rays parallel to a line through the center of the mirror are reflected off the mirror and pass through the mirror's focal point. Rays from the top of the object are reflected downward and those from the bottom are reflected upward.

 How does your distance from the mirror affect the way your image appears in a concave lens?

4.1 Review

KEY CONCEPTS

1. Explain the term *optics* in your own words.

2. How is diffuse reflection similar to regular reflection? How is it different?

3. Describe the path that light rays take when they form an image of your smile when you look into a flat mirror.

CRITICAL THINKING

4. **Infer** Imagine seeing your reflection in a polished table top. The image is blurry and hard to recognize. What can you tell about the surface of the table from your observations?

5. **Analyze** Why do images formed by concave mirrors sometimes appear upside down?

CHALLENGE

6. **Synthesizing** Draw the letter *R* below as it would appear if you held the book up to (a) a flat mirror and (b) a convex mirror.

R

ANSWERS

1. *Optics is the study of light and vision.*

2. *Diffuse reflection and regular reflection both obey the law of reflection. Diffuse reflection sends parallel light rays in many different directions; regular reflection*

sends parallel rays in the same direction.

3. *Light rays reflected from your smile strike the mirror, bounce off, and enter your eye.*

4. *The surface is not perfectly smooth, and the reflection is not perfectly regular.*

5. *When light rays pass through a focal point, they cross to the side opposite the side from which they were reflected.*

6. (a)   (b)

CHECK YOUR READING *Answer: Distance determines whether your image is enlarged and right side up, or reduced and upside down.*

Real World Example

Convex mirrors are useful to provide a wider field of vision. They collect rays from a wide area, but they produce a small image. They are used for mirrors on cars, and are sometimes mounted at difficult corners to allow drivers to see oncoming traffic. Because convex mirrors produce images that are smaller than the object, the objects appear to be farther away than they are.

EXPLORE (the **BIG** idea)

Revisit "How Does a Spoon Reflect Your Face?" on p. 111. Have students explain the reasons for their results.

Reinforce (the **BIG** idea)

Have students relate the section to the Big Idea.

 Reinforcing Key Concepts, p. 224

4.1 ASSESS & RETEACH

Assess

 Section 4.1 Quiz, p. 62

Reteach

Have students review the concepts of mirrors and reflection by making ray diagrams. Ask students to write on one side of an index card a type of mirror (flat, convex, concave with object inside focal point, concave with object outside focal point). On the other side of the card, have them draw a ray diagram, illustrating how that mirror forms an image.

Technology Resources

Have students visit **ClassZone.com** for reaching of Key Concepts.

 CONTENT REVIEW

CONTENT REVIEW CD-ROM

MATH IN SCIENCE
Math Skills Practice for Science

Set Learning Goal

To measure and calculate the angles of incidence and reflection in various situations

Present the Science

If you were trying to attract the attention of a search plane, you could use a mirror to reflect sunlight toward the plane and attract a pilot's attention.

Develop Measurement Skills

Students should measure the angles between the rays and the normal, not between the rays and the horizon.

Advise students to draw diagrams when answering the questions, especially the Challenge question.

DIFFERENTIATION TIP Below level: Review the definition of *normal* and the use of a protractor with students who need help.

Close

Ask: If you were lost in the desert but didn't have a mirror, what else could you use to signal a search plane? How could you tell if the object might work? *Anything shiny would reflect rays. If you can see your image in the object's surface, it might work as a mirror.*

- Math Support, p. 262
- Math Practice, p. 263

Technology Resources

Students can visit **ClassZone.com** for practice in measuring angles.

 MATH TUTORIAL

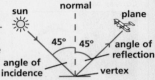

 MATH TUTORIAL
CLASSZONE.COM
Click on Math Tutorial for more help with measuring angles.

A mirror can be used to signal for help.

SKILL: MEASURING ANGLES

Send Help!

Survival kits often contain a small mirror that can be used to signal for help. If you were lost in the desert and saw a search plane overhead, you could use the mirror to reflect sunlight toward the plane and catch the pilot's attention. To aim your signal, you would use the law of reflection. The angle at which a ray of light bounces off a mirror—the angle of reflection—is always equal to the angle at which the ray strikes the mirror—the angle of incidence.

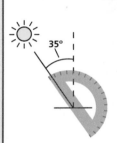

Example

Measure the angle of incidence using a protractor as follows:

(1) Place the center mark of the protractor over the vertex of the angle formed by the incident ray and the normal.

(2) Place the left 0° mark of the protractor on the incident ray.

(3) Read the number where the normal crosses the scale (35°).

(4) The angle of incidence is 35°.

ANSWER Therefore, the angle of reflection will be 35°.

Copy each of the following angles of incidence, extend its sides, and use a protractor to measure it.

1.　　2.　　3.　　4.

CHALLENGE Copy the drawing below. Use a protractor to find the angle of reflection necessary to signal the plane from point A.

• A

ANSWERS

1. 28°

2. 60°

3. 20°

4. 45°

CHALLENGE

25°　25°

KEY CONCEPT

4.2 Lenses form images by refracting light.

◀ **BEFORE, you learned**

- Waves can refract when they move from one medium to another
- Refraction changes the direction of a wave

▶ **NOW, you will learn**

- How a material medium can refract light
- How lenses control refraction
- How lenses produce images

VOCABULARY

lens p. 121
focal length p. 123

EXPLORE Refraction

How does material bend light?

PROCEDURE

1. Place the pencil in the cup, as shown in the photograph. Look at the cup from the side so that you see part of the pencil through the cup.

2. Fill the cup one third full with water and repeat your observations.

3. Gently add oil until the cup is two-thirds full. After the oil settles into a separate layer, observe.

MATERIALS

- clear plastic cup
- pencil
- water
- mineral oil

WHAT DO YOU THINK?

- How did the appearance of the pencil change when you added the water and the oil?
- What might explain these changes?

A medium can refract light.

When sunlight strikes a window, some of the light rays reflect off the surface of the glass. Other rays continue through the glass, but their direction is slightly changed. This slight change in direction is called refraction. Refraction occurs when a wave strikes a new medium—such as the window—at an angle other than 90° and keeps going forward in a slightly different direction.

Refraction occurs because one side of the wave reaches the new medium slightly before the other side does. That side changes speed, while the other continues at its previous speed, causing the wave to turn.

 CHECK YOUR READING How does the motion of a light wave change when it refracts?

Chapter 4: **Light and Optics** 119 **D**

▶ **Set Learning Goals**

Students will

- Identify how a material medium can refract light.
- Describe how lenses control refraction.
- Recognize how lenses produce images.
- Discover through experimentation how to use a convex lens to focus an image.

◀ **3-Minute Warm-Up**

Display Transparency 28 or copy this exercise on the board:

Match each definition to the correct term.

Definitions

1. a picture of an object formed by light rays *e*
2. the point where parallel light rays striking a concave mirror meet *c*
3. a surface that curves out like the back of a spoon *b*

Terms

a. concave
b. convex
c. focal point
d. optics
e. image

T 3-Minute Warm-Up, p. T28

4.2 MOTIVATE

EXPLORE Refraction

PURPOSE To show how different materials bend light rays

TIP *10 min.* Have students think about their observations in terms of the direction of light waves.

WHAT DO YOU THINK? *The pencil appears to break where the water and air meet and where the water and the oil meet. The oil bends light more than the water does.*

CHECK YOUR READING *Answer: Light waves change direction slightly when they refract.*

Chapter 4 **119 D**

RESOURCES FOR DIFFERENTIATED INSTRUCTION

Below Level

UNIT RESOURCE BOOK
- Reading Study Guide A, pp. 227–228
- Decoding Support, p. 261

AUDIO CDS

R Additional **INVESTIGATION,**
Bending Light, A, B, & C, pp. 273–281
Teacher Instructions, pp. 284–285

Advanced

UNIT RESOURCE BOOK
Challenge and Extension, p. 233

English Learners

UNIT RESOURCE BOOK
Spanish Reading Study Guide, pp. 231–232

AUDIO CDS

- Audio Readings in Spanish
- Audio Readings (English)

Teach from Visuals

To help students interpret the diagram of alight wave passing through air and glass, ask:

• Does light move faster through air or through glass? *through air*

• Is glass a thin or a dense medium? *dense*

Develop Critical Thinking

APPLY Have students do the following experiment and apply their knowledge of refraction to explain the results. Put a penny in the bottom of a teacup. Lower your head until the penny just disappears from view behind the rim of the cup. Without moving your head or the cup, have a partner fill the cup with water. The penny will appear to float into view. *When you add water, rays of light reflected from the penny refract enough to reach your eye as they pass into the air from the water.*

Ongoing Assessment

Identify how a material medium can refract light.

Ask: In what direction does light turn when it refracts? *If the new medium slows the wave, the wave will turn toward the normal. If the new medium speeds the wave up, the wave will turn away from the normal.*

Refraction of Light

COMBINATION NOTES
Sketch the ways light is refracted when it moves into a denser medium and into a thinner medium.

Recall that waves travel at different speeds in different mediums. The direction in which a light wave turns depends on whether the new medium slows the wave down or allows it to travel faster. Like reflection, refraction is described in terms of an imaginary line—called the normal—that is perpendicular to the new surface. If the medium slows the wave, the wave will turn toward the normal. If the new medium lets the wave speed up, the wave will turn away from the normal. The wave in the diagram below turns toward the normal as it slows down in the new medium.

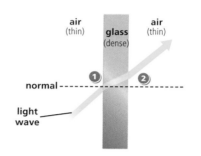

① Waves moving at an angle into a denser medium turn toward the normal.

② Waves moving at an angle into a thinner medium turn away from the normal.

READING TiP
A dense medium has more mass in a given volume than a thin medium.

? A

Light from the Sun travels toward Earth through the near vacuum of outer space. Sunlight refracts when it reaches the new medium of Earth's upper atmosphere. Earth's upper atmosphere is relatively thin and refracts light only slightly. Denser materials, such as water and glass, refract light more.

By measuring the speed of light in different materials and comparing this speed to the speed of light in a vacuum, scientists have been able to determine exactly how different materials refract light. This knowledge has led to the ability to predict and control refraction, which is the basis of much optical technology.

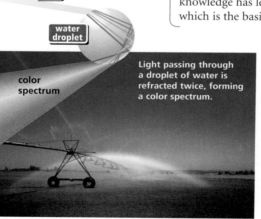

Light passing through a droplet of water is refracted twice, forming a color spectrum.

Refraction and Rainbows

You've seen rainbows in the sky after a rainstorm or hovering in the spray of a sprinkler. Rainbows are caused by refraction and reflection of light through spherical water drops, which act as prisms. Just as a prism separates the colors of white light, producing the color spectrum, each water drop separates the wavelengths of sunlight to produce a spectrum. You can see the effect in the diagram on the left.

DIFFERENTIATE INSTRUCTION

? More Reading Support

A Can Earth's atmosphere refract light? *yes*

B How do spherical water drops cause rainbows? *act as prisms to separate wavelengths of white light*

Additional Investigation To reinforce Section 4.2 learning goals, use the following full-period investigation:

R **Additional INVESTIGATION,** Bending Light, A, B, & C, pp. 273–281, 284–285 (Advanced students should complete Levels B & C.)

Shape determines how lenses form images.

When you look at yourself in a flat mirror, you see your image clearly, without distortions. Similarly, when you look through a plain glass window, you can see what is on the other side clearly. Just as curved mirrors distort images, certain transparent mediums called lenses alter what you see through them. A **lens** is a clear optical tool that refracts light. Different lenses refract light in different ways and form images useful for a variety of purposes.

READING TiP

Distort means to change the shape of something by twisting or moving the parts around.

Convex and Concave Lenses

Like mirrors, lenses can be convex or concave. A convex lens is curved outward; a concave lens is curved inward. A lens typically has two sides that are curved, as shown in the illustration below.

Convex Lens   **Concave Lens**

focal point

principal axis

A convex lens causes parallel light rays to meet at a focal point.

A concave lens causes parallel light rays to spread out.

Convex Parallel light rays passing through a convex lens are refracted inward. They meet at a focal point on the other side of the lens. The rays are actually refracted twice—once upon entering the lens and once upon leaving it. This is because both times they are entering a new medium at an angle other than 90 degrees. Rays closest to the edges of the lens are refracted most. Rays passing through the center of the lens—along the principal axis, which connects the centers of the two curved surfaces—are not refracted at all. They pass through to the same focal point as all rays parallel to them.

▼ REMINDER

The focal point is the point at which parallel light rays meet after being reflected or refracted.

Concave Parallel light rays that pass through a concave lens are refracted outward. As with a convex lens, the rays are refracted twice. Rays closest to the edges of the lens are refracted most; rays at the very center of the lens pass straight through without being deflected. Because they are refracted away from each other, parallel light rays passing through a concave lens do not meet.

⌃ CHECK YOUR READiNG   Compare what happens to parallel light rays striking a concave mirror with those striking a concave lens.

Chapter 4: **Light and Optics** 121   **D**

Teach Difficult Concepts

Students often have a hard time understanding how light changes speed when it moves into a different medium. Use the analogy of a wagon with its two right wheels moving on a sidewalk and its two left wheels moving in mud. The wheels moving in mud will turn at a slower speed than the wheels on the sidewalk, and the wagon will turn toward the left.

Teach from Visuals

To help students interpret the lens diagram, ask:

• What happens at the focal point? *The refracted rays of light meet.*

• Do the light rays that pass through a concave lens ever meet? *no*

History of Science

The curved surface of a lens must be carefully shaped if it is to focus accurately. Early lens makers copied the shape of a mold onto a piece of glass. They ground and polished the lens using rouge, an abrasive powder made of iron oxide.

CHECK YOUR READiNG   *Answer: A concave mirror causes parallel light waves to move towards each other so they meet at a focal point. Parallel light waves passing through a concave lens move away from each other and do not meet.*

DIFFERENTIATE INSTRUCTION

(?) More Reading Support

C How does a convex lens refract light? *It bends parallel light rays so that they meet at a focal point on the other side of the lens.*

English Learners Words that have multiple meanings may be confusing for English learners. For example, words like *vacuum* and *medium* are presented in a new context in this chapter. Have students write down each of these words and all their possible meanings. Then have them decide which meaning makes the most sense in the context of the sentence where it is found. To illustrate the concept of concave and convex lenses, put these terms on the Science Word Wall with visual reminders of how each lens refracts light.

Teach from Visuals

To help students interpret the lens diagrams, ask:

- Do the light waves passing through a convex lens move toward or away from each other? *toward each other*

- What is a focal length? *the distance from the center of a lens to its focal point*

- The images formed by a convex lens are similar to the images formed by what type of mirror? *concave*

 This visual is also available as T30 in the Unit Transparency Book.

Address Misconceptions

IDENTIFY Ask: Do magnifying glasses always enlarge images of an object, or can they form smaller images as well? If students say "enlarged images only," they may think that a magnifying glass is made with a special kind of lens that only magnifies.

CORRECT Pass out magnifying glasses and let students experiment with them. Students should discover that when the object is more than one focal length from the lens, a magnifying glass forms upside down images that can be enlarged, the same size as the object, or reduced.

REASSESS Ask students what a magnifying glass is. *A magnifying glass is simply a convex lens that is used in a particular way.*

Ongoing Assessment

Recognize how lenses form images.

Ask: Which types of lens forms images by causing light rays to meet at a focal point? *convex*

 Answer: at the middle of the image of the penguin

How a Convex Lens Forms an Image

A convex lens forms an image by refracting light rays. Light rays reflected from an object are refracted when they enter the lens and again when they leave the lens. They meet to form the image.

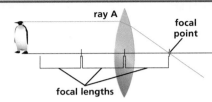

① Light rays reflect off the penguin in all directions and many enter the lens. Here a single ray (A) from the top of the penguin enters the lens and is refracted downward.

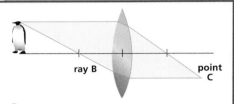

② Another light ray (B) from the top of the penguin passes through the lens at the bottom and meets the first ray at point C. All of the rays from the top of the penguin passing through the lens meet at this point.

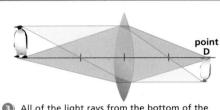

③ All of the light rays from the bottom of the penguin meet at a different point (D). Light rays from all parts of the penguin meet at corresponding points on the image.

READING VISUALS Where do light rays reflected from the middle of the penguin meet?

DIFFERENTIATE INSTRUCTION

Advanced If light is traveling in a direction that is perpendicular to the boundary between mediums, no refraction occurs even though the light's speed changes. Ask students to explain why this is so. *The light strikes the boundary at a right angle, so all sides of the wave enter the medium at the same time, all sides change speed at the same time, and no refraction occurs.*

 Challenge and Extension, p. 233

Images Formed by Lenses

When light rays from an object pass through a lens, an image of the object is formed. The type of image depends on the lens and, for convex lenses, on the distance between the lens and the object.

Notice the distance between the penguin and the lens in the illustration on page 122. The distance is measured in terms of a **focal length,** which is the distance from the center of the lens to the lens's focal point. The penguin is more than two focal lengths from the camera lens, which means the image formed is upside down and smaller.

If the penguin were between one and two focal lengths away from a convex lens, the image formed would be upside down and larger. Overhead projectors form this type of image, which is then turned right side up by a mirror and projected onto a screen for viewing.

Finally, if an object is less than one focal length from a convex lens, it will appear right side up and larger. In order to enlarge an object so that you can see details, you hold a magnifying lens close to the object. In the photograph, you see a face enlarged by a magnifying lens. The boy's face is less than one focal length from the lens.

If you look at an object through a concave lens, you'll see an image of the object that is right side up and smaller than the object normally appears. In the case of concave lenses, the distance between the object and the lens does not make a difference in the type of image that is formed. In the next section you'll see how the characteristics of the images formed by different lenses play a role in complex optical tools.

SIMULATION
CLASSZONE.COM
Work with convex and concave lenses to form images.

CHECK YOUR READING When will an image formed by a convex lens be upside down?

4.2 Review

KEY CONCEPTS

1. What quality of a material affects how much it refracts light?

2. How does the curve in a lens cause it to refract light differently from a flat piece of glass?

3. How does a camera lens form an image?

CRITICAL THINKING

4. **Infer** You look through a lens and see an image of a building upside down. What type of lens are you looking through?

5. **Make a Model** Draw the path of a light ray moving at an angle from air into water. Write a caption to explain the process.

◯ CHALLENGE

6. Study the diagram on the opposite page. Describe the light rays that would pass through the labeled focal point. Where are they coming from, and how are they related to each other?

Chapter 4: **Light and Optics 123** **D**

ANSWERS

1. the speed of light in the material

2. Because the angle at which light strikes a curved surface varies across the surface, the amount of refraction for different light rays varies also.

3. A camera lens refracts light waves inward toward a focal point.

4. a convex lens

5. Diagrams should show the refracted light beam turning toward the normal. Sample

caption: Because water slows the light wave, the wave bends toward the normal.

6. All light rays reflected off the penguin that are parallel to the principal axis of the lens will pass through the focal point of the lens.

Ongoing Assessment

CHECK YOUR READING *Answer: The image will be upside down when the object is one or more focal lengths from the lens.*

Teacher Demo

Students often learn best when they are allowed to make informal observations on their own. Pass out various lenses and let students discover for themselves how the image changes with the shape of the lens and with the focal length.

EXPLORE (the BIG idea)

Revisit "Why Do Things Look Different Through Water?" on p. 111. Have students explain the reasons for their results.

Reinforce (the BIG idea)

Have students relate the section to the Big Idea.

R Reinforcing Key Concepts, p. 234

4.2 ASSESS & RETEACH

Assess

A Section 4.2 Quiz, p. 63

Reteach

Have students make a table listing the shape of lenses and mirrors and the type, orientation, and size of images formed. Ask students to use the table to compare the images from mirrors to those from lenses.

Technology Resources

Have students visit **ClassZone.com** for reaching of Key Concepts.

 CONTENT REVIEW

 CONTENT REVIEW CD-ROM

Chapter 4 **123** **D**

CHAPTER INVESTIGATION

Focus

PURPOSE Students will focus images using a convex lens and determine the conditions that produce different kinds of images

OVERVIEW Students will examine the images formed by a convex lens as they vary the distance between the lens and the object. Students will see

- an enlarged, upright, virtual image when the distance between the object and the lens is less than one focal length
- an enlarged, inverted, real image when the distance between the object and the lens is between one and two focal lengths
- a reduced, inverted, real image when the distance between the object and the lens is more than two focal lengths

Lab Preparation

- Review the concepts of focal point and focal length.
- Prior to the investigation, have students read through the investigation and prepare their data tables. Or you may wish to copy and distribute datasheets and rubrics.

 UNIT RESOURCE BOOK, pp. 264–272

 SCIENCE TOOLKIT, F15

Lab Management

- The lens can be mounted in an upright position in some modeling clay on the table surface.
- Students may need help setting up the lens and the light.
- Darken the room as much as possible. If the tape arrow does not produce clear images, have students make a light arrow by masking most of the flashlight cover instead.

SAFETY Caution students not to look directly into the flashlight.

INCLUSION Have students with visual impairments use large lenses and meter sticks with large numbers.

Looking at Lenses

OVERVIEW AND PURPOSE Optical tools such as microscopes, telescopes, and eyeglasses use lenses to create images of objects. In this lab, you will use what you have learned about light and lenses to

- experiment with a convex lens to focus images of objects
- determine what makes it possible to focus images of objects.

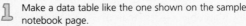 **Procedure**

PART A

1. Make a data table like the one shown on the sample notebook page.

2. Draw a stick figure on one index card. Assemble the cards, clay, and lens as shown in the photograph.

3. Position the convex lens so that you can see an enlarged, right-side up image of the stick figure. Measure the distances between the lens and the card, and between the lens and your eye. Record the distances in your data table.

4. Position the lens so that you can see an enlarged, upside down image of the stick figure. Measure the distances between the lens and the object, and between the lens and your eye. Record the distances in your data table.

5. Position the lens so that you can see a reduced, upside down image of the stick figure. Measure the distances between the lens and the object, and between the lens and your eye. Record the distances in your data table.

MATERIALS
- index card
- marker
- modeling clay
- convex lens
- meter stick
- flashlight
- masking tape
- white poster board

INVESTIGATION RESOURCES

 CHAPTER INVESTIGATION, Looking at Lenses
- Level A, pp. 264–267
- Level B, pp. 268–271
- Level C, p. 272

Advanced students should complete Levels B & C.

 Writing a Lab Report, D12–13

Technology Resources

Customize this student lab as needed or look for an alternative. Print rubrics to assess student lab reports.

 **Lab Generator CD-ROM**

PART B

6 Put an arrow made of tape on the lens of the flashlight as shown.

step 6

7 Assemble poster board and clay to make a screen. Arrange the flashlight, lens, and screen as shown below right.

8 Shine the beam from the flashlight through the lens to form an enlarged, upside down image on the screen. Measure the distances between the lens and the flashlight and between the lens and the screen.

9 Position the light and screen to produce a reduced, upside down image. Measure the distances between the lens and the flashlight and between the lens and the screen.

10 Position the light and screen to produce an enlarged right-side up image.

▶ Observe and Analyze — Write It Up

1. **RECORD OBSERVATIONS** Draw pictures of each setup in steps 3–9 to show what happened. Be sure your data table is complete.

2. **ANALYZE** What was the distance from the lens to the object in step 3? Answer this question for each of the other steps. How do the distances compare?

3. **ANALYZE** What happened when you tried to form the three types of images on the screen? How can you explain these results?

▶ Conclude — Write It Up

1. **ANALYZE** What conclusions can you draw about the relationship between the distances you measured and the type of image that was produced?

2. **IDENTIFY LIMITS** Describe possible sources of error in your procedure or any places where errors might have occurred.

3. **APPLY** What kind of lenses are magnifying glasses? When a magnifying glass produces a sharp clear image, where is the object located in relation to the lens?

step 7

▶ INVESTIGATE Further

CHALLENGE If you were to repeat steps 8 and 9 with a concave lens, you would not be able to focus an image on the piece of paper. Why not?

Looking at Lenses

Problem How can you use a convex lens to focus different types of images of an object?

Hypothesize

Observe and Analyze

Table 1. Distances from Lens

Image	Object	Eye
Object enlarged and right-side up		
Object enlarged and upside down		
Object reduced and upside down		
	Flashlight	Screen
Object enlarged and right-side up		
Object enlarged and upside down		
Object reduced and upside down		

Conclude

Chapter 4: **Light and Optics** 125 **D**

Post-Lab Discussion

- Have students make ray diagrams to summarize their results. Diagrams should clearly show why images can be enlarged or reduced and inverted or upright when the object is at different distances.

- Describe this scenario: You are camping and want to start a fire, but you have no matches. You have a lens. Can you start a fire in a pile of dead leaves? What kind of lens would you need? *yes; convex* Where would the lens have to be? *exactly one focal length from the leaves. The Sun's rays would focus on the leaves, and the concentration of light could start a fire.*

▶ Observe and Analyze — Write It Up

SAMPLE DATA *Object enlarged, right-side up: 0–1 focal length from lens; Object enlarged, upside down: 1–2 focal lengths from lens; Object reduced, upside down: more than 2 focal lengths from lens; Lightbulb enlarged, right-side up: no image; Lightbulb image enlarged, upside down: 1–2 focal lengths from lens; Lightbulb image reduced, upside down: more than 2 focal lengths from lens.*

1. See students' diagrams.

2. Step 3: less than one focal length; Steps 4 and 8: between one and two focal lengths; Steps 5 and 9: more than two focal lengths

3. Step 8: enlarged, upside down image appears on screen when distance from lens to screen is about twice the distance from light to lens. Step 9: reduced, upside down image is visible on screen when distance from lens to screen is about equal to the distance from light to lens. Step 10: enlarged, right side up image appears on screen when lens is very close to screen and the flashlight very close to the lens.

   Explanation: Different images are formed when the screen is within one focal length, about one focal length, and outside one focal length from the lens.

▶ Conclude — Write It Up

1. When the distance between an object and a convex lens is: 1. less than one focal length, an enlarged, upright image forms. 2. between one and two focal lengths, an enlarged, upside down image forms. 3. more than two focal lengths, a reduced, upside down image forms.

2. misidentifying the type of image, measuring incorrectly, and not having the lens in the correct position

3. convex lens; an object must be located less than one focal length in front of the lens.

▶ INVESTIGATE Further

CHALLENGE Answer: They do not cause light rays to meet at a focal point.

4.3 FOCUS

▶ Set Learning Goals

Students will

- Recognize how the eye depends on natural lenses.
- Explain how artificial lenses can be used to correct vision problems.
- Observe and describe through an experiment how the eye focuses an image.

◀ 3-Minute Warm-Up

Display Transparency 29 or copy this exercise on the board:

Imagine a woman is spearfishing at a lake. She sees a fish through the water ahead of her and aims her spear directly at the image. Will she hit the fish? Explain your answer. *No; the image and the actual location of the fish are different because the light is refracted.*

If the woman is spearfishing underwater while scuba diving and she aims directly at a fish, will she hit it? Why? *Yes; because both the woman and the fish are underwater, there is no refraction.*

 3-Minute Warm-Up, p. T29

4.3 MOTIVATE

EXPLORE Focusing Vision

PURPOSE To investigate how the human eye focuses

TIP *10 min.* Students who cannot keep one eye closed should hold one hand over the eye.

WHAT DO YOU THINK? *The nearby object looks out of focus. The eye changes its focal point.*

KEY CONCEPT

4.3 The eye is a natural optical tool.

◀ **BEFORE, you learned**

- Mirrors and lenses focus light to form images
- Mirrors and lenses can alter images in useful ways

▶ **NOW, you will learn**

- How the eye depends on natural lenses
- How artificial lenses can be used to correct vision problems

VOCABULARY

cornea p. 127
pupil p. 127
retina p. 127

EXPLORE Focusing Vision

How does the eye focus an image?

PROCEDURE

① Position yourself so you can see an object about 6 meters (20 feet) away.

② Close one eye, hold up your index finger, and bring it as close to your open eye as you can while keeping the finger clearly in focus.

③ Keeping your finger in place, look just to the side at the more distant object and focus your eye on it.

④ Without looking away from the more distant object, observe your finger.

WHAT DO YOU THINK?

- How does the nearby object look when you are focusing on something distant?
- What might be happening in your eye to cause this change in the nearby object?

The eye gathers and focuses light.

The eyes of human beings and many other animals are natural optical tools that process visible light. Eyes transmit light, refract light, and respond to different wavelengths of light. Eyes contain natural lenses that focus images of objects. Eyes convert the energy of light waves into signals that can be sent to the brain. The brain interprets these signals as shape, brightness, and color. Altogether, these processes make vision possible.

In this section, you will learn how the eye works. You will also learn how artificial lenses can be used to improve vision.

D 126 Unit: Waves, Sound, and Light

RESOURCES FOR DIFFERENTIATED INSTRUCTION

Below Level
UNIT RESOURCE BOOK
- Reading Study Guide A, pp. 237–238
- Decoding Support, p. 261

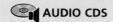 **AUDIO CDS**

Advanced
UNIT RESOURCE BOOK
- Challenge and Extension, p. 243
- Challenge Reading, pp. 257–258

English Learners
UNIT RESOURCE BOOK
Spanish Reading Study Guide, pp. 241–242

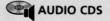 **AUDIO CDS**

- Audio Readings in Spanish
- Audio Readings (English)

How Light Travels Through the Human Eye

1. Light enters the eye through the **cornea** (KAWR-nee-uh), a transparent membrane that covers the eye. The cornea acts as a convex lens and does most of the refracting in the eye.

2. The light then continues through the **pupil,** a circular opening that controls how much light enters the eye. The pupil is surrounded by the iris, which opens and closes to change the size of the pupil.

3. Next the light passes through the part of the eye called the lens. The lens is convex on both sides. It refracts light to make fine adjustments for near and far objects. Unlike the cornea, the lens is attached to tiny muscles that contract and relax to control the amount of refraction that occurs and to move the focal point.

READING TiP
The word *lens* can refer both to an artificial optical tool and to a specific part of the eye.

4. The light passes through the clear center of the eye and strikes the **retina** (REHT-uhn-uh). The retina contains specialized cells that respond to light. Some of these cells send signals through the optic nerve to the brain. The brain interprets these signals as images.

How the Human Eye Forms an Image

The cornea and lens together focus a reduced, inverted image on the retina.

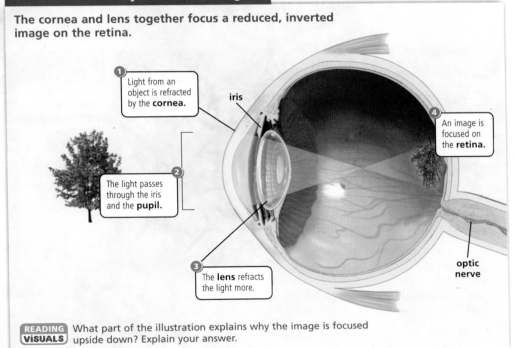

1. Light from an object is refracted by the **cornea.**

iris

2. The light passes through the iris and the **pupil.**

3. The **lens** refracts the light more.

4. An image is focused on the **retina.**

optic nerve

READING VISUALS What part of the illustration explains why the image is focused upside down? Explain your answer.

Chapter 4: **Light and Optics** 127 **D**

4.3 INSTRUCT

Address Misconceptions

IDENTIFY Ask: What is the main focusing part of the eye? If students say the lens, they may hold the misconception that most or all refraction occurs in the lens.

CORRECT Call students' attention to the diagram on this page. Point out that the cornea is shaped like a convex lens and does most of the refracting for the eye. Because the cornea covers the eye, most refraction actually takes place in the cornea rather than in the lens.

REASSESS Have students draw a diagram of light rays approaching an eye and being refracted by the cornea. The light should be further refracted by the lens.

Technology Resources

Visit **ClassZone.com** for background on common student misconceptions.

MISCONCEPTION DATABASE

Teach from Visuals

To help students interpret the eye visual, ask:

- What is the shape of the cornea? *convex*
- Where is the focal point? *where the lines cross*
- What does the optic nerve do? *sends signals to the brain*

Ongoing Assessment

Recognize how the eye depends on natural lenses.

Ask: What two lenses are part of the eye? *the cornea and the lens*

READING VISUALS *Answer: The part that shows light rays crossing at a focal point so that the ones on top go to the bottom of the image and vice versa.*

DIFFERENTIATE INSTRUCTION

? More Reading Support

A What part of the eye does most light refraction? *the cornea*

B Where does the eye focus an image? *on the retina*

English Learners Developing a Science Word Wall in the classroom will help English learners learn new vocabulary. Offering a visual reminder with the word can reinforce meaning as well. Place the words *pupil, cornea,* and *retina* on the Science Word Wall and include a diagram of the eye showing the location of each part.

INVESTIGATE Vision

PURPOSE To use a magnifying glass to observe how an image becomes focused

TIPS *10 min.*
- Use a meter stick to measure all distances.
- The distance between the plate and the glass will vary, depending on the focal length of the lens used.

WHAT DO YOU THINK? *The distance gets smaller. Instead of moving the position of the retina, the eye changes the focal point of the lens.*

CHALLENGE *Use a lens with a different focal point to refocus the image.*

 Datasheet, Vision, p. 244

Technology Resources

Customize this student lab as needed or look for an alternative. Print rubrics to assess student lab reports.

 Lab Generator CD-ROM

Metacognitive Strategy

Ask students to summarize in their own words how images are focused.

Ongoing Assessment

 CHECK YOUR READING *Answer: cone cell*

How the Eye Forms Images

For you to see an object clearly, your eye must focus an image of the object on your retina. The light reflected from each particular spot on the object must converge on a matching point on your retina. Many such points make up an image of an entire object. Because the light rays pass through the lens's focal point, the image is upside down. The brain interprets this upside down image as an object that is right-side up.

For a complete image to be formed in the eye and communicated to the brain, more than the lens and the cornea are needed. The retina also plays an important role. The retina contains specialized cells that detect brightness and color and other qualities of light.

COMBINATION NOTES Make a chart showing how light interacts with different parts of the eye.

? C

Rod Cells Rod cells distinguish between white and black and shades of gray. Rods respond to faint light, so they help with night vision.

? D

Cone Cells Cone cells respond to different wavelengths of light, so they detect color. There are three types of cones, one for each of the colors red, blue, and green. Cones respond to other colors with combinations of these three, as the screen of a color monitor does. The brain interprets these combinations as the entire color spectrum.

CHECK YOUR READING Which type of cell in the retina detects color?

INVESTIGATE Vision

How does distance affect vision?

PROCEDURE

① Arrange the materials as shown so that the lamp shines through the lens onto the plate. The lens should be about $\frac{2}{3}$ a meter from the lamp.

② Adjust the distance between the plate and the lens until you see a focused image of the bulb on the plate. Measure this distance.

③ Move the lens until it is about a meter and a half from the lamp. Adjust the plate once again to get a focused image, then measure the distance between the plate and the lens.

WHAT DO YOU THINK?
- How does the distance needed between the plate and the lens change when the lamp is farther from the lens?
- How is what happens in the eye different from what you did to refocus the image?

CHALLENGE How could you change the model to make it more like what happens in the eye?

SKILL FOCUS Observing

MATERIALS
- convex lens
- index card
- modeling clay
- white paper plate
- lamp

TIME 10 minutes

DIFFERENTIATE INSTRUCTION

? More Reading Support

C The eye forms what type of image? *upside down*

D What do rod cells do? *distinguish between black and white and shades of gray*

English Learners Some English learners may have never had a vision test or seen an optometrist. Go over the different ways a doctor can correct a person's vision. Explain how eyeglasses correct nearsightedness or farsightedness. Talk about how contact lenses are used. Encourage students to use the Internet to find out more about surgical methods of vision correction.

Advanced

 Challenge and Extension, p. 243

Corrective lenses can improve vision.

What happens when the image formed by the lens of the eye does not fall exactly on the retina? The result is that the image appears blurry. This can occur either because of the shape of the eye or because of how the lens works. Artificial lenses can be used to correct this problem.

Corrective Lenses

A person who is nearsighted cannot see objects clearly unless they are near. Nearsightedness occurs when the lens of the eye focuses the image in front of the retina. The farther away the object is, the farther in front of the retina the image forms. This problem can be corrected with glasses made with concave lenses. The concave lenses spread out the rays of light before they enter the eye. The point at which the rays meet then falls on the retina.

 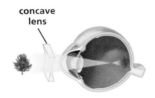

nearsighted eye   **concave lens**

① image in front of retina   ② image at retina

Objects are clearer to a farsighted person when the objects are farther away. Farsightedness occurs when the lens of the eye focuses an object's image behind the retina. This condition can result from aging, which may make the lens less flexible. The closer the object is, the farther behind the retina the image forms. Farsightedness can be corrected with glasses made from convex lenses. The convex lenses bend the light rays inward before they enter the eye. The point at which the rays meet then falls on the retina.

READING TiP

Nearsighted people can see objects near to them best. *Farsighted* people can see objects better when the objects are farther away.

farsighted eye   **convex lens**

① image behind retina   ② image at retina

 CHECK YOUR READING What kind of lens is used for correcting nearsightedness?

Teach from Visuals

To help students understand corrective lenses, ask:

- Where does the image form in a nearsighted eye? *in front of the retina*
- Where does the image form in a farsighted eye? *behind the retina*

Integrate the Sciences

Perfect vision is measured at 20/20, which means that your eye sees at 20 feet what a normal eye can see at the same distance. The larger the second number, the blurrier the image. So 20/40 means that your eye sees at 20 feet what a normal eye can see at 40 feet. Legal blindness is 20/200 or worse in both eyes.

Ongoing Assessment

Explain how artificial lenses can be used to correct vision problems.

Ask: How does a concave lens correct nearsightedness? *By spreading out the light before it enters the eye, the lens lengthens the focal length in the eye, and the image focuses on the retina.*

CHECK YOUR READING *Answer: Nearsightedness can be corrected with a concave lens.*

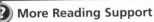

CHECK YOUR READING   *Answer: surgery and contact lenses*

Reinforce (the **BIG** idea)

Have students relate the section to the Big Idea.

 R   Reinforcing Key Concepts, p. 245

 4.3 ASSESS & RETEACH

Assess

A   Section 4.3 Quiz, p. 64

Reteach

Ask students the following questions to extend their understanding of sight:

- Where are the images projected on the retina interpreted? *in the brain*
- What happens to the lens of the eye when light passes through? *It refracts light to make adjustments for near and far objects.*

Have students find their blind spot—the place where the retina joins the optic nerve.

- Make a diagram of an X and a black spot located about 6 cm apart.
- Close your left eye, look at the X, and gradually move the diagram away from the eye.

At a distance of about 25 cm, the spot will disappear.

Technology Resources

Have students visit **ClassZone.com** for reteaching of Key Concepts.

 CONTENT REVIEW

CONTENT REVIEW CD-ROM

Surgery and Contact Lenses

Wearing glasses is an effective way to correct vision. It is also possible to change the shape of the cornea to make the eye refract properly. The cornea is responsible for two-thirds of the refraction that takes place inside the eye. As you know, the eye's lens changes shape to focus an image, but the shape of the cornea does not ordinarily change.

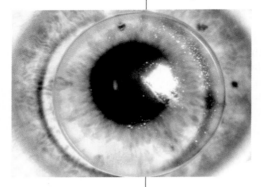

Contact lenses fit directly onto the cornea, changing the way light is refracted as it enters a person's eye.

However, using advanced surgical technology, doctors can change the shape of the cornea. By doing this, they change the way light rays focus in the eye so that the image lines up with the retina. To correct for nearsightedness, surgeons remove tissue from the center of the cornea. This flattens the cornea and makes it less convex so that it will refract less. To correct for farsightedness, surgeons remove tissue from around the edges of the cornea. This increases the cornea's curvature to make it refract more. Surgery changes the shape of the cornea permanently and can eliminate the need for eyeglasses.

Contact lenses also correct vision by changing the way the cornea refracts light. Contact lenses are corrective lenses that fit directly onto the cornea. The lenses actually float on a thin layer of tears. The moisture, the contact lens, and the cornea all function together as a single lens. Because the change is temporary, contacts, like eyeglasses, can be adapted to new changes in the eye.

 CHECK YOUR READING   What are two ways of changing the way the cornea refracts light to correct vision?

4.3 Review

KEY CONCEPTS
1. Where are images focused in an eye with perfect vision?
2. What causes people with nearsightedness to see blurry images of objects at a distance?
3. What kind of lens is used for correcting farsightedness? Why?

CRITICAL THINKING
4. **Make a Model** Draw a diagram to answer the following question: How does a convex lens affect the way a nearsighted eye focuses an image?
5. **Analyze** What distance would an eye doctor need to measure to correct a problem with nearsightedness or farsightedness?

CHALLENGE
6. **Apply** A person alternates between wearing glasses and wearing contact lenses to correct farsightedness. Are the contact lenses more or less convex than the lenses of the glasses? Explain the reasoning behind your response.

D   130   Unit: **Waves, Sound, and Light**

ANSWERS

1. on the retina

2. The eye focuses the image in front of the retina.

3. Convex lens; it bends light inward, which moves the focal point of the image forward toward the retina.

4. Student diagrams should show that the focal point has moved even farther away from the retina, making the eye even more nearsighted.

5. the distance between the image and the retina

6. more convex, because they are closer to the eye and must do the same amount of refraction in less space

KEY CONCEPT
Optical technology makes use of light waves.

BEFORE, you learned	**NOW,** you will learn
• Mirrors are optical tools that use reflection • Lenses are optical tools that use refraction • The eye is a natural optical tool • Lenses can correct vision	• How mirrors and lenses can be combined to make complex optical tools • How optical tools are used to extend natural vision • How laser light is made and used in optical technology

VOCABULARY

laser p. 135
fiber optics p. 137

EXPLORE Combining Lenses

How can lenses be combined?

PROCEDURE

1. Assemble the lenses, clay, and index cards as shown in the photograph.

2. Line the lenses up so that you have a straight line of sight through them.

3. Experiment with different distances between
   • the lenses
   • the far lens and an object
   • the near lens and your eye
   Find an arrangement that allows you to see a clear image of an object through both lenses.

MATERIALS
• 2 convex lenses
• modeling clay
• 2 index cards

WHAT DO YOU THINK?
• What kind of image could you see? What arrangement or arrangements work best to produce an image?
• How do you think the lenses are working together to focus the image?

Mirrors and lenses can be combined to make more powerful optical tools.

COMBINATION NOTES
As you read this section, make a list of optical tools. Add sketches to help you remember important concepts.

If you know about submarines, then you know how much they depend on their periscopes to see above the water. Periscopes are made by combining mirrors. Lenses can also be combined. In the eye, for example, the cornea and the eye's lens work together to focus an image. Mirrors and lenses can be combined with each other, as they are in an overhead projector. Many of the most powerful and complex optical tools are based on different combinations of mirrors and lenses.

Chapter 4: **Light and Optics** 131 **D**

RESOURCES FOR DIFFERENTIATED INSTRUCTION

Below Level
UNIT RESOURCE BOOK
• Reading Study Guide A, pp. 248–249
• Decoding Support, p. 261

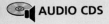

Advanced
UNIT RESOURCE BOOK
Challenge and Extension, p. 254

English Learners
UNIT RESOURCE BOOK
Spanish Reading Study Guide, pp. 252–253

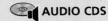

• Audio Readings in Spanish
• Audio Readings (English)

4.4 FOCUS

○ Set Learning Goals
Students will

• Describe how mirrors and lenses can be combined to make complex optical tools.

• Explain how optical tools are used to extend natural vision.

• Recognize how laser light is made and used in optical technology.

• Design their own experiment by making a model of a telescope and explaining how it works.

○ 3-Minute Warm-Up

Display Transparency 29 or copy this exercise on the board:

Decide if these statements are true. If not true, correct them.

1. The lens of the human eye is concave on both sides. *The lens of the human eye is convex on both sides.*

2. Rods and cones, located in the retina, are important in night and color vision. *true*

3. The eyes convert the energy of light waves into nerve signals that travel to the brain. *true*

T 3-Minute Warm-Up, p. T29

4.4 MOTIVATE

EXPLORE Combining Lenses

PURPOSE To determine how two lenses work together to focus an image

TIP *10 min.* Use a protractor to ensure that the lenses are exactly vertical.

WHAT DO YOU THINK? *An enlarged image can be seen when the lenses are the distance of their combined focal lengths apart, the object is more than one focal length from the far lens, and the viewer is about one focal length from the near lens.*

History of Science

In the 17th century, Robert Hooke and Anton van Leeuwenhoek both developed microscopes and discovered the world of tiny living things. Although Hooke's microscope had two lenses and was similar to modern compound microscopes, poor-quality lenses provided little detail. Van Leeuwenhoek's microscope had only one lens, but it was good enough to show many details in cells and organisms.

Ongoing Assessment

Describe how mirrors and lenses can be combined to make complex optical tools.

Ask: What type of telescope combines mirrors and lenses? *a reflecting telescope*

CHECK YOUR READING *Answer: The objective lens forms an enlarged real image. The eyepiece forms an enlarged virtual image of the first image.*

CHECK YOUR READING *Answer: A reflecting telescope uses two mirrors and one lens to focus an image. A refracting telescope uses two lenses to focus an image.*

Microscopes

Microscopes are used to see objects that are too small to see well with the naked eye. An ordinary microscope works by combining convex lenses. The lens closer to the object is called the objective. The object is between one and two focal lengths from this lens, so the lens focuses an enlarged image of the object inside the microscope.

The other microscope lens—the one you look through—is called the eyepiece. You use this lens to look at the image formed by the objective. Like a magnifying glass, the eyepiece lens forms an enlarged image of the first image.

Very small objects do not reflect much light. Most microscopes use a lamp or a mirror to shine more light on the object.

CHECK YOUR READING Which types of images do the lenses in a microscope form?

Telescopes

RESOURCE CENTER
CLASSZONE.COM
Find out more about microscopes and telescopes.

Telescopes are used to see objects that are too far away to see well with the naked eye. One type of telescope, called a refracting telescope, is made by combining lenses. Another type of telescope, called a reflecting telescope, is made by combining lenses and mirrors.

Refracting telescopes combine convex lenses, just as microscopes do. However, the objects are far away from the objective lens instead of near to it. The object is more than two focal lengths from the objective lens, so the lens focuses a reduced image of the object inside the telescope. The eyepiece of a telescope then forms an enlarged image of the first image, just as a microscope does. This second image enlarges the object.

Reflecting telescopes work in the same way that refracting telescopes do. However, there is no objective lens where light enters the telescope. Instead, a concave mirror at the opposite end focuses an image of the object. A small flat mirror redirects the image to the side of the telescope. With this arrangement, the eyepiece does not interfere with light on its way to the concave mirror. The eyepiece then forms an enlarged image of the first image.

Both refracting and reflecting telescopes must adjust for the small amount of light received from distant objects. The amount of light gathered can be increased by increasing the diameter of the objective lens or mirror. Large mirrors are easier and less expensive to make than large lenses. So reflecting telescopes can produce brighter images more cheaply than refracting telescopes.

CHECK YOUR READING How is a reflecting telescope different from a refracting telescope?

DIFFERENTIATE INSTRUCTION

? More Reading Support

A What does the eyepiece of a microscope do? *forms an enlarged image of the first image*

B What type of telescope has two convex lenses? *refracting*

English Learners Label any refracting telescopes, reflecting telescopes, or microscopes in the classroom. English learners might be confused by phrasal verbs such as *made up* (p. 135). Tell students that in this context, "up" is not a literal direction, but rather a part of the verb. If students still have trouble, offer synonyms such as "composed" or "consists of."

Microscopes and Telescopes

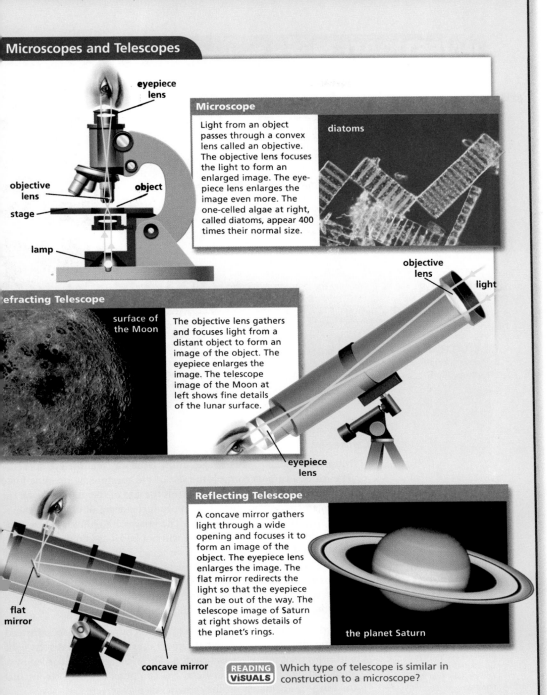

eyepiece lens

Microscope

Light from an object passes through a convex lens called an objective. The objective lens focuses the light to form an enlarged image. The eyepiece lens enlarges the image even more. The one-celled algae at right, called diatoms, appear 400 times their normal size.

diatoms

objective lens

object

stage

lamp

Refracting Telescope

surface of the Moon

The objective lens gathers and focuses light from a distant object to form an image of the object. The eyepiece enlarges the image. The telescope image of the Moon at left shows fine details of the lunar surface.

objective lens

light

eyepiece lens

Reflecting Telescope

A concave mirror gathers light through a wide opening and focuses it to form an image of the object. The eyepiece lens enlarges the image. The flat mirror redirects the light so that the eyepiece can be out of the way. The telescope image of Saturn at right shows details of the planet's rings.

the planet Saturn

flat mirror

concave mirror

READING VISUALS Which type of telescope is similar in construction to a microscope?

Chapter 4: **Light and Optics** 133 **D**

DIFFERENTIATE INSTRUCTION

Advanced Have students investigate the focal lengths of the lenses in microscopes and refracting telescopes. Students should understand why lenses with long or short focal lengths are used. Students can make ray diagrams tracing the path of light through these instruments. *Microscopes use an objective lens with a very short focal length because the object being examined is very close to this lens. Telescopes use an objective lens with a very long focal length because the object being viewed is very far away.*

R Challenge and Extension, p. 254

Teach from Visuals

To help students interpret the microscope and telescope visual:

- Both microscopes and refracting telescopes contain two lenses. Ask: How are these two instruments different? *The image in a microscope is enlarged twice. The refracting telescope enlarges the image once.*

- Ask: What is the advantage to having the eyepiece out of the way in a reflecting telescope? *The advantage of having the eyepiece out of the way is that it will not block light from entering the telescope.*

Address Misconceptions

IDENTIFY Ask: What is the function of each lens in a telescope? If students respond that both lenses produce enlarged images, they may hold the misconception that the only purpose of a telescope is to enlarge an image.

CORRECT Have students reexamine the visual on page 122, which shows that the image of an object located more than two focal lengths from a convex lens is reduced and inverted. Drawing a ray diagram will help reinforce this concept.

REASSESS Ask: What is the main function of an objective lens? *to collect as much light from the distant object as possible*

Technology Resources

Visit **ClassZone.com** for background on common student misconceptions.

MISCONCEPTION DATABASE

Ongoing Assessment

READING VISUALS *Answer: a refracting telescope*

INVESTIGATE Optical Tools

PURPOSE To design and build a telescope to find out how two lenses work together

TIPS *30 min.* Let students explore for a few minutes. Offer these suggestions if necessary.

- The two lenses should be at a distance equal to the sum of the focal lengths of the two lenses.
- Use a large convex lens with a long focal length for the objective and a smaller convex lens with a short focal length for the eyepiece.

WHAT DO YOU THINK? *The object should be more than two focal lengths from the objective lens. Students will see an enlarged inverted image if their telescope works.*

CHALLENGE *The image is first upside down, because the objective lens forms an inverted image. The eyepiece then forms an upright image of the inverted image.*

 Datasheet, Optical Tools, p. 255

Technology Resources

Customize this student lab as needed or look for an alternative. Print rubrics to assess student lab reports.

 Lab Generator CD-ROM

Teaching with Technology

If you have access to a telescope, let students use it to compare its images with the images created by their homemade telescope.

Ongoing Assessment

Explain how optical tools are used to extend natural vision.

Ask: How do microscopes and telescopes extend natural vision? *They enlarge objects that are either too small or too far away to be seen with the naked eye.*

INVESTIGATE Optical Tools

How can you make a simple telescope?

Use what you have learned about how a telescope works to build one. Figure out how far apart the two lenses need to be and use that information to construct a working model.

SKILL FOCUS
Making models

MATERIALS
- 2 convex lenses
- 2 cardboard tubes
- duct tape

TIME
30 minutes

PROCEDURE

1. Decide how the lenses should be positioned in relation to an object you select to view.
2. Adjust the lenses until you get a clear image.
3. Use the other materials to fix the lenses into place and to make it possible to adjust the distance between them.

WHAT DO YOU THINK?

- How did you end up positioning the lenses in relation to the object?
- Did your telescope work? Why do you think you got this result?

CHALLENGE Is your telescope image upside down or right-side up? How can you explain this observation?

Cameras

Most film cameras focus images in the same way that the eye does. The iris of a camera controls the size of the aperture, an opening for light, just as the iris of an eye controls the size of the pupil. Like an eye, a camera uses a convex lens to produce images of objects that are more than two focal lengths away. The images are reduced in size and upside down. In the eye, an image will not be focused unless it falls exactly on the retina. In a camera, an image will not be focused unless it falls exactly on the film. The camera does not change the shape of its lens as the eye does to change the focal point. Instead, the camera allows you to move the lens nearer to or farther away from the film until the object you want to photograph is in focus.

READING TiP
The term *digital* is often used to describe technology involving computers. Computers process information digitally, that is, using numbers.

A digital camera focuses images just as a film camera does. Instead of using film, though, the digital camera uses a sensor that detects light and converts it into electrical charges. These charges are recorded by a small computer inside the camera. The computer can then reconstruct the image immediately on the camera's display screen.

DIFFERENTIATE INSTRUCTION

 More Reading Support

C What kind of an image is produced in a camera? *reduced in size and upside down*

Alternative Assessment Have students make diagrams showing how the lenses in a refracting telescope have to be positioned in relation to each other. Ask them to explain why the lenses have to be positioned in this way.

How Cameras Work

A camera focuses an image in the same way as an eye.

film camera

 light

 lens

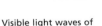 iris

 aperture film

READING VISUALS What part of a camera corresponds to the pupil of an eye?

Eye and Camera

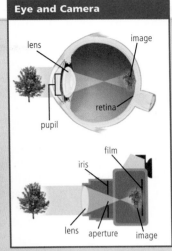

image
lens
retina
pupil

film
iris
lens aperture image

Digital Camera

A **digital camera** records images digitally, that is, using a computer.

Lasers use light in new ways.

A **laser** (LAY-zuhr) is a device that produces an intense, concentrated beam of light that is brighter than sunlight. The word *laser* means "light amplification by stimulated emission of radiation." Laser light has many uses. It carries a lot of energy and can be controlled precisely.

Ordinary visible light is made up of many different wavelengths. Even colored light usually contains many different wavelengths. But a laser beam is made up of light waves with a single wavelength and a pure color. In addition, the crests and troughs of the waves are in phase, which means that they are lined up so they match exactly.

REMINDER
The crest of a wave is its highest point. The trough of a wave is its lowest point.

Visible light waves of different wavelengths

Light waves of a single wavelength

Single wavelength waves in phase

Chapter 4: **Light and Optics** 135 **D**

Teach from Visuals

To help students interpret the visual of how cameras work, ask:

• What part of a camera corresponds to the retina of the eye? *the film*

• What replaces the film in a digital camera? *A small computer stores light waves from the object in the form of a digitized image.*

Teach Difficult Concepts

Some students may still be having trouble understanding where a lens focuses a real image. Remind them that the focal point of a lens is the point where parallel rays passing through the lens meet. Rays from different points on an object are not parallel and so meet a little beyond the focal point of the lens. Every point on the object has a corresponding point where light rays meet on the image.

Teacher Demo

Light a candle in a darkened room, and hold a magnifying glass between the candle and the wall. An inverted image of the candle will appear on the wall. If the light waves from a particular point on the object don't quite converge at the wall, the real image of the candle will not fall exactly on the wall, and it will be blurry. To focus the image, move the magnifying glass closer or farther away from the candle. When you focus a camera, you turn the lens to move it closer or farther away from the film surface. As you move the lens, you line up the focused real image of an object so it falls directly on the surface of the film.

Ongoing Assessment

READING VISUALS *Answer: the iris*

DIFFERENTIATE INSTRUCTION

More Reading Support

D What is a laser? *a device that produces an intense, concentrated beam of light*

Below Level Ask students who have a hard time reading to describe and compare the path of light in a camera and in the human eye. Ask them to bring in a camera to show the class the different parts.

Chapter 4 **135** **D**

To help students interpret the laser visual, ask:

- What happens to the light as the mirrors reflect it back and forth? *It becomes stronger and more concentrated.*
- What do the light waves in a laser look like? *They are all one wavelength and parallel, and their troughs and crests line up.*

Real World Example

A bar code is a specific arrangement of bars and spaces. When a cashier moves a bar-coded item in front of a scanner, a laser scans the bar code. The white spaces between the bars reflect light in bursts through the scanner window, through a partial mirror, and onto a detector. The detector changes these light bursts into digital signals, which travel to a central computer. The computer processes the signals and sends information about the price of the item to the cash register.

Ongoing Assessment

Recognize how laser light is made and used in optical technology.

Ask: How is a laser beam made? *An energy source stimulates the atoms in a material to give off light waves of a single wavelength. Mirrors concentrate the light waves to produce a laser.*

Light waves in a laser beam are highly concentrated and exactly parallel. Ordinary light spreads out, growing more faint as it gets farther from its source. Laser light spreads out very little. After traveling 1 kilometer (0.6 mi), a laser beam may have a diameter of only one meter.

Making Laser Light

RESOURCE CENTER
CLASSZONE.COM
Learn more about lasers.

A laser is made in a special tube called an optical cavity. A material that is known to give off a certain wavelength of light, such as a ruby crystal, is placed inside the tube. Next, an energy source, such as a bright flash of light, stimulates the material, causing it to emit, or give off, light waves. Both ends of the crystal are mirrored so that they reflect light back and forth between them. One end is mirrored more than the other. As the light waves pass through the crystal, they cause the material to give off more light waves—all perfectly parallel, all with the same wavelength, and all with their crests and troughs lined up. Eventually the beam becomes concentrated and strong enough to penetrate the less-mirrored end of the crystal. What comes out of the end is a laser beam.

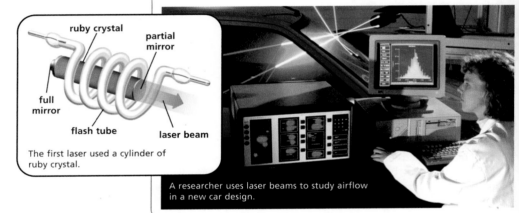

ruby crystal
partial mirror
full mirror
flash tube
laser beam

The first laser used a cylinder of ruby crystal.

A researcher uses laser beams to study airflow in a new car design.

Visual Uses of Lasers

Lasers are used today in an amazing variety of ways. One of these ways is to create devices that do the kind of work the human eye does—detecting and interpreting light waves. For example, surveyors once used telescopes to measure distances and angles. Now lasers can be used to take these measurements more precisely. Lasers are used to read bar codes, to scan images and pages of text, and to create holograms—three-dimensional images that appear to hover in the air. Holograms, which are hard to reproduce, are sometimes used in important documents so that the documents cannot be duplicated.

D 136   Unit: Waves, Sound, and Light

DIFFERENTIATE INSTRUCTION

? More Reading Support

E How are the light waves in a laser beam arranged? *exactly parallel*

F How are lasers used today? *to detect and interpret light waves*

English Learners English learners may not have prior knowledge of some concepts in this section. Explain holograms on p. 136 and aquariums on p. 137. Be sure students understand the difference between the film camera and the digital camera on p. 134.

iber Optics

Some laser applications use visible light in ways that have nothing to do with vision. One of the fastest growing technologies is fiber optics. **Fiber optics** is technology based on the use of laser light to send signals through transparent wires called optical fibers. Fiber optics makes use of a light behavior called total internal reflection. Total internal reflection occurs when all of the light inside a medium reflects off the inner surface of the medium.

When light strikes the inner surface of a transparent medium, it may pass through the surface or it may be reflected back into the medium. Which one occurs depends on the angle at which the light hits the surface. For example, if you look through the sides of an aquarium, you can see what is behind it. But if you look at the surface of the water from below, it will act like a mirror, reflecting the inside of the aquarium.

Laser light is very efficient at total internal reflection. It can travel long distances inside clear fibers of glass or other materials. Light always travels in a straight line; however, by reflecting off the sides of the fibers, laser light inside fibers can go around corners and even completely reverse direction.

CHECK YOUR READING What is total internal reflection? What questions do you have about this light behavior?

Fiber optics is important in communications, because it can be used to transmit information very efficiently. Optical fibers can carry more signals than a corresponding amount of electrical cable. Optical cables can be used in place of electrical wires for telephone lines, cable television, and broadband Internet connections.

Fiber optics also has visual uses. For example, fiber optics is used in medicine to look inside the body. Using optical cable, doctors can examine organs and diagnose illnesses without surgery or x-rays. Optical fibers can also deliver laser light to specific points inside the body to help surgeons with delicate surgery.

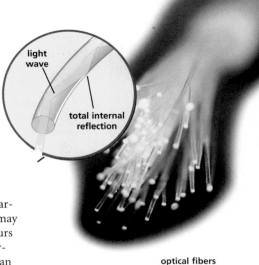

light wave

total internal reflection

optical fibers

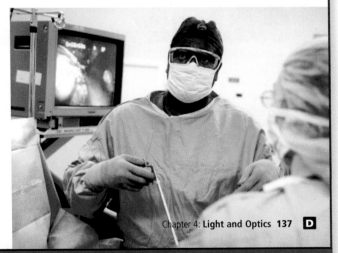

This surgeon uses fiber optics to see inside a patient's body.

Chapter 4: **Light and Optics** 137 **D**

Integrate the Sciences

Because optical fibers are so flexible and can transmit and receive light, they are used in flexible digital cameras for medical imaging. An endoscope uses fiber optics and powerful lens systems to provide lighting and visualization of the interior of a body part. The endoscope uses two fiber-optic lines. One carries light into the body cavity, and the other carries the image of the body cavity back to the physician's viewing lens.

Reinforce (the BIG idea)

Have students relate the section to the Big Idea.

 Reinforcing Key Concepts, p. 256

ASSESS & RETEACH

Assess

 Section 4.4 Quiz, p. 65

Reteach

Have groups each make a poster of an optical tool of their choosing. They should give a short history of the development of the tool, show a ray diagram explaining the optics of the tool, and include pictures of various types of the tool. If students choose optical tools that have not been covered in the text, you may want them to give reports to the class.

Technology Resources

Have students visit **ClassZone.com** for reteaching of Key Concepts.

 CONTENT REVIEW

 CONTENT REVIEW CD-ROM

Drawing power from a laser beam, the space elevator climbs to an orbiting space station.

Future Uses of Lasers

Research involving new uses of lasers continues at an amazing pace. Many new discoveries and developments in science and technology today are possible only because of lasers.

One area of research in which lasers have made a big impact is nanotechnology—the development of super-tiny machines and tools. Laser light can be controlled very precisely, so scientists can use it to perform extremely fine operations. For example, lasers could be used to cut out parts to make molecule-size motors. Lasers can also be used as "optical tweezers" to handle extremely small objects such as molecules. Scientists are even beginning to use lasers to change the shape of molecules. They do this by varying the laser's wavelength.

Future applications of lasers are also sure to involve new ways of transferring energy. Remember that a wave is a disturbance that transfers energy. Laser light is made up of EM waves. EM waves can move energy over great distances without losing any of it. When EM waves encounter a material medium, their energy can then be converted into other forms and put to use.

One possible future use of lasers is to supply energy to spacecraft. Scientists imagine a day when orbiting space stations will make rockets unnecessary. A cable between the ground and the station will make it possible for a "space elevator" to escape Earth's gravity by climbing up the cable. The elevator will be powered by an Earth-based laser. A device on board the elevator will convert the laser's energy into electrical power.

4.4 Review

KEY CONCEPTS

1. How do refracting and reflecting telescopes use convex lenses and mirrors?

2. What is different about the way a camera focuses images from the way an eye focuses images?

3. How is laser light different from ordinary light?

CRITICAL THINKING

4. **Predict** What would happen to laser light if it passed through a prism?

5. **Analyze** What are two ways reflection is involved in fiber optics?

CHALLENGE

6. **Apply** How could the speed of light and a laser beam be used to measure the distance between two satellites?

ANSWERS

1. Refracting telescopes use one lens to focus and another to magnify the image; reflecting telescopes use a lens to magnify an image focused by a mirror.

2. The eye focuses by changing shape of the lens. A camera focuses by changing distance between the lens and the film.

3. Laser light is made up of a single wavelength and a pure color. Ordinary light is made up of many different wavelengths.

4. It would not produce a color spectrum, because all of it would be refracted the same amount.

5. to make a laser and to keep the laser light within optical fibers

6. One satellite could shine the beam at the other satellite. Measure the time it takes for the beam to reflect and return from the other satellite. Use the speed of light to calculate the distance.

Optics in Photography

Photographers use the science of optics to help them make the best photographs possible. For example, a portrait photographer chooses the right equipment and lighting to make each person look his or her best. A photographer needs to understand how light reflects, refracts, and diffuses to achieve just the right effect.

Using Reflection

A gold-colored reflector reflects only gold-colored wavelengths of light onto the subject. Photographers use these to fill in shadows and add warmth.

without gold reflector

with gold reflector

Using Diffusion

When light is directed toward a curved reflective surface, the light scatters in many directions. This diffused light produces a softer appearance than direct light.

direct light

diffused light

Using Refraction

Lenses refract light in different ways. A long lens makes the subject appear closer. A wide-angle lens includes more space around the subject.

long lens

wide-angle lens

EXPLORE

1. **COMPARE** Find photos of people and compare them to the photos above. Which would have been improved by the use of a gold reflector? a long lens? diffused light?

2. **CHALLENGE** Using a disposable camera and a desk lamp, experiment with photography yourself. Try using a piece of paper as a reflector and observe its effects on the photograph. What happens if you use more than one reflector? What happens if you use a different color of paper?

EXPLORE

COMPARE *Sample Answer: A gold reflector will improve photographs by filling in shadows and adding warmth. A long lens enlarges details. Diffused light softens images.*

CHALLENGE *Multiple reflectors scatter the light coming from all directions and eliminate shadows. Colored reflectors absorb some wavelengths of light, changing the color of the light reflected off the subject.*

SCIENCE ON THE JOB
Relevance of Science to Non-science Jobs

Set Learning Goal

To understand why photographers need a knowledge of optics

Present the Science

Photographers can choose from a wide variety of lenses.

- A **wide-angle lens** has a very short focal length, which takes in a wide field of view. Because of its short focal length, it must be close to the film to form a sharp image. Wide-angle lenses produce a relatively small image of the subject and include much of the background.

- **Telephoto lenses,** also called **long lenses,** have a long focal length and a narrow field of view. They must be relatively far from the film. They produce an enlarged image that seems closer than it really is.

- A **zoom lens** has a variable focal length. Many photographers like using this lens so they don't have to carry several lenses.

Discussion Question

Ask: Why must photographers understand how colors form from the interaction of different wavelengths of light? *Color affects the appearance of a photograph in subtle ways. Photographers can insert color filters in front of the camera lens to soften the image, create shadows and contrast, or tone down harsh colors.*

Close

Ask: When would a photographer want to use a wide-angle lens? a long lens? *Sample Answer: A wide-angle lens could be used for panoramic photographs of scenery or crowd scenes. Long lenses are useful for close-ups of small details or for blow-ups.*

BACK TO

 the BIG idea

All optical technology uses reflection or refraction of light waves. Give students a list of optical tools and ask them to state which of these two phenomena are important for the operation of the tool. *refraction: microscope, camera, refracting telescope, contact lenses, corrective eyeglasses; reflection: lasers, periscope, fiber optics; both: reflecting telescope*

◐ KEY CONCEPTS SUMMARY

SECTION 4.1
Ask: How does regular reflection help mirrors form images? *Regular reflection causes the rays reflected from the mirror to keep the same arrangement as the rays coming from the object.*

SECTION 4.2
Ask: If the penguin is more than two focal lengths from the lens, what kind of image is formed? *a reduced inverted image*

Ask: If the penguin is between one and two focal lengths from the lens, what kind of image is formed? *an inverted image that is larger than the penguin*

SECTION 4.3
Ask: Does the eye shown need a corrective lens for normal sight? Why? *No; the image is focused directly on the retina.*

SECTION 4.4
Ask: Why does the tube have mirrors at both ends? *The mirrors reflect the light back and forth. One of the mirrors allows some of the light waves through, producing a laser beam.*

Review Concepts

- Big Idea Flow Chart, p. T25
- Chapter Outline, pp. T31–T32

 # Chapter Review

the BIG idea
Optical tools depend on the wave behavior of light.

 **CONTENT REVIEW**
CLASSZONE.COM

◐ KEY CONCEPTS SUMMARY

4.1 Mirrors form images by reflecting light.

flat mirror

- Light rays obey the law of reflection.
- Mirrors work by regular reflection.
- Curved mirrors can form images that are distorted in useful ways.

VOCABULARY
optics p. 113
image p. 114
diffuse reflection p. 114
regular reflection p. 114
law of reflection p. 114
concave p. 116
focal point p. 116
convex p. 117

4.2 Lenses form images by refracting light.

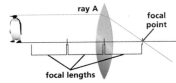
ray A
focal point
focal lengths

- Lenses have curved surfaces that refract parallel light waves in different amounts.
- Convex lenses bend light inward toward a focal point.
- Concave lenses spread light out.
- Lenses form a variety of useful images.

VOCABULARY
lens p. 121
focal length p. 123

4.3 The eye is a natural optical tool.

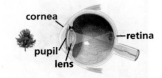
cornea
retina
pupil
lens

- The eyes of humans and many animals use lenses to focus images on the retina.
- The retina detects images and sends information about them to the brain.

VOCABULARY
cornea p. 127
pupil p. 127
retina p. 127

4.4 Optical technology makes use of light waves.

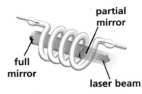
partial mirror
full mirror
laser beam

- Many optical tools are made by combining mirrors and lenses.
- Examples of optical tools include telescopes, microscopes, cameras, and lasers.
- Lasers have a wide variety of uses.

VOCABULARY
laser p. 135
fiber optics p. 137

D 140 Unit: Waves, Sound, and Light

Technology Resources

Have students visit **ClassZone.com** or use the CD-ROM for a cumulative review of concepts.

Engage students in a whole-class interactive review of Key Concepts. Edit content as you wish.

 CONTENT REVIEW

CONTENT REVIEW CD-ROM

POWER PRESENTATIONS

Reviewing Vocabulary

For each item below, fill in the blank. If the left column is blank, give a brief description or definition. If the right column is blank, give the correct term.

Term	Description
1.	shape like the inside of a bowl
2. convex	
3.	science of light, vision, and related technology
4.	picture of object formed by light rays
5. focal point	
6.	controls the amount of light entering the eye
7.	distance between mirror or lens and place where light rays meet
8. fiber optics	
9. law of reflection	
10.	concentrated, parallel light waves of a single wavelength

Reviewing Key Concepts

Multiple Choice *Choose the letter of the best answer.*

11. What shape is a mirror that reflects parallel light rays toward a focal point?

- **a.** convex
- **b.** flat
- **c.** concave
- **d.** regular

12. According to the law of reflection, a light ray striking a mirror

- **a.** continues moving through the mirror in the same direction
- **b.** moves into the mirror at a slightly different angle
- **c.** bounces off the mirror toward the direction it came from
- **d.** bounces off the mirror at the same angle it hits

13. Reflecting telescopes focus images using

- **a.** several mirrors
- **b.** several lenses
- **c.** both mirrors and lenses
- **d.** either a mirror or a lens, but not both

14. Ordinary light differs from laser light in that ordinary light waves

- **a.** all have the same wavelength
- **b.** tend to spread out
- **c.** stay parallel to each other
- **d.** all have their crests and troughs lined up

15. Nearsighted vision is corrected when lenses

- **a.** reflect light away from the eye
- **b.** allow light rays to focus on the retina
- **c.** allow light to focus slightly past the retina
- **d.** help light rays reflect regularly

16. Lasers do work similar to that of human vision when they are used to

- **a.** perform surgery
- **b.** send phone signals over optical cable
- **c.** scan bar codes at the grocery store
- **d.** change the shape of molecules

Short Answer *Write a short answer to each question.*

17. Name one optical tool, describe how it works, and explain some of its uses.

18. How are the images that are produced by a convex mirror different from those produced by a concave mirror?

19. Describe what typically happens to a ray of light from the time it enters the eye until it strikes the retina.

20. How do lenses correct nearsightedness and farsightedness?

21. What does a refracting telescope have in common with a simple microscope?

22. Describe two ways the distance of an object from a lens can affect the appearance of the object's image.

Reviewing Vocabulary

1. concave
2. curved outward like the bottom of a spoon
3. optics
4. image
5. the point where a concave mirror focuses light rays
6. pupil
7. focal length
8. technology based on the use of laser light
9. states that the angle of reflection equals the angle of incidence
10. laser beam

Reviewing Key Concepts

11. c
12. d
13. c
14. b
15. b
16. c
17. Answers will vary. For example, students might name a microscope and describe how it uses lenses to make images of small objects.
18. Convex mirrors produce reduced, right-side up images. Concave mirrors can produce images that are reduced, enlarged, right-side up, or upside down.
19. Light first enters the eye through the cornea, where it is refracted. It then continues through the pupil, which controls how much light enters the eye. After the pupil, the light passes through the lens of the eye, where the light is refracted to make adjustments for nearby and distant objects.
20. by spreading out the rays of light for nearsightedness and bending light rays inward for farsightedness; in both cases, they cause the image to fall on the retina
21. Both use a convex lens to focus and enlarge an image.
22. The distance can change the image's size and determine whether the image is upside down or right side up.

ASSESSMENT RESOURCES

UNIT ASSESSMENT BOOK
- Chapter Test A, pp. 66–69
- Chapter Test B, pp. 70–73
- Chapter Test C, pp. 74–77
- Alternative Assessment, pp. 78–79
- Unit Test, A, B, & C, pp. 80–91

SPANISH ASSESSMENT BOOK
- Spanish Chapter Test, pp. 293–296
- Spanish Unit Test, pp. 297–300

Technology Resources

Edit test items and answer choices.

Test Generator CD-ROM

Visit **ClassZone.com** to extend test practice.

Test Practice

Thinking Critically

23. C   24. D   25. A   26. B

For 27–32, sample answers are given.

27. reflect light; curved changes direction of light

28. bend light rays; convex bends rays inward, concave bends rays outward

29. involved in refraction of light by lenses; focal point is a point, focal length is a distance.

30. both are vision problems; nearsighted people can't see far objects clearly, farsighted people can't see near objects clearly

31. contain convex lenses; microscope focuses enlarged image, telescope focuses reduced image

32. involve reflection of light rays; in regular reflection parallel light rays reflect in the same direction, while in total internal reflection light inside a transparent medium reflects off the inner surface

33. distance from the lens to retina; image forms on the retina.

34. Laser light waves don't spread out; they stay parallel as they are reflected through the fiber.

35. more curved to shorten focal length

36. Lasers used in surgery have narrow beams that can be precisely controlled and carry energy to cut tissues and seal blood vessels.

the BIG idea

37. The refractor is used to test a person's eyesight. Light waves bend when they pass through the lens of a refractor. The patient benefits by having vision problems corrected with eyeglasses.

38. Sketches should show two mirrors at right angles and standing on a third mirror to form a half cube. Light rays bounce off each mirror at the same angle that they hit. Light striking the corner where mirrors meet is reflected from mirror to mirror, then reflected back parallel to original path.

UNIT PROJECTS

Have students present their projects. Use the appropriate rubrics from the URB to evaluate their work.

 Unit Projects, pp. 5–10

Thinking Critically

In the four diagrams below, light rays are shown interacting with a material medium. For the next four questions, choose the letter of the diagram that answers the question.

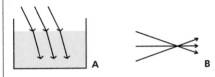

23. **INTERPRET** Which diagram shows regular reflection?

24. **INTERPRET** Which diagram shows diffuse reflection?

25. **INTERPRET** Which diagram shows refraction?

26. **INTERPRET** Which diagram shows light rays converging at a focal point?

Copy the chart below. For each pair of terms, write down one way they are alike (compare) and one way they are different (contrast).

Terms	Compare	Contrast
27. flat mirror, curved mirror		
28. convex lens, concave lens		
29. focal point, focal length		
30. nearsighted, farsighted		
31. simple microscope, refracting telescope		
32. regular reflection, total internal reflection		

33. **INFER** What is the approximate focal length of the eye's lens? How do you know?

34. **ANALYZE** Why is laser light used in fiber optics?

35. **APPLY** In order to increase the magnification of a magnifying glass, would you need to make the convex surfaces of the lens more or less curved?

36. **APPLY** Describe a possible use for laser light not mentioned in the chapter. What characteristics of laser light does this application make use of?

the BIG idea

37. **SYNTHESIZE** Using what you have learned in this chapter, describe two possible uses of an optical tool like the one shown on pages 110–111. Explain what wave behaviors of light would be involved in these uses. Then explain how these uses could benefit the person in the photo.

38. **APPLY** Make a sketch of an optical tool that would use 3 mirrors to make a beam of light return to its source. Your sketch should include:
   - the path of light waves through the tool
   - labels indicating the names of parts and how they affect the light.
   - 2 or 3 sentences at the bottom describing one possible use of the tool.

UNIT PROJECTS

Evaluate all the data, results, and information from your project folder. Prepare to present your project.

MONITOR AND RETEACH

If students have trouble applying the concepts in items 23–26, conduct a demonstration using a mirror, a flashlight, a prism, and a convex lens. Use the mirror to redirect a beam of light by reflection. Use the prism to redirect the beam by refraction. Use the lens to focus the beam into a concentrated spot. Then have students diagram the path of the light rays in each of the three events.

Students may benefit from summarizing sections of the chapter.

 Summarizing the Chapter, pp. 282–283

Standardized Test Practice

For practice on your state test, go to . . .
TEST PRACTICE
CLASSZONE.COM

Interpreting Diagrams

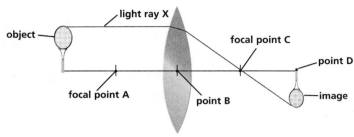

Study the diagram above and then answer the questions that follow.

1. What kind of lens is shown in the diagram?

  a. concave   **c.** flat

  b. convex   **d.** prism

2. What happens to parallel light rays passing through this type of lens?

  a. They become polarized.

  b. They form a rainbow.

  c. They bend inward.

  d. They bend outward.

3. All light rays parallel to light ray X will pass through what point?

  a. point A   **c.** point C

  b. point B   **d.** point D

4. How far is the object in the diagram from the lens?

  a. less than one focal length

  b. one focal length

  c. about two focal lengths

  d. more than three focal lengths

5. Where would you position a screen in order to see the image in focus on the screen?

  a. at point A

  b. at point B

  c. at point C

  d. at point D

Extended Response

Answer the two questions below in detail. Include some of the terms from the word box. Underline each term you use in your answer.

concave	focal point	real image
convex	refraction	virtual image
flat mirror	reflection	magnifying glass

6. What kind of mirror would you use to see what is happening over a broad area? Why?

7. Choose one of the following optical tools and explain how it uses mirrors and/or lenses to form an image: camera, telescope, periscope, microscope

---

METACOGNITIVE ACTIVITY

Have students answer the following questions in their **Science Notebook:**

1. How does what you learned about the eye relate to your life?

2. Describe a scenario in which you might use an optical tool.

3. Summarize the key concepts and main ideas of this chapter that apply to your Unit Project.

Interpreting Diagrams

1. *b*   3. *c*   5. *d*

2. *c*   4. *c*

Extended Response

6. RUBRIC

4 points for a response that thoroughly answers the question and uses the following terms accurately:

  • convex   • concave

Sample: I would use a <u>convex</u> mirror because parallel light rays move away from each other. My image in a convex mirror will appear smaller than me. You will be able to see a broad area because the images would look like they have shrunk. If I use a <u>concave</u> mirror and stand inside the focal point, my image will appear very large, and I would not be able to see a broad area.

3 points for a less thorough response that uses both terms accurately

2 points for a response that adequately answers the question and uses one term accurately

1 point for a response that adequately answers the question, but does not use the terms

7. RUBRIC

4 points for a response that correctly answers the question and uses the following terms accurately:

  • convex   • focal point   • real image

Sample: A film camera uses a <u>convex</u> lens to produce images of objects that are more than two focal lengths away. The images are reduced in size and upside down. The camera does not change the shape of its lens to change the <u>focal point</u>. Instead, you can move the lens nearer to or farther away from the film until the object you want to photograph is in focus. In a camera, the focal length is the distance between the lens and the <u>real image</u> of the object.

3 points for a response that correctly answers the question and uses two terms accurately

2 points for a response that uses one term accurately

1 point for a response that correctly answers the question, but does not use the terms

Student Resource Handbooks

Scientific Thinking Handbook

Making Observations

An **observation** is an act of noting and recording an event, characteristic, behavior, or anything else detected with an instrument or with the senses.

Observations allow you to make informed hypotheses and to gather data for experiments. Careful observations often lead to ideas for new experiments. There are two categories of observations:

- **Quantitative observations** can be expressed in numbers and include records of time, temperature, mass, distance, and volume.

- **Qualitative observations** include descriptions of sights, sounds, smells, and textures.

EXAMPLE

A student dissolved 30 grams of Epsom salts in water, poured the solution into a dish, and let the dish sit out uncovered overnight. The next day, she made the following observations of the Epsom salt crystals that grew in the dish.

> To determine the mass, the student found the mass of the dish before and after growing the crystals and then used subtraction to find the difference.

> The student measured several crystals and calculated the mean length. (To learn how to calculate the mean of a data set, see page R36.)

Table 1. Observations of Epsom Salt Crystals

Quantitative Observations	Qualitative Observations
• mass = 30 g • mean crystal length = 0.5 cm • longest crystal length = 2 cm	• Crystals are clear. • Crystals are long, thin, and rectangular. • White crust has formed around edge of dish.

> Photographs or sketches are useful for recording qualitative observations.

 Epsom salt crystals

MORE ABOUT OBSERVING

- Make quantitative observations whenever possible. That way, others will know exactly what you observed and be able to compare their results with yours.

- It is always a good idea to make qualitative observations too. You never know when you might observe something unexpected.

Predicting and Hypothesizing

A **prediction** is an expectation of what will be observed or what will happen. A **hypothesis** is a tentative explanation for an observation or scientific problem that can be tested by further investigation.

EXAMPLE

Suppose you have made two paper airplanes and you wonder why one of them tends to glide farther than the other one.

1. Start by asking a question.

2. Make an educated guess. After examination, you notice that the wings of the airplane that flies farther are slightly larger than the wings of the other airplane.

3. Write a prediction based upon your educated guess, in the form of an "If . . . , then . . ." statement. Write the independent variable after the word *if,* and the dependent variable after the word *then.*

4. To make a hypothesis, explain why you think what you predicted will occur. Write the explanation after the word *because.*

1. Why does one of the paper airplanes glide farther than the other?

2. The size of an airplane's wings may affect how far the airplane will glide.

3. Prediction: If I make a paper airplane with larger wings, then the airplane will glide farther.

To read about independent and dependent variables, see page R30.

4. Hypothesis: If I make a paper airplane with larger wings, then the airplane will glide farther, because the additional surface area of the wing will produce more lift.

Notice that the part of the hypothesis after *because* adds an explanation of why the airplane will glide farther.

MORE ABOUT HYPOTHESES

- The results of an experiment cannot prove that a hypothesis is correct. Rather, the results either support or do not support the hypothesis.

- Valuable information is gained even when your hypothesis is not supported by your results. For example, it would be an important discovery to find that wing size is not related to how far an airplane glides.

- In science, a hypothesis is supported only after many scientists have conducted many experiments and produced consistent results.

Inferring

An **inference** is a logical conclusion drawn from the available evidence and prior knowledge. Inferences are often made from observations.

EXAMPLE

A student observing a set of acorns noticed something unexpected about one of them. He noticed a white, soft-bodied insect eating its way out of the acorn.

The student recorded these observations.

Observations
- There is a hole in the acorn, about 0.5 cm in diameter, where the insect crawled out.
- There is a second hole, which is about the size of a pinhole, on the other side of the acorn.
- The inside of the acorn is hollow.

Here are some inferences that can be made on the basis of the observations.

Inferences
- The insect formed from the material inside the acorn, grew to its present size, and ate its way out of the acorn.
- The insect crawled through the smaller hole, ate the inside of the acorn, grew to its present size, and ate its way out of the acorn.
- An egg was laid in the acorn through the smaller hole. The egg hatched into a larva that ate the inside of the acorn, grew to its present size, and ate its way out of the acorn.

When you make inferences, be sure to look at all of the evidence available and combine it with what you already know.

MORE ABOUT INFERENCES

Inferences depend both on observations and on the knowledge of the people making the inferences. Ancient people who did not know that organisms are produced only by similar organisms might have made an inference like the first one. A student today might look at the same observations and make the second inference. A third student might have knowledge about this particular insect and know that it is never small enough to fit through the smaller hole, leading her to the third inference.

Identifying Cause and Effect

In a **cause-and-effect relationship,** one event or characteristic is the result of another. Usually an effect follows its cause in time.

There are many examples of cause-and-effect relationships in everyday life.

Cause	Effect
Turn off a light.	Room gets dark.
Drop a glass.	Glass breaks.
Blow a whistle.	Sound is heard.

Scientists must be careful not to infer a cause-and-effect relationship just because one event happens after another event. When one event occurs after another, you cannot infer a cause-and-effect relationship on the basis of that information alone. You also cannot conclude that one event caused another if there are alternative ways to explain the second event. A scientist must demonstrate through experimentation or continued observation that an event was truly caused by another event.

EXAMPLE

Make an Observation

Suppose you have a few plants growing outside. When the weather starts getting colder, you bring one of the plants indoors. You notice that the plant you brought indoors is growing faster than the others are growing. You cannot conclude from your observation that the change in temperature was the cause of the increased plant growth, because there are alternative explanations for the observation. Some possible explanations are given below.

- The humidity indoors caused the plant to grow faster.

- The level of sunlight indoors caused the plant to grow faster.

- The indoor plant's being noticed more often and watered more often than the outdoor plants caused it to grow faster.

- The plant that was brought indoors was healthier than the other plants to begin with.

To determine which of these factors, if any, caused the indoor plant to grow faster than the outdoor plants, you would need to design and conduct an experiment.

See pages R28–R35 for information about designing experiments.

Recognizing Bias

Television, newspapers, and the Internet are full of experts claiming to have scientific evidence to back up their claims. How do you know whether the claims are really backed up by good science?

Bias is a slanted point of view, or personal prejudice. The goal of scientists is to be as objective as possible and to base their findings on facts instead of opinions. However, bias often affects the conclusions of researchers, and it is important to learn to recognize bias.

When scientific results are reported, you should consider the source of the information as well as the information itself. It is important to critically analyze the information that you see and read.

SOURCES OF BIAS

There are several ways in which a report of scientific information may be biased. Here are some questions that you can ask yourself:

1. **Who is sponsoring the research?**

   Sometimes, the results of an investigation are biased because an organization paying for the research is looking for a specific answer. This type of bias can affect how data are gathered and interpreted.

2. **Is the research sample large enough?**

   Sometimes research does not include enough data. The larger the sample size, the more likely that the results are accurate, assuming a truly random sample.

3. **In a survey, who is answering the questions?**

   The results of a survey or poll can be biased. The people taking part in the survey may have been specifically chosen because of how they would answer. They may have the same ideas or lifestyles. A survey or poll should make use of a random sample of people.

4. **Are the people who take part in a survey biased?**

   People who take part in surveys sometimes try to answer the questions the way they think the researcher wants them to answer. Also, in surveys or polls that ask for personal information, people may be unwilling to answer questions truthfully.

SCIENTIFIC BIAS

It is also important to realize that scientists have their own biases because of the types of research they do and because of their scientific viewpoints. Two scientists may look at the same set of data and come to completely different conclusions because of these biases. However, such disagreements are not necessarily bad. In fact, a critical analysis of disagreements is often responsible for moving science forward.

Identifying Faulty Reasoning

Faulty reasoning is wrong or incorrect thinking. It leads to mistakes and to wrong conclusions. Scientists are careful not to draw unreasonable conclusions from experimental data. Without such caution, the results of scientific investigations may be misleading.

EXAMPLE

Scientists try to make generalizations based on their data to explain as much about nature as possible. If only a small sample of data is looked at, however, a conclusion may be faulty. Suppose a scientist has studied the effects of the El Niño and La Niña weather patterns on flood damage in California from 1989 to 1995. The scientist organized the data in the bar graph below.

The scientist drew the following conclusions:

1. The La Niña weather pattern has no effect on flooding in California.

2. When neither weather pattern occurs, there is almost no flood damage.

3. A weak or moderate El Niño produces a small or moderate amount of flooding.

4. A strong El Niño produces a lot of flooding.

Flood and Storm Damage in California

Estimated damage (millions of dollars)

- Weak–moderate El Niño
- Strong El Niño

Starting year of season (July 1–June 30)

SOURCE: *Governor's Office of Emergency Services, California*

For the six-year period of the scientist's investigation, these conclusions may seem to be reasonable. However, a six-year study of weather patterns may be too small of a sample for the conclusions to be supported. Consider the following graph, which shows information that was gathered from 1949 to 1997.

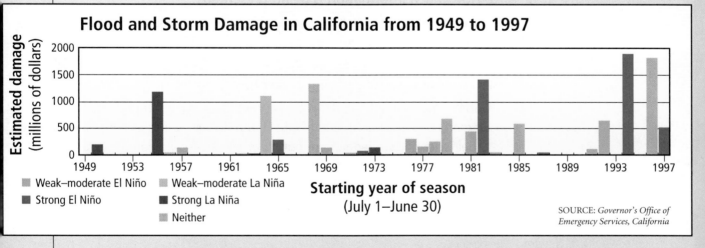

Flood and Storm Damage in California from 1949 to 1997

Estimated damage (millions of dollars)

- Weak–moderate El Niño
- Strong El Niño
- Weak–moderate La Niña
- Strong La Niña
- Neither

Starting year of season (July 1–June 30)

SOURCE: *Governor's Office of Emergency Services, California*

The only one of the conclusions that all of this information supports is number 3: a weak or moderate El Niño produces a small or moderate amount of flooding. By collecting more data, scientists can be more certain of their conclusions and can avoid faulty reasoning.

Analyzing Statements

To **analyze** a statement is to examine its parts carefully. Scientific findings are often reported through media such as television or the Internet. A report that is made public often focuses on only a small part of research. As a result, it is important to question the sources of information.

Evaluate Media Claims

To **evaluate** a statement is to judge it on the basis of criteria you've established. Sometimes evaluating means deciding whether a statement is true.

Reports of scientific research and findings in the media may be misleading or incomplete. When you are exposed to this information, you should ask yourself some questions so that you can make informed judgments about the information.

1. **Does the information come from a credible source?**

   Suppose you learn about a new product and it is stated that scientific evidence proves that the product works. A report from a respected news source may be more believable than an advertisement paid for by the product's manufacturer.

2. **How much evidence supports the claim?**

   Often, it may seem that there is new evidence every day of something in the world that either causes or cures an illness. However, information that is the result of several years of work by several different scientists is more credible than an advertisement that does not even cite the subjects of the experiment.

3. **How much information is being presented?**

   Science cannot solve all questions, and scientific experiments often have flaws. A report that discusses problems in a scientific study may be more believable than a report that addresses only positive experimental findings.

4. **Is scientific evidence being presented by a specific source?**

   Sometimes scientific findings are reported by people who are called experts or leaders in a scientific field. But if their names are not given or their scientific credentials are not reported, their statements may be less credible than those of recognized experts.

Differentiate Between Fact and Opinion

Sometimes information is presented as a fact when it may be an opinion. When scientific conclusions are reported, it is important to recognize whether they are based on solid evidence. Again, you may find it helpful to ask yourself some questions.

1. **What is the difference between a fact and an opinion?**

   A **fact** is a piece of information that can be strictly defined and proved true. An **opinion** is a statement that expresses a belief, value, or feeling. An opinion cannot be proved true or false. For example, a person's age is a fact, but if someone is asked how old they feel, it is impossible to prove the person's answer to be true or false.

2. **Can opinions be measured?**

   Yes, opinions can be measured. In fact, surveys often ask for people's opinions on a topic. But there is no way to know whether or not an opinion is the truth.

HOW TO DIFFERENTIATE FACT FROM OPINION

Human Activities and the Environment

Opinions
Notice words or phrases that express beliefs or feelings. The words *unfortunately* and *careless* show that opinions are being expressed.

Unfortunately, human use of fossil fuels is one of the most significant developments of the past few centuries. Humans rely on fossil fuels, a non-renewable energy resource, for more than 90 percent of their energy needs.

Facts
Statements that contain statistics tend to be facts. Writers often use facts to support their opinions.

Opinion
Look for statements that speculate about events. These statements are opinions, because they cannot be proved.

This careless misuse of our planet's resources has resulted in pollution, global warming, and the destruction of fragile ecosystems. For example, oil pipelines carry more than one million barrels of oil each day across tundra regions. Transporting oil across such areas can only result in oil spills that poison the land for decades.

Lab Handbook

LAB HANDBOOK

Safety Rules

Before you work in the laboratory, read these safety rules twice. Ask your teacher to explain any rules that you do not completely understand. Refer to these rules later on if you have questions about safety in the science classroom.

Directions

- Read all directions and make sure that you understand them before starting an investigation or lab activity. If you do not understand how to do a procedure or how to use a piece of equipment, ask your teacher.
- Do not begin any investigation or touch any equipment until your teacher has told you to start.
- Never experiment on your own. If you want to try a procedure that the directions do not call for, ask your teacher for permission first.
- If you are hurt or injured in any way, tell your teacher immediately.

Dress Code

goggles

apron

gloves

- Wear goggles when
  — using glassware, sharp objects, or chemicals
  — heating an object
  — working with anything that can easily fly up into the air and hurt someone's eye
- Tie back long hair or hair that hangs in front of your eyes.
- Remove any article of clothing—such as a loose sweater or a scarf—that hangs down and may touch a flame, chemical, or piece of equipment.
- Observe all safety icons calling for the wearing of eye protection, gloves, and aprons.

Heating and Fire Safety

fire safety

heating safety

- Keep your work area neat, clean, and free of extra materials.
- Never reach over a flame or heat source.
- Point objects being heated away from you and others.
- Never heat a substance or an object in a closed container.
- Never touch an object that has been heated. If you are unsure whether something is hot, treat it as though it is. Use oven mitts, clamps, tongs, or a test-tube holder.
- Know where the fire extinguisher and fire blanket are kept in your classroom.
- Do not throw hot substances into the trash. Wait for them to cool or use the container your teacher puts out for disposal.

Electrical Safety

electrical safety

- Never use lamps or other electrical equipment with frayed cords.
- Make sure no cord is lying on the floor where someone can trip over it.
- Do not let a cord hang over the side of a counter or table so that the equipment can easily be pulled or knocked to the floor.
- Never let cords hang into sinks or other places where water can be found.
- Never try to fix electrical problems. Inform your teacher of any problems immediately.
- Unplug an electrical cord by pulling on the plug, not the cord.

Chemical Safety

chemical safety

poison

fumes

- If you spill a chemical or get one on your skin or in your eyes, tell your teacher right away.
- Never touch, taste, or sniff any chemicals in the lab. If you need to determine odor, waft. Wafting consists of holding the chemical in its container 15 centimeters (6 in.) away from your nose, and using your fingers to bring fumes from the container to your nose.
- Keep lids on all chemicals you are not using.
- Never put unused chemicals back into the original containers. Throw away extra chemicals where your teacher tells you to.
- Pour chemicals over a sink or your work area, not over the floor.
- If you get a chemical in your eye, use the eyewash right away.
- Always wash your hands after handling chemicals, plants, or soil.

Wafting

Glassware and Sharp-Object Safety

sharp objects

- If you break glassware, tell your teacher right away.
- Do not use broken or chipped glassware. Give these to your teacher.
- Use knives and other cutting instruments carefully. Always wear eye protection and cut away from you.

Animal Safety

- Never hurt an animal.
- Touch animals only when necessary. Follow your teacher's instructions for handling animals.
- Always wash your hands after working with animals.

Cleanup

disposal

- Follow your teacher's instructions for throwing away or putting away supplies.
- Clean your work area and pick up anything that has dropped to the floor.
- Wash your hands.

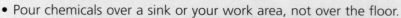

Using Lab Equipment

Different experiments require different types of equipment. But even though experiments differ, the ways in which the equipment is used are the same.

Beakers

- Use beakers for holding and pouring liquids.
- Do not use a beaker to measure the volume of a liquid. Use a graduated cylinder instead. (See page R16.)
- Use a beaker that holds about twice as much liquid as you need. For example, if you need 100 milliliters of water, you should use a 200- or 250-milliliter beaker.

Test Tubes

- Use test tubes to hold small amounts of substances.
- Do not use a test tube to measure the volume of a liquid.
- Use a test tube when heating a substance over a flame. Aim the mouth of the tube away from yourself and other people.
- Liquids easily spill or splash from test tubes, so it is important to use only small amounts of liquids.

Test-Tube Holder

- Use a test-tube holder when heating a substance in a test tube.
- Use a test-tube holder if the substance in a test tube is dangerous to touch.
- Make sure the test-tube holder tightly grips the test tube so that the test tube will not slide out of the holder.
- Make sure that the test-tube holder is above the surface of the substance in the test tube so that you can observe the substance.

Test-Tube Rack

- Use a test-tube rack to organize test tubes before, during, and after an experiment.
- Use a test-tube rack to keep test tubes upright so that they do not fall over and spill their contents.
- Use a test-tube rack that is the correct size for the test tubes that you are using. If the rack is too small, a test tube may become stuck. If the rack is too large, a test tube may lean over, and some of its contents may spill or splash.

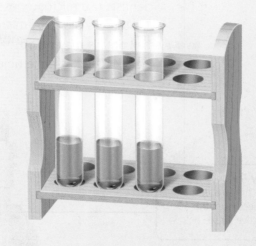

Forceps

- Use forceps when you need to pick up or hold a very small object that should not be touched with your hands.
- Do not use forceps to hold anything over a flame, because forceps are not long enough to keep your hand safely away from the flame. Plastic forceps will melt, and metal forceps will conduct heat and burn your hand.

Hot Plate

- Use a hot plate when a substance needs to be kept warmer than room temperature for a long period of time.
- Use a hot plate instead of a Bunsen burner or a candle when you need to carefully control temperature.
- Do not use a hot plate when a substance needs to be burned in an experiment.
- Always use "hot hands" safety mitts or oven mitts when handling anything that has been heated on a hot plate.

Microscope

Scientists use microscopes to see very small objects that cannot easily be seen with the eye alone. A microscope magnifies the image of an object so that small details may be observed. A microscope that you may use can magnify an object 400 times—the object will appear 400 times larger than its actual size.

Body The body separates the lens in the eyepiece from the objective lenses below.

Nosepiece The nosepiece holds the objective lenses above the stage and rotates so that all lenses may be used.

High-Power Objective Lens This is the largest lens on the nosepiece. It magnifies an image approximately 40 times.

Stage The stage supports the object being viewed.

Diaphragm The diaphragm is used to adjust the amount of light passing through the slide and into an objective lens.

Mirror or Light Source Some microscopes use light that is reflected through the stage by a mirror. Other microscopes have their own light sources.

Eyepiece Objects are viewed through the eyepiece. The eyepiece contains a lens that commonly magnifies an image 10 times.

Coarse Adjustment This knob is used to focus the image of an object when it is viewed through the low-power lens.

Fine Adjustment This knob is used to focus the image of an object when it is viewed through the high-power lens.

Low-Power Objective Lens This is the smallest lens on the nosepiece. It magnifies an image approximately 10 times.

Arm The arm supports the body above the stage. Always carry a microscope by the arm and base.

Stage Clip The stage clip holds a slide in place on the stage.

Base The base supports the microscope.

VIEWING AN OBJECT

1. Use the coarse adjustment knob to raise the body tube.

2. Adjust the diaphragm so that you can see a bright circle of light through the eyepiece.

3. Place the object or slide on the stage. Be sure that it is centered over the hole in the stage.

4. Turn the nosepiece to click the low-power lens into place.

5. Using the coarse adjustment knob, slowly lower the lens and focus on the specimen being viewed. Be sure not to touch the slide or object with the lens.

6. When switching from the low-power lens to the high-power lens, first raise the body tube with the coarse adjustment knob so that the high-power lens will not hit the slide.

7. Turn the nosepiece to click the high-power lens into place.

8. Use the fine adjustment knob to focus on the specimen being viewed. Again, be sure not to touch the slide or object with the lens.

MAKING A SLIDE, OR WET MOUNT

1 Place the specimen in the center of a clean slide.

2 Place a drop of water on the specimen.

3 Place a cover slip on the slide. Put one edge of the cover slip into the drop of water and slowly lower it over the specimen.

4 Remove any air bubbles from under the cover slip by gently tapping the cover slip.

5 Dry any excess water before placing the slide on the microscope stage for viewing.

Spring Scale (Force Meter)

- Use a spring scale to measure a force pulling on the scale.
- Use a spring scale to measure the force of gravity exerted on an object by Earth.
- To measure a force accurately, a spring scale must be zeroed before it is used. The scale is zeroed when no weight is attached and the indicator is positioned at zero.
- Do not attach a weight that is either too heavy or too light to a spring scale. A weight that is too heavy could break the scale or exert too great a force for the scale to measure. A weight that is too light may not exert enough force to be measured accurately.

Graduated Cylinder

- Use a graduated cylinder to measure the volume of a liquid.
- Be sure that the graduated cylinder is on a flat surface so that your measurement will be accurate.
- When reading the scale on a graduated cylinder, be sure to have your eyes at the level of the surface of the liquid.
- The surface of the liquid will be curved in the graduated cylinder. Read the volume of the liquid at the bottom of the curve, or meniscus (muh-NIHS-kuhs).
- You can use a graduated cylinder to find the volume of a solid object by measuring the increase in a liquid's level after you add the object to the cylinder.

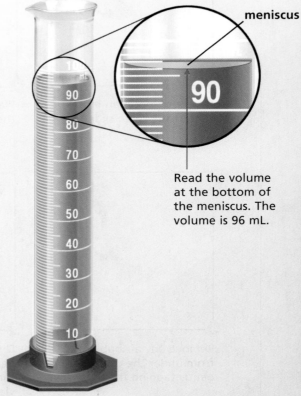

meniscus

Read the volume at the bottom of the meniscus. The volume is 96 mL.

Metric Rulers

- Use metric rulers or meter sticks to measure objects' lengths.

- Do not measure an object from the end of a metric ruler or meter stick, because the end is often imperfect. Instead, measure from the 1-centimeter mark, but remember to subtract a centimeter from the apparent measurement.

- Estimate any lengths that extend between marked units. For example, if a meter stick shows centimeters but not millimeters, you can estimate the length that an object extends between centimeter marks to measure it to the nearest millimeter.

- **Controlling Variables** If you are taking repeated measurements, always measure from the same point each time. For example, if you're measuring how high two different balls bounce when dropped from the same height, measure both bounces at the same point on the balls—either the top or the bottom. Do not measure at the top of one ball and the bottom of the other.

EXAMPLE

How to Measure a Leaf

1. Lay a ruler flat on top of the leaf so that the 1-centimeter mark lines up with one end. Make sure the ruler and the leaf do not move between the time you line them up and the time you take the measurement.

2. Look straight down on the ruler so that you can see exactly how the marks line up with the other end of the leaf.

3. Estimate the length by which the leaf extends beyond a marking. For example, the leaf below extends about halfway between the 4.2-centimeter and 4.3-centimeter marks, so the apparent measurement is about 4.25 centimeters.

4. Remember to subtract 1 centimeter from your apparent measurement, since you started at the 1-centimeter mark on the ruler and not at the end. The leaf is about 3.25 centimeters long (4.25 cm – 1 cm = 3.25 cm).

Triple-Beam Balance

This balance has a pan and three beams with sliding masses, called riders. At one end of the beams is a pointer that indicates whether the mass on the pan is equal to the masses shown on the beams.

1. Make sure the balance is zeroed before measuring the mass of an object. The balance is zeroed if the pointer is at zero when nothing is on the pan and the riders are at their zero points. Use the adjustment knob at the base of the balance to zero it.

2. Place the object to be measured on the pan.

3. Move the riders one notch at a time away from the pan. Begin with the largest rider. If moving the largest rider one notch brings the pointer below zero, begin measuring the mass of the object with the next smaller rider.

4. Change the positions of the riders until they balance the mass on the pan and the pointer is at zero. Then add the readings from the three beams to determine the mass of the object.

300 g	position of largest rider
90 g	position of middle rider
+ 3 g	position of smallest rider
393 g	mass of beaker

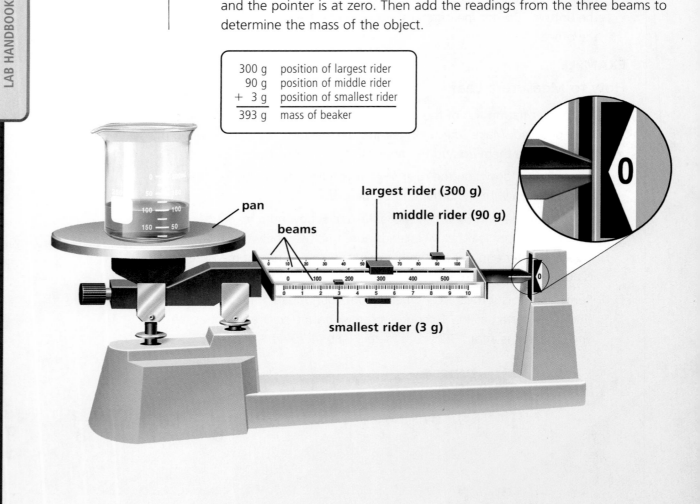

pan

beams

largest rider (300 g)

middle rider (90 g)

smallest rider (3 g)

Double-Pan Balance

This type of balance has two pans. Between the pans is a pointer that indicates whether the masses on the pans are equal.

1. Make sure the balance is zeroed before measuring the mass of an object. The balance is zeroed if the pointer is at zero when there is nothing on either of the pans. Many double-pan balances have sliding knobs that can be used to zero them.

2. Place the object to be measured on one of the pans.

3. Begin adding standard masses to the other pan. Begin with the largest standard mass. If this adds too much mass to the balance, begin measuring the mass of the object with the next smaller standard mass.

4. Add standard masses until the masses on both pans are balanced and the pointer is at zero. Then add the standard masses together to determine the mass of the object being measured.

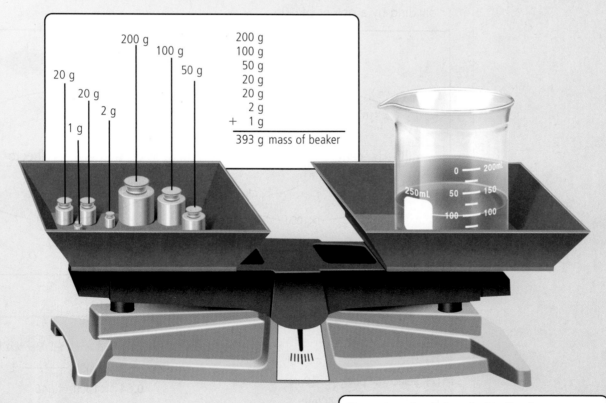

200 g	
100 g	
50 g	
20 g	
20 g	
2 g	
+ 1 g	
393 g	mass of beaker

Never place chemicals or liquids directly on a pan. Instead, use the following procedure:

1. Determine the mass of an empty container, such as a beaker.

2. Pour the substance into the container, and measure the total mass of the substance and the container.

3. Subtract the mass of the empty container from the total mass to find the mass of the substance.

The Metric System and SI Units

Scientists use International System (SI) units for measurements of distance, volume, mass, and temperature. The International System is based on multiples of ten and the metric system of measurement.

Basic SI Units		
Property	Name	Symbol
length	meter	m
volume	liter	L
mass	kilogram	kg
temperature	kelvin	K

SI Prefixes		
Prefix	Symbol	Multiple of 10
kilo-	k	1000
hecto-	h	100
deca-	da	10
deci-	d	0.1 $\left(\frac{1}{10}\right)$
centi-	c	0.01 $\left(\frac{1}{100}\right)$
milli-	m	0.001 $\left(\frac{1}{1000}\right)$

Changing Metric Units

You can change from one unit to another in the metric system by multiplying or dividing by a power of 10.

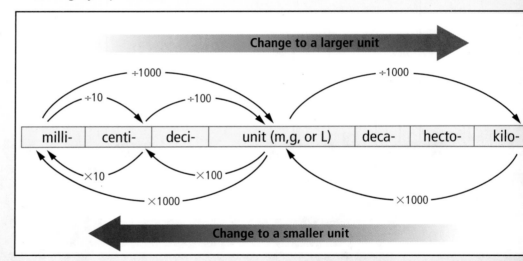

Example

Change 0.64 liters to milliliters.

(1) Decide whether to multiply or divide.

(2) Select the power of 10.

ANSWER 0.64 L = 640 mL

Change to a smaller unit by multiplying.

mL ⟵ × 1000 ⟶ L

0.64 × 1000 = **640.**

Example

Change 23.6 grams to kilograms.

(1) Decide whether to multiply or divide.

(2) Select the power of 10.

ANSWER 23.6 g = 0.0236 kg

Change to a larger unit by dividing.

g ⟶ ÷ 1000 ⟶ kg

23.6 ÷ 1000 = **0.0236**

Temperature Conversions

Even though the kelvin is the SI base unit of temperature, the degree Celsius will be the unit you use most often in your science studies. The formulas below show the relationships between temperatures in degrees Fahrenheit (°F), degrees Celsius (°C), and kelvins (K).

$$°C = \frac{5}{9}(°F - 32)$$

$$°F = \frac{9}{5}°C + 32$$

$$K = °C + 273$$

See page R42 for help with using formulas.

Examples of Temperature Conversions

Condition	Degrees Celsius	Degrees Fahrenheit
Freezing point of water	0	32
Cool day	10	50
Mild day	20	68
Warm day	30	86
Normal body temperature	37	98.6
Very hot day	40	104
Boiling point of water	100	212

Converting Between SI and U.S. Customary Units

Use the chart below when you need to convert between SI units and U.S. customary units.

SI Unit	From SI to U.S. Customary			From U.S. Customary to SI		
Length	When you know	multiply by	to find	When you know	multiply by	to find
kilometer (km) = 1000 m	kilometers	0.62	miles	miles	1.61	kilometers
meter (m) = 100 cm	meters	3.28	feet	feet	0.3048	meters
centimeter (cm) = 10 mm	centimeters	0.39	inches	inches	2.54	centimeters
millimeter (mm) = 0.1 cm	millimeters	0.04	inches	inches	25.4	millimeters
Area	When you know	multiply by	to find	When you know	multiply by	to find
square kilometer (km^2)	square kilometers	0.39	square miles	square miles	2.59	square kilometers
square meter (m^2)	square meters	1.2	square yards	square yards	0.84	square meters
square centimeter (cm^2)	square centimeters	0.155	square inches	square inches	6.45	square centimeters
Volume	When you know	multiply by	to find	When you know	multiply by	to find
liter (L) = 1000 mL	liters	1.06	quarts	quarts	0.95	liters
	liters	0.26	gallons	gallons	3.79	liters
	liters	4.23	cups	cups	0.24	liters
	liters	2.12	pints	pints	0.47	liters
milliliter (mL) = 0.001 L	milliliters	0.20	teaspoons	teaspoons	4.93	milliliters
	milliliters	0.07	tablespoons	tablespoons	14.79	milliliters
	milliliters	0.03	fluid ounces	fluid ounces	29.57	milliliters
Mass	When you know	multiply by	to find	When you know	multiply by	to find
kilogram (kg) = 1000 g	kilograms	2.2	pounds	pounds	0.45	kilograms
gram (g) = 1000 mg	grams	0.035	ounces	ounces	28.35	grams

Precision and Accuracy

When you do an experiment, it is important that your methods, observations, and data be both precise and accurate.

low precision

precision, but not accuracy

precision and accuracy

Precision

In science, **precision** is the exactness and consistency of measurements. For example, measurements made with a ruler that has both centimeter and millimeter markings would be more precise than measurements made with a ruler that has only centimeter markings. Another indicator of precision is the care taken to make sure that methods and observations are as exact and consistent as possible. Every time a particular experiment is done, the same procedure should be used. Precision is necessary because experiments are repeated several times and if the procedure changes, the results will change.

EXAMPLE

Suppose you are measuring temperatures over a two-week period. Your precision will be greater if you measure each temperature at the same place, at the same time of day, and with the same thermometer than if you change any of these factors from one day to the next.

Accuracy

In science, it is possible to be precise but not accurate. **Accuracy** depends on the difference between a measurement and an actual value. The smaller the difference, the more accurate the measurement.

EXAMPLE

Suppose you look at a stream and estimate that it is about 1 meter wide at a particular place. You decide to check your estimate by measuring the stream with a meter stick, and you determine that the stream is 1.32 meters wide. However, because it is hard to measure the width of a stream with a meter stick, it turns out that you didn't do a very good job. The stream is actually 1.14 meters wide. Therefore, even though your estimate was less precise than your measurement, your estimate was actually more accurate.

Making Data Tables and Graphs

Data tables and graphs are useful tools for both recording and communicating scientific data.

Making Data Tables

You can use a **data table** to organize and record the measurements that you make. Some examples of information that might be recorded in data tables are frequencies, times, and amounts.

EXAMPLE

Suppose you are investigating photosynthesis in two elodea plants. One sits in direct sunlight, and the other sits in a dimly lit room. You measure the rate of photosynthesis by counting the number of bubbles in the jar every ten minutes.

1. Title and number your data table.
2. Decide how you will organize the table into columns and rows.
3. Any units, such as seconds or degrees, should be included in column headings, not in the individual cells.

Table 1. Number of Bubbles from Elodea

Time (min)	Sunlight	Dim Light
0	0	0
10	15	5
20	25	8
30	32	7
40	41	10
50	47	9
60	42	9

> Always number and title data tables.

The data in the table above could also be organized in a different way.

Table 1. Number of Bubbles from Elodea

Light Condition	Time (min)						
	0	10	20	30	40	50	60
Sunlight	0	15	25	32	41	47	42
Dim light	0	5	8	7	10	9	9

> Put units in column heading.

Making Line Graphs

You can use a **line graph** to show a relationship between variables. Line graphs are particularly useful for showing changes in variables over time.

EXAMPLE

Suppose you are interested in graphing temperature data that you collected over the course of a day.

Table 1. Outside Temperature During the Day on March 7

	Time of Day						
	7:00 A.M.	9:00 A.M.	11:00 A.M.	1:00 P.M.	3:00 P.M.	5:00 P.M.	7:00 P.M.
Temp (°C)	8	9	11	14	12	10	6

1. Use the vertical axis of your line graph for the variable that you are measuring—temperature.

2. Choose scales for both the horizontal axis and the vertical axis of the graph. You should have two points more than you need on the vertical axis, and the horizontal axis should be long enough for all of the data points to fit.

3. Draw and label each axis.

4. Graph each value. First find the appropriate point on the scale of the horizontal axis. Imagine a line that rises vertically from that place on the scale. Then find the corresponding value on the vertical axis, and imagine a line that moves horizontally from that value. The point where these two imaginary lines intersect is where the value should be plotted.

5. Connect the points with straight lines.

Be sure to add a number and a title to your graph.

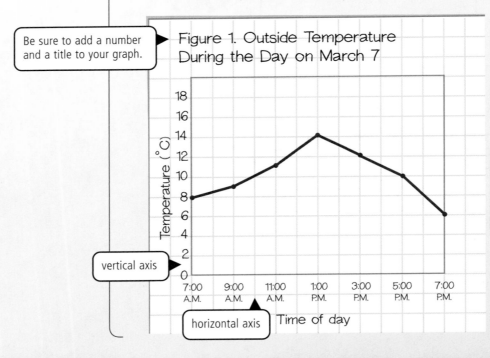

Figure 1. Outside Temperature During the Day on March 7

vertical axis

horizontal axis

Time of day

Making Circle Graphs

You can use a **circle graph,** sometimes called a pie chart, to represent data as parts of a circle. Circle graphs are used only when the data can be expressed as percentages of a whole. The entire circle shown in a circle graph is equal to 100 percent of the data.

EXAMPLE

Suppose you identified the species of each mature tree growing in a small wooded area. You organized your data in a table, but you also want to show the data in a circle graph.

1. To begin, find the total number of mature trees.

   $56 + 34 + 22 + 10 + 28 = 150$

2. To find the degree measure for each sector of the circle, write a fraction comparing the number of each tree species with the total number of trees. Then multiply the fraction by 360°.

   Oak: $\frac{56}{150} \times 360° = 134.4°$

3. Draw a circle. Use a protractor to draw the angle for each sector of the graph.

4. Color and label each sector of the graph.

5. Give the graph a number and title.

Table 1. Tree Species in Wooded Area

Species	Number of Specimens
Oak	56
Maple	34
Birch	22
Willow	10
Pine	28

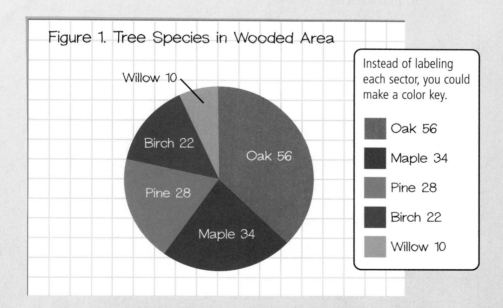

Figure 1. Tree Species in Wooded Area

Willow 10
Birch 22
Pine 28
Oak 56
Maple 34

Instead of labeling each sector, you could make a color key.

- Oak 56
- Maple 34
- Pine 28
- Birch 22
- Willow 10

Bar Graph

A **bar graph** is a type of graph in which the lengths of the bars are used to represent and compare data. A numerical scale is used to determine the lengths of the bars.

EXAMPLE

To determine the effect of water on seed sprouting, three cups were filled with sand, and ten seeds were planted in each. Different amounts of water were added to each cup over a three-day period.

Table 1. Effect of Water on Seed Sprouting

Daily Amount of Water (mL)	Number of Seeds That Sprouted After 3 Days in Sand
0	1
10	4
20	8

1. Choose a numerical scale. The greatest value is 8, so the end of the scale should have a value greater than 8, such as 10. Use equal increments along the scale, such as increments of 2.

2. Draw and label the axes. Mark intervals on the vertical axis according to the scale you chose.

3. Draw a bar for each data value. Use the scale to decide how long to make each bar.

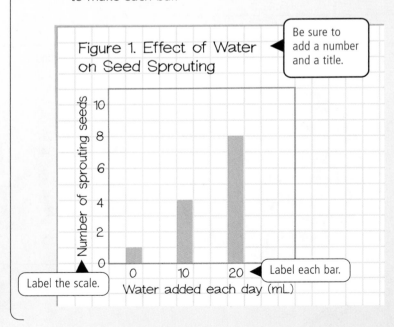

Figure 1. Effect of Water on Seed Sprouting

Be sure to add a number and a title.

Label the scale.

Label each bar.

Double Bar Graph

A **double bar graph** is a bar graph that shows two sets of data. The two bars for each measurement are drawn next to each other.

EXAMPLE

The same seed-sprouting experiment was repeated with potting soil. The data for sand and potting soil can be plotted on one graph.

1. Draw one set of bars, using the data for sand, as shown below.
2. Draw bars for the potting-soil data next to the bars for the sand data. Shade them a different color. Add a key.

Table 2. Effect of Water and Soil on Seed Sprouting

Daily Amount of Water (mL)	Number of Seeds That Sprouted After 3 Days in Sand	Number of Seeds That Sprouted After 3 Days in Potting Soil
0	1	2
10	4	5
20	8	9

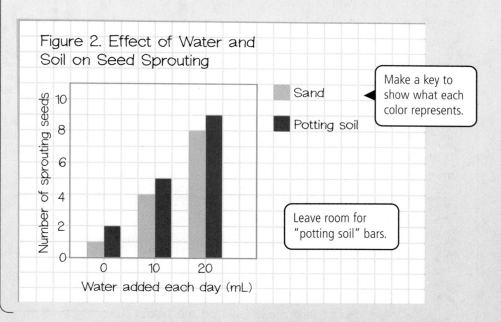

Figure 2. Effect of Water and Soil on Seed Sprouting

Make a key to show what each color represents.

Leave room for "potting soil" bars.

Designing an Experiment

Use this section when designing or conducting an experiment.

Determining a Purpose

You can find a purpose for an experiment by doing research, by examining the results of a previous experiment, or by observing the world around you. An **experiment** is an organized procedure to study something under controlled conditions.

> Don't forget to learn as much as possible about your topic before you begin.

1. Write the purpose of your experiment as a question or problem that you want to investigate.

2. Write down research questions and begin searching for information that will help you design an experiment. Consult the library, the Internet, and other people as you conduct your research.

EXAMPLE

Middle school students observed an odor near the lake by their school. They also noticed that the water on the side of the lake near the school was greener than the water on the other side of the lake. The students did some research to learn more about their observations. They discovered that the odor and green color in the lake  came from algae. They also discovered that a new fertilizer was being used on a field nearby. The students inferred that the use of the fertilizer might be related to the presence of the algae and designed a controlled experiment to find out whether they were right.

> **Problem**
>
> How does fertilizer affect the presence of algae in a lake?
>
> **Research Questions**
>
> - Have other experiments been done on this problem? If so, what did those experiments show?
> - What kind of fertilizer is used on the field? How much?
> - How do algae grow?
> - How do people measure algae?
> - Can fertilizer and algae be used safely in a lab? How?

> **Research**
> As you research, you may find a topic that is more interesting to you than your original topic, or learn that a procedure you wanted to use is not practical or safe. It is OK to change your purpose as you research.

LAB HANDBOOK

Writing a Hypothesis

A **hypothesis** is a tentative explanation for an observation or scientific problem that can be tested by further investigation. You can write your hypothesis in the form of an "If . . . , then . . . , because . . ." statement.

Hypothesis

If the amount of fertilizer in lake water is increased, then the amount of algae will also increase, because fertilizers provide nutrients that algae need to grow.

Hypotheses
For help with hypotheses, refer to page R3.

Determining Materials

Make a list of all the materials you will need to do your experiment. Be specific, especially if someone else is helping you obtain the materials. Try to think of everything you will need.

Materials

- 1 large jar or container
- 4 identical smaller containers
- rubber gloves that also cover the arms
- sample of fertilizer-and-water solution
- eyedropper
- clear plastic wrap
- scissors
- masking tape
- marker
- ruler

Determining Variables and Constants

EXPERIMENTAL GROUP AND CONTROL GROUP

An experiment to determine how two factors are related always has two groups—a control group and an experimental group.

1. Design an experimental group. Include as many trials as possible in the experimental group in order to obtain reliable results.

2. Design a control group that is the same as the experimental group in every way possible, except for the factor you wish to test.

Experimental Group: two containers of lake water with one drop of fertilizer solution added to each

Control Group: two containers of lake water with no fertilizer solution added

Go back to your materials l[ist] and make sure you have enough items listed to cove[r] both your experimental gro[up] and your control group.

VARIABLES AND CONSTANTS

Identify the variables and constants in your experiment. In a controlled experiment, a **variable** is any factor that can change. **Constants** are all of the factors that are the same in both the experimental group and the control group.

1. Read your hypothesis. The **independent variable** is the factor that you wish to test and that is manipulated or changed so that it can be tested. The independent variable is expressed in your hypothesis after the word *if*. Identify the independent variable in your laboratory report.

2. The **dependent variable** is the factor that you measure to gather results. It is expressed in your hypothesis after the word *then*. Identify the dependent variable in your laboratory report.

Hypothesis
If the amount of fertilizer in lake water is increased, then the amount of algae will also increase, because fertilizers provide nutrients that algae need to grow.

Table 1. Variables and Constants in Algae Experiment

Independent Variable	Dependent Variable	Constants
Amount of fertilizer in lake water	Amount of algae that grow	• Where the lake water is obtained • Type of container used • Light and temperature conditions where water will be stored

Set up your experiment so that you will test only one variable.

MEASURING THE DEPENDENT VARIABLE

Before starting your experiment, you need to define how you will measure the dependent variable. An **operational definition** is a description of the one particular way in which you will measure the dependent variable.

Your operational definition is important for several reasons. First, in any experiment there are several ways in which a dependent variable can be measured. Second, the procedure of the experiment depends on how you decide to measure the dependent variable. Third, your operational definition makes it possible for other people to evaluate and build on your experiment.

EXAMPLE 1

An operational definition of a dependent variable can be qualitative. That is, your measurement of the dependent variable can simply be an observation of whether a change occurs as a result of a change in the independent variable. This type of operational definition can be thought of as a "yes or no" measurement.

Table 2. Qualitative Operational Definition of Algae Growth

Independent Variable	Dependent Variable	Operational Definition
Amount of fertilizer in lake water	Amount of algae that grow	Algae grow in lake water

A qualitative measurement of a dependent variable is often easy to make and record. However, this type of information does not provide a great deal of detail in your experimental results.

EXAMPLE 2

An operational definition of a dependent variable can be quantitative. That is, your measurement of the dependent variable can be a number that shows how much change occurs as a result of a change in the independent variable.

Table 3. Quantitative Operational Definition of Algae Growth

Independent Variable	Dependent Variable	Operational Definition
Amount of fertilizer in lake water	Amount of algae that grow	Diameter of largest algal growth (in mm)

A quantitative measurement of a dependent variable can be more difficult to make and analyze than a qualitative measurement. However, this type of data provides much more information about your experiment and is often more useful.

Writing a Procedure

Write each step of your procedure. Start each step with a verb, or action word, and keep the steps short. Your procedure should be clear enough for someone else to use as instructions for repeating your experiment.

> If necessary, go back to your materials list and add any materials that you left out.

> **Controlling Variables**
> The same amount of fertilizer solution must be added to two of the four containers.

> **Controlling Variables**
> All four containers must receive the same amount of light.

Procedure

1. Put on your gloves. Use the large container to obtain a sample of lake water.

2. Divide the sample of lake water equally among the four smaller containers.

3. Use the eyedropper to add one drop of fertilizer solution to two of the containers.

4. Use the masking tape and the marker to label the containers with your initials, the date, and the identifiers "Jar 1 with Fertilizer," "Jar 2 with Fertilizer," "Jar 1 without Fertilizer," and "Jar 2 without Fertilizer."

5. Cover the containers with clear plastic wrap. Use the scissors to punch ten holes in each of the covers.

6. Place all four containers on a window ledge. Make sure that they all receive the same amount of light.

7. Observe the containers every day for one week.

8. Use the ruler to measure the diameter of the largest clump of algae in each container, and record your measurements daily.

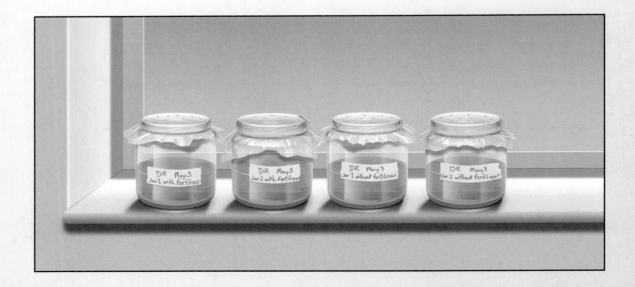

Recording Observations

Once you have obtained all of your materials and your procedure has been approved, you can begin making experimental observations. Gather both quantitative and qualitative data. If something goes wrong during your procedure, make sure you record that too.

> **Observations**
> For help with making qualitative and quantitative observations, refer to page R2.

> For more examples of data tables, see page R23.

Table 4. Fertilizer and Algae Growth

Date and Time	Experimental Group		Control Group		Observations
	Jar 1 with Fertilizer (diameter of algae in mm)	Jar 2 with Fertilizer (diameter of algae in mm)	Jar 1 without Fertilizer (diameter of algae in mm)	Jar 2 without Fertilizer (diameter of algae in mm)	
5/3 4:00 P.M.	0	0	0	0	condensation in all containers
5/4 4:00 P.M.	0	3	0	0	tiny green blobs in jar 2 with fertilizer
5/5 4:15 P.M.	4	5	0	3	green blobs in jars 1 and 2 with fertilizer and jar 2 without fertilizer
5/6 4:00 P.M.	5	6	0	4	water light green in jar 2 with fertilizer
5/7 4:00 P.M.	8	10	0	6	water light green in jars 1 and 2 with fertilizer and in jar 2 without fertilizer
5/8 3:30 P.M.	10	18	0	6	cover off jar 2 with fertilizer
5/9 3:30 P.M.	14	23	0	8	drew sketches of each container

> Notice that on the sixth day, the observer found that the cover was off one of the containers. It is important to record observations of unintended factors because they might affect the results of the experiment.

> Use technology, such as a microscope, to help you make observations when possible.

Drawings of Samples Viewed Under Microscope on 5/9 at 100x

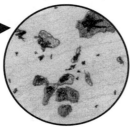

Jar 1 with Fertilizer

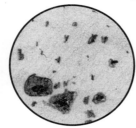

Jar 2 with Fertilizer

Jar 1 without Fertilizer

Jar 2 without Fertilizer

Summarizing Results

To summarize your data, look at all of your observations together. Look for meaningful ways to present your observations. For example, you might average your data or make a graph to look for patterns. When possible, use spreadsheet software to help you analyze and present your data. The two graphs below show the same data.

EXAMPLE 1

Always include a number and a title with a graph.

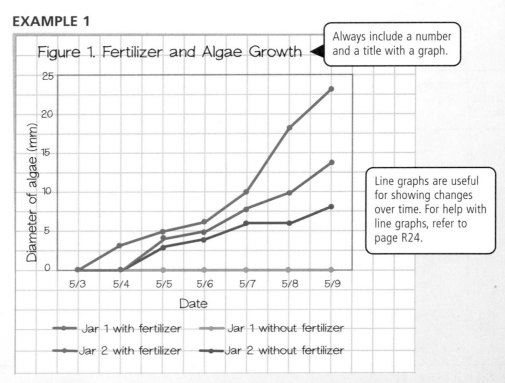

Figure 1. Fertilizer and Algae Growth

Line graphs are useful for showing changes over time. For help with line graphs, refer to page R24.

Bar graphs are useful for comparing different data sets. This bar graph has four bars for each day. Another way to present the data would be to calculate averages for the tests and the controls, and to show one test bar and one control bar for each day.

EXAMPLE 2

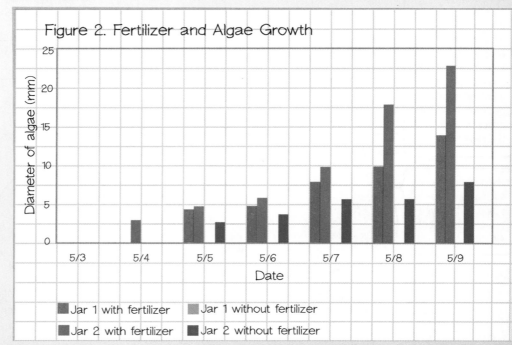

Figure 2. Fertilizer and Algae Growth

Drawing Conclusions

RESULTS AND INFERENCES

To draw conclusions from your experiment, first write your results. Then compare your results with your hypothesis. Do your results support your hypothesis? Be careful not to make inferences about factors that you did not test.

> For help with making inferences, see page R4.

Results and Inferences

The results of my experiment show that more algae grew in lake water to which fertilizer had been added than in lake water to which no fertilizer had been added. My hypothesis was supported. I infer that it is possible that the growth of algae in the lake was caused by the fertilizer used on the field.

> Notice that you cannot conclude from this experiment that the presence of algae in the lake was due only to the fertilizer.

QUESTIONS FOR FURTHER RESEARCH

Write a list of questions for further research and investigation. Your ideas may lead you to new experiments and discoveries.

Questions for Further Research

- What is the connection between the amount of fertilizer and algae growth?
- How do different brands of fertilizer affect algae growth?
- How would algae growth in the lake be affected if no fertilizer were used on the field?
- How do algae affect the lake and the other life in and around it?
- How does fertilizer affect the lake and the life in and around it?
- If fertilizer is getting into the lake, how is it getting there?

Math Handbook

Describing a Set of Data

Means, medians, modes, and ranges are important math tools for describing data sets such as the following widths of fossilized clamshells.

13 mm   25 mm   14 mm   21 mm   16 mm   23 mm   14 mm

Mean

The **mean** of a data set is the sum of the values divided by the number of values.

Example

To find the mean of the clamshell data, add the values and then divide the sum by the number of values.

$$\frac{13 \text{ mm} + 25 \text{ mm} + 14 \text{ mm} + 21 \text{ mm} + 16 \text{ mm} + 23 \text{ mm} + 14 \text{ mm}}{7} = \frac{126 \text{ mm}}{7} = 18 \text{ mm}$$

ANSWER The mean is 18 mm.

Median

The **median** of a data set is the middle value when the values are written in numerical order. If a data set has an even number of values, the median is the mean of the two middle values.

Example

To find the median of the clamshell data, arrange the values in order from least to greatest. The median is the middle value.

13 mm   14 mm   14 mm   16 mm   21 mm   23 mm   25 mm

ANSWER The median is 16 mm.

Mode

The **mode** of a data set is the value that occurs most often.

Example

To find the mode of the clamshell data, arrange the values in order from least to greatest and determine the value that occurs most often.

13 mm   14 mm   14 mm   16 mm   21 mm   23 mm   25 mm

ANSWER The mode is 14 mm.

A data set can have more than one mode or no mode. For example, the following data set has modes of 2 mm and 4 mm:

2 mm   2 mm   3 mm   4 mm   4 mm

The data set below has no mode, because no value occurs more often than any other.

2 mm   3 mm   4 mm   5 mm

Range

The **range** of a data set is the difference between the greatest value and the least value.

Example

To find the range of the clamshell data, arrange the values in order from least to greatest.

13 mm   14 mm   14 mm   16 mm   21 mm   23 mm   25 mm

Subtract the least value from the greatest value.

13 mm is the least value.

25 mm is the greatest value.

25 mm − 13 mm = 12 mm

ANSWER The range is 12 mm.

Using Ratios, Rates, and Proportions

You can use ratios and rates to compare values in data sets. You can use proportions to find unknown values.

Ratios

A **ratio** uses division to compare two values. The ratio of a value a to a nonzero value b can be written as $\frac{a}{b}$.

Example

The height of one plant is 8 centimeters. The height of another plant is 6 centimeters. To find the ratio of the height of the first plant to the height of the second plant, write a fraction and simplify it.

$$\frac{8 \text{ cm}}{6 \text{ cm}} = \frac{4 \times \overset{1}{\cancel{2}}}{3 \times \underset{1}{\cancel{2}}} = \frac{4}{3}$$

ANSWER The ratio of the plant heights is $\frac{4}{3}$.

You can also write the ratio $\frac{a}{b}$ as "a to b" or as $a:b$. For example, you can write the ratio of the plant heights as "4 to 3" or as 4:3.

Rates

A **rate** is a ratio of two values expressed in different units. A unit rate is a rate with a denominator of 1 unit.

Example

A plant grew 6 centimeters in 2 days. The plant's rate of growth was $\frac{6 \text{ cm}}{2 \text{ days}}$. To describe the plant's growth in centimeters per day, write a unit rate.

Divide numerator and denominator by 2: $\quad \dfrac{6 \text{ cm}}{2 \text{ days}} = \dfrac{6 \text{ cm} \div 2}{2 \text{ days} \div 2}$

You divide 2 days by 2 to get 1 day, so divide 6 cm by 2 also.

Simplify: $\quad = \dfrac{3 \text{ cm}}{1 \text{ day}}$

ANSWER The plant's rate of growth is 3 centimeters per day.

Proportions

A **proportion** is an equation stating that two ratios are equivalent. To solve for an unknown value in a proportion, you can use cross products.

Example

If a plant grew 6 centimeters in 2 days, how many centimeters would it grow in 3 days (if its rate of growth is constant)?

Write a proportion:	$\dfrac{6 \text{ cm}}{2 \text{ days}} = \dfrac{x \text{ cm}}{3 \text{ days}}$
Set cross products:	$6 \cdot 3 = 2x$
Multiply 6 and 3:	$18 = 2x$
Divide each side by 2:	$\dfrac{18}{2} = \dfrac{2x}{2}$
Simplify:	$9 = x$

ANSWER The plant would grow 9 centimeters in 3 days.

Using Decimals, Fractions, and Percents

Decimals, fractions, and percentages are all ways of recording and representing data.

Decimals

A **decimal** is a number that is written in the base-ten place value system, in which a decimal point separates the ones and tenths digits. The values of each place is ten times that of the place to its right.

Example

A caterpillar traveled from point *A* to point *C* along the path shown.

A   **36.9 cm**   **B**   **52.4 cm**   C

ADDING DECIMALS To find the total distance traveled by the caterpillar, add the distance from *A* to *B* and the distance from *B* to *C*. Begin by lining up the decimal points. Then add the figures as you would whole numbers and bring down the decimal point.

```
  36.9 cm
+ 52.4 cm
  89.3 cm
```

ANSWER The caterpillar traveled a total distance of 89.3 centimeters.

Example continued

SUBTRACTING DECIMALS To find how much farther the caterpillar traveled on the second leg of the journey, subtract the distance from *A* to *B* from the distance from *B* to *C*.

$$\begin{array}{r} 52.4 \text{ cm} \\ - 36.9 \text{ cm} \\ \hline 15.5 \text{ cm} \end{array}$$

ANSWER The caterpillar traveled 15.5 centimeters farther on the second leg of the journey.

Example

A caterpillar is traveling from point *D* to point *F* along the path shown. The caterpillar travels at a speed of 9.6 centimeters per minute.

D   E   33.6 cm   F

MULTIPLYING DECIMALS You can multiply decimals as you would whole numbers. The number of decimal places in the product is equal to the sum of the number of decimal places in the factors.

For instance, suppose it takes the caterpillar 1.5 minutes to go from *D* to *E*. To find the distance from *D* to *E*, multiply the caterpillar's speed by the time it took.

9.6	1 decimal place
× 1.5	+ 1 decimal place
480	
96	
14.40	2 decimal places

Align as shown.

ANSWER The distance from *D* to *E* is 14.4 centimeters.

DIVIDING DECIMALS When you divide by a decimal, move the decimal points the same number of places in the divisor and the dividend to make the divisor a whole number.

For instance, to find the time it will take the caterpillar to travel from *E* to *F*, divide the distance from *E* to *F* by the caterpillar's speed.

$$9.6 \overline{)33.6}$$

Move each decimal point one place to the right.

$$\begin{array}{r} 3.5 \\ 96 \overline{)336.} \\ \underline{288} \\ 480 \\ \underline{480} \\ 0 \end{array}$$

Line up decimal points.

ANSWER The caterpillar will travel from *E* to *F* in 3.5 minutes.

Fractions

A **fraction** is a number in the form $\frac{a}{b}$, where b is not equal to 0. A fraction is in **simplest form** if its numerator and denominator have a greatest common factor (GCF) of 1. To simplify a fraction, divide its numerator and denominator by their GCF.

Example

A caterpillar is 40 millimeters long. The head of the caterpillar is 6 millimeters long. To compare the length of the caterpillar's head with the caterpillar's total length, you can write and simplify a fraction that expresses the ratio of the two lengths.

$$\text{Write the ratio of the two lengths:} \quad \frac{\text{Length of head}}{\text{Total length}} = \frac{6 \text{ mm}}{40 \text{ mm}}$$

$$\text{Write numerator and denominator as products of numbers and the GCF:} \quad = \frac{3 \times 2}{20 \times 2}$$

$$\text{Divide numerator and denominator by the GCF:} \quad = \frac{3 \times \overset{1}{\cancel{2}}}{20 \times \underset{1}{\cancel{2}}}$$

$$\text{Simplify:} \quad = \frac{3}{20}$$

ANSWER In simplest form, the ratio of the lengths is $\frac{3}{20}$.

Percents

A **percent** is a ratio that compares a number to 100. The word *percent* means "per hundred" or "out of 100." The symbol for *percent* is %.

For instance, suppose 43 out of 100 caterpillars are female. You can represent this ratio as a percent, a decimal, or a fraction.

Percent	Decimal	Fraction
43%	0.43	$\frac{43}{100}$

Example

In the preceding example, the ratio of the length of the caterpillar's head to the caterpillar's total length is $\frac{3}{20}$. To write this ratio as a percent, write an equivalent fraction that has a denominator of 100.

$$\text{Multiply numerator and denominator by 5:} \quad \frac{3}{20} = \frac{3 \times 5}{20 \times 5}$$

$$= \frac{15}{100}$$

$$\text{Write as a percent:} \quad = 15\%$$

ANSWER The caterpillar's head represents 15 percent of its total length.

Using Formulas

A mathematical **formula** is a statement of a fact, rule, or principle. It is usually expressed as an equation.

In science, a formula often has a word form and a symbolic form. The formula below expresses Ohm's law.

The term *variable* is also used in science to refer to a factor that can change during an experiment.

Word Form

$$\text{Current} = \frac{\text{voltage}}{\text{resistance}}$$

Symbolic Form

$$I = \frac{V}{R}$$

In this formula, I, V, and R are variables. A mathematical **variable** is a symbol or letter that is used to represent one or more numbers.

MATH HANDBOOK

Example

Suppose that you measure a voltage of 1.5 volts and a resistance of 15 ohms. You can use the formula for Ohm's law to find the current in amperes.

Write the formula for Ohm's law: $\quad I = \dfrac{V}{R}$

Substitute 1.5 volts for V *and 15 ohms for* R: $\quad I = \dfrac{1.5 \text{ volts}}{15 \text{ ohms}}$

Simplify: $\quad I = 0.1 \text{ amp}$

ANSWER The current is 0.1 ampere.

If you know the values of all variables but one in a formula, you can solve for the value of the unknown variable. For instance, Ohm's law can be used to find a voltage if you know the current and the resistance.

Example

Suppose that you know that a current is 0.2 amperes and the resistance is 18 ohms. Use the formula for Ohm's law to find the voltage in volts.

Write the formula for Ohm's law: $\quad I = \dfrac{V}{R}$

Substitute 0.2 amp for I *and 18 ohms for* R: $\quad 0.2 \text{ amp} = \dfrac{V}{18 \text{ ohms}}$

Multiply both sides by 18 ohms: $\quad 0.2 \text{ amp} \cdot 18 \text{ ohms} = V$

Simplify: $\quad 3.6 \text{ volts} = V$

ANSWER The voltage is 3.6 volts.

Finding Areas

The area of a figure is the amount of surface the figure covers.

Area is measured in square units, such as square meters (m^2) or square centimeters (cm^2). Formulas for the areas of three common geometric figures are shown below.

Area = (side length)^2
$A = s^2$

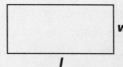

Area = length × width
$A = lw$

Area = $\frac{1}{2}$ × base × height
$A = \frac{1}{2} bh$

Example

Each face of a halite crystal is a square like the one shown. You can find the area of the square by using the steps below.

Write the formula for the area of a square: $A = s^2$

Substitute 3 mm for s: $= (3\text{ mm})^2$

Simplify: $= 9\text{ mm}^2$

3 mm

3 mm

ANSWER The area of the square is 9 square millimeters.

Finding Volumes

The volume of a solid is the amount of space contained by the solid.

Volume is measured in cubic units, such as cubic meters (m^3) or cubic centimeters (cm^3). The volume of a rectangular prism is given by the formula shown below.

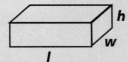

Volume = length × width × height
$V = lwh$

Example

A topaz crystal is a rectangular prism like the one shown. You can find the volume of the prism by using the steps below.

Write the formula for the volume of a rectangular prism: $V = lwh$

Substitute dimensions: $= 20\text{ mm} \times 12\text{ mm} \times 10\text{ mm}$

Simplify: $= 2400\text{ mm}^3$

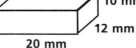

10 mm

12 mm

20 mm

ANSWER The volume of the rectangular prism is 2400 cubic millimeters.

Using Significant Figures

The **significant figures** in a decimal are the digits that are warranted by the accuracy of a measuring device.

When you perform a calculation with measurements, the number of significant figures to include in the result depends in part on the number of significant figures in the measurements. When you multiply or divide measurements, your answer should have only as many significant figures as the measurement with the fewest significant figures.

Example

Using a balance and a graduated cylinder filled with water, you determined that a marble has a mass of 8.0 grams and a volume of 3.5 cubic centimeters. To calculate the density of the marble, divide the mass by the volume.

Write the formula for density: $\text{Density} = \dfrac{\text{mass}}{\text{Volume}}$

Substitute measurements: $= \dfrac{8.0 \text{ g}}{3.5 \text{ cm}^3}$

Use a calculator to divide: $\approx 2.285714286 \text{ g/cm}^3$

ANSWER Because the mass and the volume have two significant figures each, give the density to two significant figures. The marble has a density of 2.3 grams per cubic centimeter.

Using Scientific Notation

Scientific notation is a shorthand way to write very large or very small numbers. For example, 73,500,000,000,000,000,000,000 kg is the mass of the Moon. In scientific notation, it is 7.35×10^{22} kg.

Example

You can convert from standard form to scientific notation.

Standard Form	Scientific Notation
720,000	7.2×10^5
5 decimal places left	Exponent is 5.
0.000291	2.91×10^{-4}
4 decimal places right	Exponent is −4.

You can convert from scientific notation to standard form.

Scientific Notation	Standard Form
4.63×10^7	46,300,000
Exponent is 7.	7 decimal places right
1.08×10^{-6}	0.00000108
Exponent is −6.	6 decimal places left

Note-Taking Handbook

Note-Taking Strategies

Taking notes as you read helps you understand the information. The notes you take can also be used as a study guide for later review. This handbook presents several ways to organize your notes.

Content Frame

1. Make a chart in which each column represents a category.
2. Give each column a heading.
3. Write details under the headings.

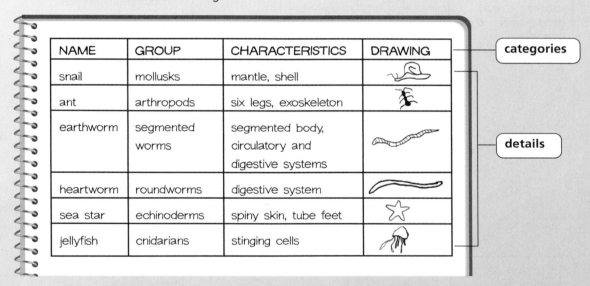

NAME	GROUP	CHARACTERISTICS	DRAWING
snail	mollusks	mantle, shell	
ant	arthropods	six legs, exoskeleton	
earthworm	segmented worms	segmented body, circulatory and digestive systems	
heartworm	roundworms	digestive system	
sea star	echinoderms	spiny skin, tube feet	
jellyfish	cnidarians	stinging cells	

categories

details

Combination Notes

1. For each new idea or concept, write an informal outline of the information.
2. Make a sketch to illustrate the concept, and label it.

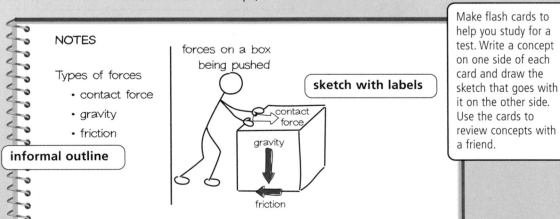

NOTES

Types of forces
- contact force
- gravity
- friction

informal outline

forces on a box being pushed

sketch with labels

contact force

gravity

friction

Make flash cards to help you study for a test. Write a concept on one side of each card and draw the sketch that goes with it on the other side. Use the cards to review concepts with a friend.

Main Idea and Detail Notes

1. In the left-hand column of a two-column chart, list main ideas. The blue headings express main ideas throughout this textbook.

2. In the right-hand column, write details that expand on each main idea.

You can shorten the headings in your chart. Be sure to use the most important words.

When studying for tests, cover up the detail notes column with a sheet of paper. Then use each main idea to form a question—such as "How does latitude affect climate?" Answer the question, and then uncover the detail notes column to check your answer.

MAIN IDEAS	DETAIL NOTES
1. Latitude affects climate. *main idea 1*	1. Places close to the equator are usually warmer than places close to the poles. 1. Latitude has the same effect in both hemispheres. *details about main idea*
2. Altitude affects climate. *main idea 2*	2. Temperature decreases with altitude. 2. Altitude can overcome the effect of latitude on temperature. *details about main idea*

Main Idea Web

1. Write a main idea in a box.

2. Add boxes around it with related vocabulary terms and important details.

You can find definitions near highlighted terms.

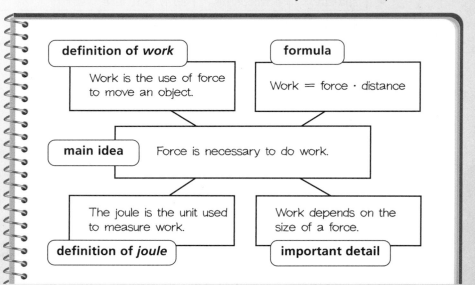

definition of *work*
Work is the use of force to move an object.

formula
Work = force · distance

main idea
Force is necessary to do work.

The joule is the unit used to measure work.

Work depends on the size of a force.

definition of *joule*

important detail

NOTE-TAKING HANDBOOK

Mind Map

1. Write a main idea in the center.
2. Add details that relate to one another and to the main idea.

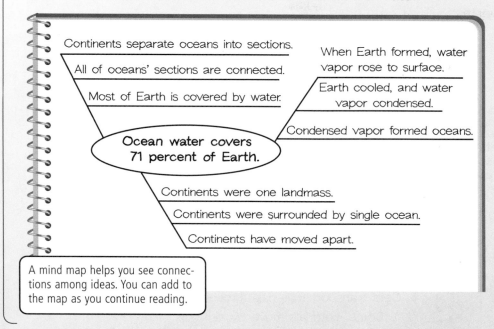

Continents separate oceans into sections.

All of oceans' sections are connected.

Most of Earth is covered by water.

When Earth formed, water vapor rose to surface.

Earth cooled, and water vapor condensed.

Condensed vapor formed oceans.

Ocean water covers 71 percent of Earth.

Continents were one landmass.

Continents were surrounded by single ocean.

Continents have moved apart.

A mind map helps you see connections among ideas. You can add to the map as you continue reading.

Supporting Main Ideas

1. Write a main idea in a box.
2. Add boxes underneath with information—such as reasons, explanations, and examples—that supports the main idea.

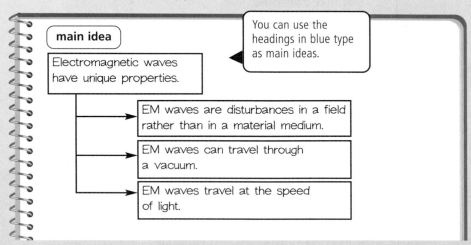

main idea

Electromagnetic waves have unique properties.

You can use the headings in blue type as main ideas.

EM waves are disturbances in a field rather than in a material medium.

EM waves can travel through a vacuum.

EM waves travel at the speed of light.

Outline

1. Copy the chapter title and headings from the book in the form of an outline.

2. Add notes that summarize in your own words what you read.

Cell Processes

I. Cells capture and release energy. — 1st key idea

   A. All cells need energy. — 1st subpoint of I

   B. Some cells capture light energy. — 2nd subpoint of I

      1. Process of photosynthesis — 1st detail about B

      2. Chloroplasts (site of photosynthesis) — 2nd detail about B

      3. Carbon dioxide and water as raw materials

      4. Glucose and oxygen as products

   C. All cells release energy.

      1. Process of cellular respiration

      2. Fermentation of sugar to carbon dioxide

      3. Bacteria that carry out fermentation

II. Cells transport materials through membranes.

   A. Some materials move by diffusion.

      1. Particle movement from higher to lower concentrations

      2. Movement of water through membrane (osmosis)

   B. Some transport requires energy.

      1. Active transport

      2. Examples of active transport

Correct Outline Form

Include a title.

Arrange key ideas, subpoints, and details as shown.

Indent the divisions of the outline as shown.

Use the same grammatical form for items of the same rank. For example, if A is a sentence, B must also be a sentence.

You must have at least two main ideas or subpoints. That is, every A must be followed by a B, and every 1 must be followed by a 2.

Concept Map

1. Write an important concept in a large oval.

2. Add details related to the concept in smaller ovals.

3. Write linking words on arrows that connect the ovals.

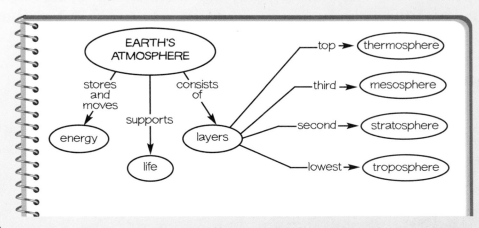

The main ideas or concepts can often be found in the blue headings. An example is "The atmosphere stores and moves energy." Use nouns from these concepts in the ovals, and use the verb or verbs on the lines.

Venn Diagram

1. Draw two overlapping circles, one for each item that you are comparing.

2. In the overlapping section, list the characteristics that are shared by both items.

3. In the outer sections, list the characteristics that are peculiar to each item.

4. Write a summary that describes the information in the Venn diagram.

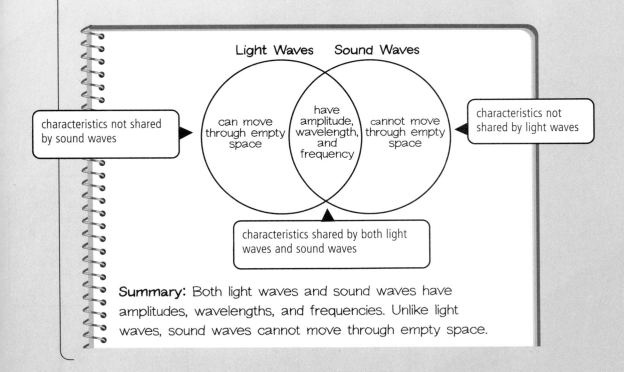

Summary: Both light waves and sound waves have amplitudes, wavelengths, and frequencies. Unlike light waves, sound waves cannot move through empty space.

Vocabulary Strategies

Important terms are highlighted in this book. A definition of each term can be found in the sentence or paragraph where the term appears. You can also find definitions in the Glossary. Taking notes about vocabulary terms helps you understand and remember what you read.

Description Wheel

1. Write a term inside a circle.
2. Write words that describe the term on "spokes" attached to the circle.

> When studying for a test with a friend, read the phrases on the spokes one at a time until your friend identifies the correct term.

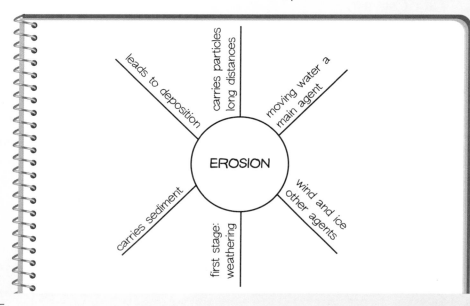

(Description Wheel, center: **EROSION**; spokes: leads to deposition, carries particles long distances, moving water a main agent, wind and ice other agents, first stage: weathering, carries sediment)

Four Square

1. Write a term in the center.
2. Write details in the four areas around the term.

Definition	Characteristics
any living thing	needs food, water, air; needs energy; grows, develops, reproduces

ORGANISM

Examples	Nonexamples
dogs, cats, birds, insects, flowers, trees	rocks, water, dirt

> Include a definition, some characteristics, and examples. You may want to add a formula, a sketch, or examples of things that the term does *not* name.

Frame Game

1. Write a term in the center.
2. Frame the term with details.

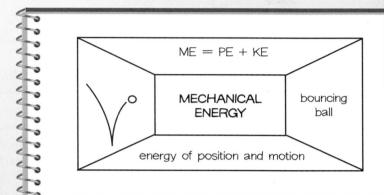

Include examples, descriptions, sketches, or sentences that use the term in context. Change the frame to fit each new term.

Magnet Word

1. Write a term on the magnet.
2. On the lines, add details related to the term.

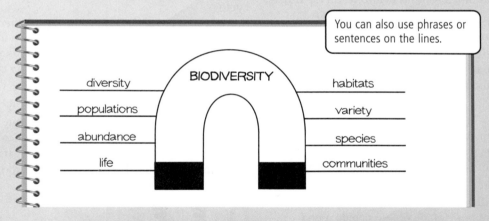

You can also use phrases or sentences on the lines.

Word Triangle

1. Write a term and its definition in the bottom section.
2. In the middle section, write a sentence in which the term is used correctly.
3. In the top section, draw a small picture to illustrate the term.

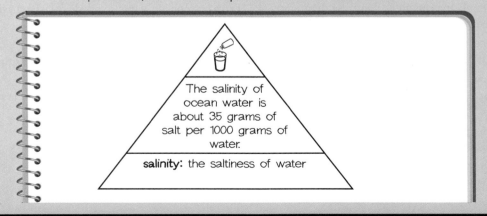

Glossary

A, B, C

absorption (uhb-SAWRP-shuhn)
The disappearance of a wave into a medium. When a wave is absorbed, the energy transferred by the wave is converted into another form of energy, usually thermal energy. (p. 93)

> **absorción** La desaparición de una onda dentro de un medio. Cuando se absorbe una onda, la energía transferida por la onda se convierte a otra forma de energía, normalmente a energía térmica.

acoustics (uh-KOO-stihks)
The scientific study of sound; the behavior of sound waves inside a space. (p. 55)

> **acústica** El estudio científico del sonido; el comportamiento de las ondas sonoras dentro de un espacio.

amplification
The strengthening of an electrical signal, often used to increase the intensity of a sound wave. (p. 55)

> **amplificación** El fortalecimiento de una señal eléctrica, a menudo se usa para aumentar la intensidad de una onda sonora.

amplitude
The maximum distance that a disturbance causes a medium to move from its rest position; the distance between a crest or trough of a wave and line through the center of a wave. (p. 17)

> **amplitud** La distancia máxima que se mueve un medio desde su posición de reposo debido a una perturbación; la distancia entre una cresta o valle de una onda y una línea que pasa por el centro de la onda.

atom
The smallest particle of an element that has the chemical properties of that element. (p. xv)

> **átomo** La partícula más pequeña de un elemento que tiene las propiedades químicas de ese elemento.

bioluminescence
The production of light by living organisms. (p. 89)

> **bioluminiscencia** La producción de luz por parte de organismos vivos.

compound
A substance made up of two or more different types of atoms bonded together.

> **compuesto** Una sustancia formada por dos o más diferentes tipos de átomos enlazados.

concave
Curved inward toward the center, like the inside of a spoon. (p. 116)

> **cóncavo** Dicho de una superficie con curvatura hacia dentro, como la parte interna de una cuchara.

convex
Curved outward, like the underside of a spoon. (p. 116)

> **convexo** Dicho de una superficie con curvatura hacia afuera, como la parte externa de una cuchara.

cornea (KAWR-nee-uh)
A transparent membrane that covers the eye. (p. 127)

> **córnea** Una membrana transparente que cubre el ojo.

crest
The highest point, or peak, of a wave. (p. 17)

> **cresta** El punto más alto, o el pico, de una onda.

cycle
n. A series of events or actions that repeat themselves regularly; a physical and/or chemical process in which one material continually changes locations and/or forms. Examples include the water cycle, the carbon cycle, and the rock cycle.

v. To move through a repeating series of events or actions.

> **ciclo** *s.* Una serie de eventos o acciones que se repiten regularmente; un proceso físico y/o químico en el cual un material cambia continuamente de lugar y/o forma. Ejemplos: el ciclo del agua, el ciclo del carbono y el ciclo de las rocas.

D, E, F

data
Information gathered by observation or experimentation that can be used in calculating or reasoning. *Data* is a plural word; the singular is datum.

> **datos** Información reunida mediante observación o experimentación y que se puede usar para calcular o para razonar.

decibel dB
The unit used to measure the intensity of a sound wave. (p. 52)

> **decibel** La unidad que se usa para medir la intensidad de una onda sonora.

density
A property of matter representing the mass per unit volume.

densidad Una propiedad de la materia que representa la masa por unidad de volumen.

diffraction
The spreading out of waves as they pass through an opening or around the edges of an obstacle. (p. 26)

difracción La dispersión de las ondas al pasar por una apertura o alrededor de los bordes de un obstáculo.

diffuse reflection
The reflection of parallel light rays in many different directions. (p. 114)

reflexión difusa La reflexión de rayos de luz paralelos en muchas direcciones diferentes.

Doppler effect
The change in perceived pitch that occurs when the source or the one who hears the sound is moving. (p. 50)

efecto Doppler El cambio en el tono percibido que ocurre cuando la fuente o el receptor de un sonido está en movimiento.

echolocation
The sending out of high-pitched sound waves and the interpretation of the returning echoes. (p. 59)

ecolocación El envío de ondas sonoras de tono alto y la interpretación de los ecos que regresan.

electromagnetic spectrum EM spectrum
The range of all electromagnetic frequencies, including the following types (from lowest to highest frequency): radio waves, microwaves, infrared light, visible light, ultraviolet light, x-rays, and gamma rays. (p. 80)

espectro electromagnético La escala de todas las frecuencias electromagnéticas, incluyendo los siguientes tipos (de la frecuencia más baja a la más alta): ondas de radio, microondas, luz infrarroja, luz visible, luz ultravioleta, rayos X y rayos gamma.

electromagnetic wave EM wave
A type of wave, such as a light wave or radio wave, that does not require a medium to travel; a disturbance that transfers energy through a field. (p. 73)

onda electromagnética Un tipo de onda, como una onda luminosa o de radio, que no requiere un medio para propagarse; una perturbación que transfiere energía a través de un campo.

element
A substance that cannot be broken down into a simpler substance by ordinary chemical changes. An element consists of atoms of only one type. (p. xv)

elemento Una sustancia que no puede descomponerse en otra sustancia más simple por medio de cambios químicos normales. Un elemento consta de átomos de un solo tipo.

energy
The ability to do work or to cause a change. For example, the energy of a moving bowling ball knocks over pins; energy from food allows animals to move and to grow; and energy from the Sun heats Earth's surface and atmosphere, which causes air to move. (p. xix)

energía La capacidad para trabajar o causar un cambio. Por ejemplo, la energía de una bola de boliche en movimiento tumba los pinos; la energía proveniente de su alimento permite a los animales moverse y crecer; la energía del Sol calienta la superficie y la atmósfera de la Tierra, lo que ocasiona que el aire se mueva.

experiment
An organized procedure to study something under controlled conditions. (p. xxiv)

experimento Un procedimiento organizado para estudiar algo bajo condiciones controladas.

fiber optics
Technology based on the use of laser light to send signals through transparent wires called optical fibers. This technology is often used in communications. (p. 137)

fibra óptica Tecnología basada en el uso de luz de láser para mandar señales por alambres transparentes llamados fibras ópticas. Esta tecnología se usa a menudo en comunicaciones.

field
An area around an object where the object can apply a force—such as gravitational force, magnetic force, or electrical force—on another object without touching it.

campo Un área alrededor de un objeto donde el objeto puede aplicar una fuerza, como fuerza gravitacional, fuerza magnética o fuerza eléctrica, sobre otro objeto sin tocarlo.

fluorescence (flu-REHS-uhns)
A phenomenon in which a material absorbs electromagnetic radiation of one wavelength and gives off electromagnetic radiation of a different wavelength. (p. 91)

fluorescencia Un fenómeno en el cual un material absorbe radiación electromagnética de una longitud de onda y emite radiación electromagnética de longitud de onda diferente.

focal length
The distance from the center of a convex lens to its focal point. (p. 123)

distancia focal La distancia del centro de un lente convexo a su punto focal.

focal point
The point at which parallel light rays reflected from a concave mirror come together; the point at which parallel light rays refracted by a convex lens come together. (p. 117)

punto focal El punto en el cual se unen los rayos paralelos de luz reflejados por un espejo cóncavo; el punto en el cual se unen los rayos paralelos de luz refractados por un lente convexo.

force
A push or a pull; something that changes the motion of an object. (p. xxi)

fuerza Un empuje o un jalón; algo que cambia el movimiento de un objeto.

friction
A force that resists the motion between two surfaces in contact. (p. xxi)

fricción Una fuerza que resiste el movimiento entre dos superficies en contacto.

frequency
The number of wavelengths (or wave crests) that pass a fixed point in a given amount of time, usually one second. (p. 17)

frecuencia El número de longitudes de onda (o crestas de onda) que pasan un punto fijo en un período de tiempo determinado, normalmente un segundo.

G, H, I

gamma rays
Part of the electromagnetic spectrum that consists of waves with the highest frequencies; electromagnetic waves with frequencies ranging from more than 10^{19} hertz to more than 10^{24} hertz. (p. 86)

rayos gamma Parte del espectro electromagnético que consiste de ondas con las frecuencias más altas; las ondas electromagnéticas con frecuencias de más de 10^{19} hertzios hasta más de 10^{24} hertzios.

gravity
The force that objects exert on each other because of their mass. (p. xxi)

gravedad La fuerza que los objetos ejercen entre sí debido a su masa.

hertz Hz
The unit used to measure frequency. One hertz is equal to one complete wavelength per second. (p. 46)

hercio La unidad usada para medir frecuencia. Un hercio es igual a una longitud de onda completa por segundo.

hypothesis
A tentative explanation for an observation or phenomenon. A hypothesis is used to make testable predictions. (p. xxiv)

hipótesis Una explicación provisional de una observación o de un fenómeno. Una hipótesis se usa para hacer predicciones que se pueden probar.

image
A picture of an object formed by rays of light. (p. 115)

imagen Reproducción de la figura de un objeto formada por rayos de luz.

incandescence (IHN-kuhn-DEHS-uhns)
1. The production of light by materials having high temperatures. (p. 89) 2. Light produced by an incandescent object.

incandescencia 1. La producción de luz por parte de materiales a altas temperaturas. 2. La luz producida por un objeto incandescente.

infrared light
Part of the electromagnetic spectrum that consists of waves with frequencies between those of microwaves and visible light. (p. 84)

luz infrarroja Parte del espectro electromagnético que consiste de ondas con frecuencias entre las de las microondas y las de la luz visible.

intensity
The amount of energy of a wave, per wavelength. Intensity is associated with the amplitude of a sound wave and with the quality of loudness produced by the sound wave. (p. 52)

intensidad La cantidad de energía de una onda sonora, por longitud de onda. La intensidad está asociada con la amplitud de una onda sonora y con la calidad del volumen producido por la onda sonora.

interference
The meeting and combining of waves; the adding or subtracting of wave amplitudes that occurs as waves overlap. (p. 27)

interferencia El encuentro y la combinación de ondas; la suma o la resta de amplitudes de onda que ocurre cuando las ondas se traslapan.

J, K, L

kinetic energy
The energy of motion. A moving object has the most kinetic energy at the point where it moves the fastest.

energía cinética La energía del movimiento. Un objeto que se mueve tiene su mayor energía cinética en el punto en el cual se mueve con mayor rapidez.

laser (LAY-zuhr)
A device that produces an intense, concentrated beam of light that can be brighter than sunlight. Lasers are often used in medicine and communications. (p. 135)

> **láser** Un aparato que produce un intenso rayo de luz concentrado que es más brillante que la luz del Sol. Los láseres se usan a menudo en la medicina y las comunicaciones.

law
In science, a rule or principle describing a physical relationship that always works in the same way under the same conditions. The law of conservation of energy is an example.

> **ley** En las ciencias, una regla o un principio que describe una relación física que siempre funciona de la misma manera bajo las mismas condiciones. La ley de la conservación de la energía es un ejemplo.

law of conservation of energy
A law stating that no matter how energy is transferred or transformed, it continues to exist in one form or another. (p. xix)

> **ley de la conservación de la energía** Una ley que establece que no importa cómo se transfiere o transforma la energía, toda la energía sigue presente en alguna forma u otra.

law of reflection
A law of physics stating that the angle at which light strikes a surface (the angle of incidence) equals the angle at which it reflects off the surface (the angle of reflection). (p. 114)

> **ley de la reflexión** Una ley de la física que establece que el ángulo al cual la luz incide sobre una superficie (el ángulo de incidencia) es igual al ángulo al cual se refleja (ángulo de reflexión) de la superficie.

lens
A transparent optical tool that refracts light. (p. 121)

> **lente** Una herramienta óptica transparente que refracta la luz.

longitudinal wave (LAHN-jih-TOOD-uhn-uhl)
A type of wave in which the disturbance moves in the same direction that the wave travels. (p. 14)

> **onda longitudinal** Un tipo de onda en la cual la perturbación se mueve en la misma dirección en la que viaja la onda.

luminescence
The production of light without the high temperatures needed for incandescence. (p. 89)

> **luminiscencia** La producción de luz sin las altas temperaturas necesarias para la incandescencia.

M, N, O

mass
A measure of how much matter an object is made of. (p. xv)

> **masa** Una medida de la cantidad de materia de la que está compuesto un objeto.

matter
Anything that has mass and volume. Matter exists ordinarily as a solid, a liquid, or a gas. (p. xv)

> **materia** Todo lo que tiene masa y volumen. Generalmente la materia existe como sólido, líquido o gas.

mechanical wave
A wave, such as a sound wave or a seismic wave, that transfers kinetic energy through matter. (p. 11)

> **onda mecánica** Una onda, como una onda sonora o una onda sísmica, que transfiere energía cinética a través de la materia.

medium
A substance through which a wave moves. (p. 11)

> **medio** Una sustancia a través de la cual se mueve una onda.

microwaves
Part of the electromagnetic spectrum that consists of waves with higher frequencies than radio waves, but lower frequencies than infrared waves. (p. 83)

> **microondas** Parte del espectro electromagnético que consiste de ondas con frecuencias mayores a las ondas de radio, pero menores a las de las ondas infrarrojas.

molecule
A group of atoms that are held together by covalent bonds so that they move as a single unit. (p. xv)

> **molécula** Un grupo de átomos que están unidos mediante enlaces covalentes de tal manera que se mueven como una sola unidad.

optics (AHP-tihks)
The study of light, vision, and related technology. (p. 113)

> **óptica** El estudio de la luz, la visión y la tecnología relacionada a ellas.

P, Q, R

pitch
The quality of highness or lowness of a sound. Pitch is associated with the frequency of a sound wave—the higher the frequency, the higher the pitch. (p. 45)

> **tono** La cualidad de un sonido de ser alto o bajo. El tono está asociado con la frecuencia de una onda sonora: entre más alta sea la frecuencia, más alto es el tono.

polarization (POH-luhr-ih-ZAY-shuhn)
A way of filtering light so that all of the waves vibrate in the same direction. (p. 96)

polarización Una manera de filtrar la luz para que todas las ondas vibren en la misma dirección.

primary colors
Three colors of light—red, green, and blue—that can be mixed to produce all possible colors. (p. 98)

colores primarios Tres colores de luz, rojo, verde y azul, que se pueden mezclar para producir todos los colores posibles.

primary pigments
Three colors of substances—cyan, yellow, and magenta—that can be mixed to produce all possible colors. (p. 99)

pigmentos primarios Tres colores de sustancias, cian, amarillo y magenta, que se pueden mezclar para producir todos los colores posibles.

prism
An optical tool that uses refraction to separate the different wavelengths that make up white light. (p. 97)

prisma Una herramienta óptica que usa la refracción para separar las diferentes longitudes de onda que componen la luz blanca.

pupil
The circular opening in the iris of the eye that controls how much light enters the eye. (p. 127)

pupila La apertura circular en el iris del ojo que controla cuánta luz entra al ojo.

radiation (RAY-dee-AY-shuhn)
Energy that travels across distances in the form of electromagnetic waves. (p. 75)

radiación Energía que viaja a través de la distancia en forma de ondas electromagnéticas.

radio waves
The part of the electromagnetic spectrum that consists of waves with the lowest frequencies. (p. 82)

ondas de radio La parte del espectro electromagnético que consiste de las ondas con las frecuencias más bajas.

reflection
The bouncing back of a wave after it strikes a barrier. (p. 25)

reflexión El rebote de una onda después de que incide sobre una barrera.

refraction
The bending of a wave as it crosses the boundary between two mediums at an angle other than 90 degrees. (p. 25)

refracción El doblamiento de una onda a medida que cruza el límite entre dos medios a un ángulo distinto a 90 grados.

regular reflection
The reflection of parallel light rays in the same direction. (p. 114)

reflexión especular La reflexión de rayos de luz paralelos en la misma dirección.

resonance
The strengthening of a sound wave when it combines with an object's natural vibration. (p. 48)

resonancia El fortalecimiento de una onda sonora cuando se combina con la vibración natural de un objeto.

retina (REHT-uhn-uh)
A light-sensitive membrane at the back of the inside of the eye. (p. 127)

retina Una membrana sensible a la luz en la parte trasera del interior del ojo.

S, T, U

scattering
The spreading out of light rays in all directions as particles reflect and absorb the light. (p. 95)

dispersión La disipación de los rayos de luz en todas las direcciones a medida que las partículas reflejan y absorben la luz.

sonar
Instruments that use echolocation to locate objects underwater; acronym for "sound navigation and ranging." (p. 59)

sonar Instrumentos que usan la ecolocación para localizar objetos bajo agua; acrónimo en inglés para "navegación y determinación de distancias por sonido".

sound
A type of wave that is produced by a vibrating object and that travels through matter. (p. 37)

sonido Un tipo de onda que es producida por un objeto que vibra y que viaja a través de la materia.

system
A group of objects or phenomena that interact. A system can be as simple as a rope, a pulley, and a mass. It also can be as complex as the interaction of energy and matter in the four spheres of the Earth system.

sistema Un grupo de objetos o fenómenos que interactúan. Un sistema puede ser algo tan sencillo como una cuerda, una polea y una masa. También puede ser algo tan complejo como la interacción de la energía y la materia en las cuatro esferas del sistema de la Tierra.

technology
The use of scientific knowledge to solve problems or engineer new products, tools, or processes.

tecnología El uso de conocimientos científicos para resolver problemas o para diseñar nuevos productos, herramientas o procesos.

theory
In science, a set of widely accepted explanations of observations and phenomena. A theory is a well-tested explanation that is consistent with all available evidence.

teoría En las ciencias, un conjunto de explicaciones de observaciones y fenómenos que es ampliamente aceptado. Una teoría es una explicación bien probada que es consecuente con la evidencia disponible.

transmission (trans-MIHSH-uhn)
The passage of a wave through a medium. (p. 93)

transmisión El paso de una onda a través de un medio.

transverse wave
A type of wave in which the disturbance moves at right angles, or perpendicular, to the direction in which the wave travels. (p. 13)

onda transversal Un tipo de onda en el cual la perturbación se mueve en ángulo recto, o perpendicularmente, a la dirección en la cual viaja la onda.

trough (trawf)
The lowest point, or valley, of a wave (p. 17)

valle El punto más bajo de una onda.

ultrasound
Sound waves with frequencies above 20,000 hertz, the upper limit of typical hearing levels in humans, used for medical purposes, among other things. (p. 46)

ultrasonido Ondas sonoras con frecuencias superiores a 20,000 hertzios, el límite superior de los niveles auditivos típicos de los humanos. Estas ondas tienen usos médicos, entre otros.

ultraviolet light
The part of the electromagnetic spectrum that consists of waves with frequencies higher than those of visible light and lower than those of x-rays. (p. 85)

luz ultravioleta La parte del espectro electromagnético que consiste de ondas con frecuencias superiores a las de luz visible y menores a las de los rayos X.

V, W

vacuum
A space containing few or no particles of matter. (p. 41)

vacío Un espacio que no contiene partículas de materia o bien contiene muy pocas.

variable
Any factor that can change in a controlled experiment, observation, or model. (p. R30)

variable Cualquier factor que puede cambiar en un experimento controlado, en una observación o en un modelo.

vibration
A rapid, back-and-forth motion. (p. 37)

vibración Un movimiento rápido hacia delante y hacia atrás.

visible light
The part of the electromagnetic spectrum that consists of waves detectable by the human eye. (p. 84)

luz visible La parte del espectro electromagnético que consiste de ondas detectables por el ojo humano.

volume
An amount of three-dimensional space, often used to describe the space that an object takes up. (p. xv)

volumen Una cantidad de espacio tridimensional; a menudo se usa este término para describir el espacio que ocupa un objeto.

wave
A disturbance that transfers energy from one place to another without requiring matter to move the entire distance. (p. 9)

onda Una perturbación que transfiere energía de un lugar a otro sin que sea necesario que la materia se mueva toda la distancia.

wavelength
The distance from one wave crest to the next crest; the distance from any part of one wave to the identical part of the next wave. (p. 17)

longitud de onda La distancia de una cresta de onda a la siguiente cresta; la distancia de cualquier parte de una onda a la parte idéntica de la siguiente onda.

X, Y, Z

x-rays
The part of the electromagnetic spectrum that consists of waves with high frequencies and high energies; electromagnetic waves with frequencies ranging from more than 10^{16} hertz to more than 10^{21} hertz. (p. 86)

rayos X La parte del espectro electromagnético que consiste de las ondas con altas frecuencias y altas energías; las ondas electromagnéticas con frecuencias de más de 10^{16} hertzios hasta más de 10^{21} hertzios.

Index

Page numbers for definitions are printed in **boldface** type.
Page numbers for illustrations, maps, and charts are printed in *italics*.

INDEX

E

ear, 39, *39*
ear canal, 39
eardrum, 39, *39*
earthquakes, 10, *10*, 11, *11*
echolocation, *59*, **59**
Edison, Thomas, 62, 63
electric field, 73–74, *74*
electromagnetic (EM) waves, 70–101, **73**
    See also laser, light
    artificial light, 90–92
    as a disturbance, 73–74
    formation of, 74, *74*
    frequencies, 78, 79, 81
    gamma rays, 81, *81*, **86**
    infrared light, 80, *80*, **84**, 84–85, *85*
    Internet activity, 71
    laser light and, 138
    light waves and materials, 93–99, 102
    measuring, 81
    microwaves, 77, *77*, 80, *80*, 83, **83**
    radio waves, 79, 80, *80*, 82, **82**
    sources of, 74
    spectrum, 80–81
    sunlight, 88–89, 102
    traits, 73–77, 102
    travel of, 75
    ultraviolet light, 81, *81*, 85, **85**
    uses of, 79–86, 102
    visible light, 81, *81*, **84**
    x-rays, 81, *81*, *86*, **86**
electromagnetic spectrum, *80–81*, **80–81**
elements, xv
energy, xviii, **xix**
    chemical, xviii
    conservation of, **xix**
    electrical, xviii
    forms, of, xviii
    kinetic, 11
    sources of, xix
energy transfer, 9–12
    EM waves, 76, 77
    laser, 138
    sound, 37, 41
evaluating, **R8**
    media claims, R8
evidence, collection of, xxiv
experimental group, R30
experiments, **xxiv**. *See also* lab
    conclusions, drawing, R35
    constants, determining, R30
    controlled, **R28**, R30
    designing, R28–R35
    hypothesis, writing, R29
    materials, determining, R29
    observations, recording, R33
    procedure, writing, R32

    purpose, determining, R28
    results, summarizing, R34
    variables, R30–R31, R32
exponents, 78, **R44**
eye. *See* human eye
eyepiece lens, 132, *133*
eyesight. *See* vision

F

fact, **R9**
    different from opinion, R9
farsightedness, 129, *129*, 130
faulty reasoning, **R7**
fiber optics, **137**, *137*
field, 73–74
filament, 91, *91*
fluorescence, **91**
fluorescent light bulbs, *91*, 91–92
FM waves, 82, *82*
focal length, *122*, **123**
focal point, *116*, **117**, *121*, *122*
forces, **xxi**
    contact, xxi
    electrical, xxi
    friction, **xxi**
    gravity, **xxi**
    magnetic, xxi
    physical, xx–xxi
    waves and, 10
formulas, **R42**
    wave speed, 20–21
fractions, **R41**
frequency, 16, *17*, **17**, 18, 30, 66
    Doppler effect and, 51
    electromagnetic, 78, 79
    natural, 48
    of sound waves, 45–51, *46*
    wavelength and, 18, *18*, 46
    fundamental tone, 49
friction, **xxi**
fuel cell, hydrogen, xxvii, *xxvii*

G, H

gamma rays, 80, *81*, **86**
graphs
    bar, R26, R31
    circle, R25
    double bar, R27
    line, R24, R34
    using, 57
    wave properties, 18, *19*
gravity, **xxi**, xxiv
hair cells, ear, 39, *39*, 56, *56*
halogen light bulbs, 91, *91*
hammer, 39, *39*

transparent materials, 94, *94*
transverse waves, **13,** 30
   graph of, 18, *19*
trough, **17,** 30
tsunamis, 29, *29*
tungsten, 91, *91*

U, V

ultrasound, 2–5, **46,** 47, 58–60
ultrasound scanner, 2–5, 60, *60*
ultraviolet light, 81, *81,* **85,** *85*
units of measurement
   decibel, **52**
   hertz, **46,** 47, 81
   wavelength, **17,** 30
vacuum, **41,** 75
   EM waves and, 75
variables, **R30,** R31, R32
   controlling, R17
   dependent, **R30,** R31
   independent, **R30**
vibration, **37,** 38–39, 40
visible light, 81, *81,* **84,** *135. See also* optics
vision, 126–130
   correction of, 129–130
   farsightedness, 129, *129,* 130
   formation of images, *127,* 128
   nearsightedness, 129, *129,* 130
vocabulary strategies, R50–R51
   description wheel, 36, *36,* 112, *112,* R50, *R50*
   four square, 8, *8,* 112, *112,* R50, *R50*
   frame game, 72, *72,* 112, *112,* R51, *R51*
   magnet word, R51, *R51*
   word triangle, R51, *R51*
vocal cords, 38, *38*
volume, xv, **R43**

W, X, Y, Z

water, **xvii**
   vapor, xvii
water waves. *See* waves, ocean
wavelength, 16, *17,* **17,** 30
   color and, 96–99, 100–101
   frequency and, 18, *18,* 46
waves, 6–29, **9**
   AM (amplitude modulation), 82, *82*
   amplitude, 16, *17,* **17**
   behavior, 24–28
   classification, 12–14
   compressional, 14
   diffraction, **26,** *26,* 26–27, *27,* 30
   electromagnetic, 70–101 (*see also* electromagnetic waves)
   energy and, 9–14
   FM (frequency modulation), 82, *82*

frequency, 16, *17,* **17,** 18, 30 (*see also* frequency)
height, 15, 16
Internet activity, 7
light (*see* light)
longitudinal, **14,** 18, *18,* 30
mechanical, **11,** 37, 41
medium (*see* medium)
model, *12*
ocean, 10, 12, 13, 15, 20–21
properties, 16–21
reflection, *25,* **25,** 30, 114–117, *114, 116, 117*
refraction, *25,* **25,** 30, 120, *120, 122*
rope, 9, 10, *10,* 13, *13*
sound, 14, 34–65 (*see also* sound; sound waves)
speed of, 16, 20–21, 44
transverse, **13,** 18, *19,* 30
wavelength, 16, *17,* **17,** 18 (*see also* wavelength)
wet mount, making a, R15, *R15*
white light, 96
wide-angle lens, 139, *139*
x-rays, 81, *81,* **86,** **86**

Acknowledgments

Photography

Cover © David Pu'u/Corbis; **i** © David Pu'u/Corbis; **iii** *left (top to bottom)* Photograph of James Trefil by Evan Cantwell; Photograph of Rita Ann Calvo by Joseph Calvo; Photograph of Linda Carnine by Amilcar Cifuentes; Photograph of Sam Miller by Samuel Miller; *right (top to bottom)* Photograph of Kenneth Cutler by Kenneth A. Cutler; Photograph of Donald Steely by Marni Stamm; Photograph of Vicky Vachon by Redfern Photographics; **vi** © Chip Simons/Getty Images; **vii** © Alan Kearney/Getty Images; **ix** Photographs by Sharon Hoogstraten; **xiv–xv** © Larry Hamill/age fotostock america, inc.; **xvi–xvii** © Fritz Poelking/age fotostock america, inc.; **xviii–xix** © Galen Rowell/Corbis; **xx–xxi** © Jack Affleck/SuperStock; **xxii** AP/Wide World Photos; **xxiii** © David Parker/IMI/University of Birmingham High, TC Consortium/Photo Researchers; **xxiv** *left* AP/Wide World Photos; *right Washington University Record;* **xxv** *top* © Kim Steele/Getty Images; *bottom* Reprinted with permission from S. Zhou et al., *SCIENCE* 291:1944–47. Copyright 2001 AAAS; **xxvi–xxvii** © Mike Fiala/Getty Images; **xxvii** *left* © Derek Trask/Corbis; *right* AP/Wide World Photos; **xxxii** © The Chedd-Angier Production Company; **2–3** © Paul Kuroda/SuperStock; **3** *left* © B. Benoit/Photo Researchers; *right* © Powerstock/SuperStock; **4** *top* © Stephen Frink/Corbis; *bottom* © The Chedd-Angier Production Company; **5** © George Stetten, M.D., Ph.D; **6–7** © Peter Sterling/Getty Images; **7, 9** Photographs by Sharon Hoogstraten; **11** Photograph courtesy of Earthquake Engineering Research Institute Reconnaissance Team; **12** © Michael Krasowitz/Getty Images; **13** Photograph by Sharon Hoogstraten; **15** © John Lund/Getty Images; **16** © Greg Huglin/Superstock; **17** © Arnulf Husmo/Getty Images; **19** Richard Olsenius/National Geographic Image Collection; **20** Photograph by Sharon Hoogstraten; **22** *top* © 1990 Robert Mathena/Fundamental Photographs, NYC; *bottom* Photographs by Sharon Hoogstraten; **23, 24** Photographs by Sharon Hoogstraten; **25** © 2001 Richard Megna/Fundamental Photographs, NYC; **26** *top* © 1972 FP/Fundamental Photographs, NYC; *bottom* Photograph by Sharon Hoogstraten; **27** © 1998 Richard Megna/Fundamental Photographs, NYC; **28** © Hiroshi Hara/Photonica; **29** Takaaki Uda, Public Works Research Institute, Japan/NOAA; **30** *bottom center* © 2001 Richard Megna/Fundamental Photographs, NYC; *bottom right* © 1972 FP/Fundamental Photographs, NYC; **34–35** © Chip Simons/Getty Images; **35, 37** Photographs by Sharon Hoogstraten; **39** © Susumu Nishinaga/Photo Researchers; **41** Photographs by Sharon Hoogstraten; **42** © Jeff Rotman/Getty Images; **43** © John Terence Turner/Getty Images; **44** *left* © Reuters NewMedia Inc./Corbis; *background* © Jason Hindley/Getty Images; **45** Photograph by Sharon Hoogstraten; **47** *left (top to bottom)* © Will Crocker/Getty Images; © Dorling Kindersley; © Photodisc/Getty Images; © Dorling Kindersley; © Photodisc/Getty Images; © Stephen Dalton/Animals Animals; © Steve Bloom/Getty Images; *top right* © Don Smetzer/Getty Images; *bottom right* Brian Gordon Green/National Geographic Image Collection; **48** Photograph by Sharon Hoogstraten; **49** © Dorling Kindersley; **50** © Michael Melford/Getty Images; **52** © Tom Main/Getty Images; **53** Photograph by Sharon Hoogstraten; **55** *left* © Roger Ressmeyer/Corbis; *right* Symphony Center, Home of the Chicago Symphony Orchestra; **56** © Yehoash Raphael, Kresge Hearing Research Institute, The University of Michigan; **57** © Chris Shinn/Getty Images; **58** Photograph by Sharon Hoogstraten; **59** *top left* © Stephen Dalton/OSF/Animals Animals; *top right* © Paulo de Oliveira/Getty Images; *bottom left* © AFP/Corbis; *bottom right* U.S. Navy photo by Photographer's Mate 3rd Class Lawrence Braxton/Department of Defense; **60** © Fetal Fotos; **63** © Andrew Syred/Photo Researchers; **64** *top left* © Reuters NewMedia Inc./Corbis; *bottom* Photographs by Sharon Hoogstraten; **65** Photograph by Sharon Hoogstraten; **66** *bottom left* © Stephen Dalton/OSF/Animals Animals; *bottom right* © Paulo de Oliveira/Getty Images; **68** © Photodisc/Getty Images; **70–71** © Alan Kearney/Getty Images; **71** *top, center* Photographs by Sharon Hoogstraten; *bottom* The EIT Consortium/NASA; **73** Photograph by Sharon Hoogstraten; **75** NASA, The Hubble Heritage Team, STScl, AURA; **76** Photograph by Sharon Hoogstraten; **78** *top* Palomar Observatory/Caltech; SAO; *center* NASA/MSFC/SAO; *bottom* NASA/CXC/ASU/J. Hester et al; *background* NASA/JHU/AUI/R. Giacconi et al.; **79** Photograph by Sharon Hoogstraten; **80** *left* © China Tourism Press/Getty Images; *center* © David Nunuk/Photo Researchers; *right* © Dr. Arthur Tucker/Photo Researchers; **81** *left to right* © Jeremy Woodhouse/Getty Images; © Sinclair Stammers/Photo Researchers; © Hugh Turvey/Photo Researchers; © Alfred Pasieka/Photo Researchers; **84** Photograph by Sharon Hoogstraten; **85** *top* © Dr. Arthur Tucker/Photo Researchers; *bottom* © Thomas Eisner, Cornell University; **86** © Martin Spinks; **87** © Photodisc/Getty Images; *inset* © David Young-Wolff/Getty Images; **88** Robert F. Sisson/National Geographic Image Collection; **89** © George D. Lepp/Corbis; **90** *top* © Raymond Blythe/OSF/Animals Animals; *bottom* Photograph by Sharon Hoogstraten; **92** © Traffic Technologies; **93** Photograph by Sharon Hoogstraten; **94** © Jeff Greenberg/Visuals Unlimited; **95** © Raymond Gehman/Corbis; **96** © Charles Swedlund; **97** *top* © Ace Photo Agency/Phototake; *bottom* © Dorling Kindersley; **98** Photograph by Sharon Hoogstraten; **100** *top* © Michael Newman/PhotoEdit; *bottom* Photographs by Sharon Hoogstraten; **101** Photographs by Sharon Hoogstraten; **102** *center right* Robert F. Sisson/National Geographic Image Collection; *bottom* © Ace Photo Agency/Phototake; **106** *top* The Granger Collection, New York; *bottom* © Jack and Beverly Wilgus; **107** *top* The Granger Collection, New York; *center left* Diagram of the eye from the *Opticae thesaurus. Alhazeni Arabis libri septem, nunc primum editi* by Ibn al-Haytham (Alhazen). Edited by Federico Risnero (Basleae, 1572), p. 6. Private collection, London; *center right* Courtesy of NASA/JPL/Caltech; *bottom* © Royal Greenwich Observatory/Photo Researchers; **108** *top* © Stock Connection/Alamy; *center* © Florian Marquardt; *bottom* © Museum of Holography, Chicago; **109** *top* © Bettmann/Corbis; *bottom* © Bob Masini/Phototake; **110–111** © Tom Raymond/Getty Images; **111** *top, center* Photographs by Sharon Hoogstraten; *bottom* © Philippe Plaily/Photo Researchers; **112** Photograph by Sharon Hoogstraten; **114** © Laura Dwight/Corbis; **115** Photograph by Sharon Hoogstraten; **116** © Michael Newman/PhotoEdit; **117** Photographs by Sharon Hoogstraten; **118** Peter McBride/Aurora; **119** Photograph by Sharon Hoogstraten; **120** © Richard H. Johnston/Getty Images; **122** © Kim Heacox/Getty Images; *background* © Photodisc/Getty Images; **123** © T. R. Tharp/Corbis; **124** *top* © Ruddy Gold/age photostock america, inc.; *bottom* Photograph by Sharon Hoogstraten; **125** Photographs by Sharon Hoogstraten; **126** © CMCD, 1994; **128** Photograph by Sharon Hoogstraten; **130** © Argentum/Photo Researchers; **131** Photograph by Sharon Hoogstraten; **133** *top* © Andrew Syred/Photo Researchers; *center, bottom* NASA; **134** Photograph by Sharon Hoogstraten; **135** Use of Canon Powershot S45 courtesy of Canon USA; **136** © Philippe Psaila/Photo Researchers; **137** *top* © Photodisc/Getty Images; *bottom* © Tom Stewart/corbisstockmarket.com; **138** Bradley C. Edwards, Ph.D.; **139** *top* © Photodisc/Getty Images; *center* © PhotoFlex.com; *bottom* © Michael Goldman/Photis/PictureQuest; **140** © Michael Newman/PhotoEdit; **R28** © Photodisc/Getty Images.

Illustrations

Accurate Art, Inc. 33; Ampersand Design Group 139; Argosy 10, 13, 14, 18, 19, 25, 30, 55, 61; Eric Chadwick 98, 99; Steve Cowden 12, 51, 54, 82; Dan Stuckenschneider 62, 77, 91, 102, 133, 135, 136, 140, R11–R19, R22, R32; Bart Vallecoccia 38, 39, 126, 127, 129, 135, 140.

Content Standards: 5–8

A. Science as Inquiry

As a result of activities in grades 5–8, all students should develop

Abilities Necessary to do Scientific Inquiry

A.1   Identify questions that can be answered through scientific investigations. Students should develop the ability to refine and refocus broad and ill-defined questions. An important aspect of this ability consists of students' ability to clarify questions and inquiries and direct them toward objects and phenomena that can be described, explained, or predicted by scientific investigations. Students should develop the ability to identify their questions with scientific ideas, concepts, and quantitative relationships that guide investigation.

A.2   Design and conduct a scientific investigation. Students should develop general abilities, such as systematic observation, making accurate measurements, and identifying and controlling variables. They should also develop the ability to clarify their ideas that are influencing and guiding the inquiry, and to understand how those ideas compare with current scientific knowledge. Students can learn to formulate questions, design investigations, execute investigations, interpret data, use evidence to generate explanations, propose alternative explanations, and critique explanations and procedures.

A.3   Use appropriate tools and techniques to gather, analyze, and interpret data. The use of tools and techniques, including mathematics, will be guided by the question asked and the investigations students design. The use of computers for the collection, summary, and display of evidence is part of this standard. Students should be able to access, gather, store, retrieve, and organize data, using hardware and software designed for these purposes.

A.4   Develop descriptions, explanations, predictions, and models using evidence. Students should base their explanation on what they observed, and as they develop cognitive skills, they should be able to differentiate explanation from description—providing causes for effects and establishing relationships based on evidence and logical argument. This standard requires a subject matter knowledge base so the students can effectively conduct investigations, because developing explanations establishes connections between the content of science and the contexts within which students develop new knowledge.

A.5   Think critically and logically to make the relationships between evidence and explanations. Thinking critically about evidence includes deciding what evidence should be used and accounting for anomalous data. Specifically, students should be able to review data from a simple experiment, summarize the data, and form a logical argument about the cause-and-effect relationships in the experiment. Students should begin to state some explanations in terms of the relationship between two or more variables.

A.6   Recognize and analyze alternative explanations and predictions. Students should develop the ability to listen to and respect the explanations proposed by other students. They should remain open to and acknowledge different ideas and explanations, be able to accept the skepticism of others, and consider alternative explanations.

A.7 Communicate scientific procedures and explanations. With practice, students should become competent at communicating experimental methods, following instructions, describing observations, summarizing the results of other groups, and telling other students about investigations and explanations.

A.8 Use mathematics in all aspects of scientific inquiry. Mathematics is essential to asking and answering questions about the natural world. Mathematics can be used to ask questions; to gather, organize, and present data; and to structure convincing explanations.

Understandings about Scientific Inquiry

A.9.a Different kinds of questions suggest different kinds of scientific investigations. Some investigations involve observing and describing objects, organisms, or events; some involve collecting specimens; some involve experiments; some involve seeking more information; some involve discovery of new objects and phenomena; and some involve making models.

A.9.b Current scientific knowledge and understanding guide scientific investigations. Different scientific domains employ different methods, core theories, and standards to advance scientific knowledge and understanding.

A.9.c Mathematics is important in all aspects of scientific inquiry.

A.9.d Technology used to gather data enhances accuracy and allows scientists to analyze and quantify results of investigations.

A.9.e Scientific explanations emphasize evidence, have logically consistent arguments, and use scientific principles, models, and theories. The scientific community accepts and uses such explanations until displaced by better scientific ones. When such displacement occurs, science advances.

A.9.f Science advances through legitimate skepticism. Asking questions and querying other scientists' explanations is part of scientific inquiry. Scientists evaluate the explanations proposed by other scientists by examining evidence, comparing evidence, identifying faulty reasoning, pointing out statements that go beyond the evidence, and suggesting alternative explanations for the same observations.

A.9.g Scientific investigations sometimes result in new ideas and phenomena for study, generate new methods or procedures for an investigation, or develop new technologies to improve the collection of data. All of these results can lead to new investigations.

B. Physical Science

As a result of their activities in grades 5–8, all students should develop an understanding of

Properties and Changes of Properties in Matter

B.1.a   A substance has characteristic properties, such as density, a boiling point, and solubility, all of which are independent of the amount of the sample. A mixture of substances often can be separated into the original substances using one or more of the characteristic properties.

B.1.b   Substances react chemically in characteristic ways with other substances to form new substances (compounds) with different characteristic properties. In chemical reactions, the total mass is conserved. Substances often are placed in categories or groups if they react in similar ways; metals is an example of such a group.

B.1.c   Chemical elements do not break down during normal laboratory reactions involving such treatments as heating, exposure to electric current, or reaction with acids. There are more than 100 known elements that combine in a multitude of ways to produce compounds, which account for the living and nonliving substances that we encounter.

Motions and Forces

B.2.a   The motion of an object can be described by its position, direction of motion, and speed. That motion can be measured and represented on a graph.

B.2.b   An object that is not being subjected to a force will continue to move at a constant speed and in a straight line.

B.2.c   If more than one force acts on an object along a straight line, then the forces will reinforce or cancel one another, depending on their direction and magnitude. Unbalanced forces will cause changes in the speed or direction of an object's motion.

Transfer of Energy

B.3.a   Energy is a property of many substances and is associated with heat, light, electricity, mechanical motion, sound, nuclei, and the nature of a chemical. Energy is transferred in many ways.

B.3.b   Heat moves in predictable ways, flowing from warmer objects to cooler ones, until both reach the same temperature.

B.3.c   Light interacts with matter by transmission (including refraction), absorption, or scattering (including reflection). To see an object, light from that object—emitted by or scattered from it—must enter the eye.

B.3.d   Electrical circuits provide a means of transferring electrical energy when heat, light, sound, and chemical changes are produced.

B.3.e   In most chemical and nuclear reactions, energy is transferred into or out of a system. Heat, light, mechanical motion, or electricity might all be involved in such transfers.

B.3.f   The sun is a major source of energy for changes on the earth's surface. The sun loses energy by emitting light. A tiny fraction of that light reaches the earth, transferring energy from the sun to the earth. The sun's energy arrives as light with a range of wavelengths, consisting of visible light, infrared, and ultraviolet radiation.

C. Life Science

As a result of their activities in grades 5–8, all students should develop understanding of

Structure and Function in Living Systems

C.1.a Living systems at all levels of organization demonstrate the complementary nature of structure and function. Important levels of organization for structure and function include cells, organs, tissues, organ systems, whole organisms, and ecosystems.

C.1.b All organisms are composed of cells—the fundamental unit of life. Most organisms are single cells; other organisms, including humans, are multicellular.

C.1.c Cells carry on the many functions needed to sustain life. They grow and divide, thereby producing more cells. This requires that they take in nutrients, which they use to provide energy for the work that cells do and to make the materials that a cell or an organism needs.

C.1.d Specialized cells perform specialized functions in multicellular organisms. Groups of specialized cells cooperate to form a tissue, such as a muscle. Different tissues are in turn grouped together to form larger functional units, called organs. Each type of cell, tissue, and organ has a distinct structure and set of functions that serve the organism as a whole.

C.1.e The human organism has systems for digestion, respiration, reproduction, circulation, excretion, movement, control, and coordination, and for protection from disease. These systems interact with one another.

C.1.f Disease is a breakdown in structures or functions of an organism. Some diseases are the result of intrinsic failures of the system. Others are the result of damage by infection by other organisms.

Reproduction and Heredity

C.2.a Reproduction is a characteristic of all living systems; because no individual organism lives forever, reproduction is essential to the continuation of every species. Some organisms reproduce asexually. Other organisms reproduce sexually.

C.2.b In many species, including humans, females produce eggs and males produce sperm. Plants also reproduce sexually—the egg and sperm are produced in the flowers of flowering plants. An egg and sperm unite to begin development of a new individual. That new individual receives genetic information from its mother (via the egg) and its father (via the sperm). Sexually produced offspring never are identical to either of their parents.

C.2.c Every organism requires a set of instructions for specifying its traits. Heredity is the passage of these instructions from one generation to another.

C.2.d Hereditary information is contained in genes, located in the chromosomes of each cell. Each gene carries a single unit of information. An inherited trait of an individual can be determined by one or by many genes, and a single gene can influence more than one trait. A human cell contains many thousands of different genes.

C.2.e The characteristics of an organism can be described in terms of a combination of traits. Some traits are inherited and others result from interactions with the environment.

Regulation and Behavior

C.3.a   All organisms must be able to obtain and use resources, grow, reproduce, and maintain stable internal conditions while living in a constantly changing external environment.

C.3.b   Regulation of an organism's internal environment involves sensing the internal environment and changing physiological activities to keep conditions within the range required to survive.

C.3.c   Behavior is one kind of response an organism can make to an internal or environmental stimulus. A behavioral response requires coordination and communication at many levels, including cells, organ systems, and whole organisms. Behavioral response is a set of actions determined in part by heredity and in part from experience.

C.3.d   An organism's behavior evolves through adaptation to its environment. How a species moves, obtains food, reproduces, and responds to danger are based in the species' evolutionary history.

Populations and Ecosystems

C.4.a   A population consists of all individuals of a species that occur together at a given place and time. All populations living together and the physical factors with which they interact compose an ecosystem.

C.4.b   Populations of organisms can be categorized by the function they serve in an ecosystem. Plants and some microorganisms are producers—they make their own food. All animals, including humans, are consumers, which obtain food by eating other organisms. Decomposers, primarily bacteria and fungi, are consumers that use waste materials and dead organisms for food. Food webs identify the relationships among producers, consumers, and decomposers in an ecosystem.

C.4.c   For ecosystems, the major source of energy is sunlight. Energy entering ecosystems as sunlight is transferred by producers into chemical energy through photosynthesis. That energy then passes from organism to organism in food webs.

C.4.d   The number of organisms an ecosystem can support depends on the resources available and abiotic factors, such as quantity of light and water, range of temperatures, and soil composition. Given adequate biotic and abiotic resources and no disease or predators, populations (including humans) increase at rapid rates. Lack of resources and other factors, such as predation and climate, limit the growth of populations in specific niches in the ecosystem.

Diversity and Adaptations of Organisms

C.5.a   Millions of species of animals, plants, and microorganisms are alive today. Although different species might look dissimilar, the unity among organisms becomes apparent from an analysis of internal structures, the similarity of their chemical processes, and the evidence of common ancestry.

C.5.b   Biological evolution accounts for the diversity of species developed through gradual processes over many generations. Species acquire many of their unique characteristics through biological adaptation, which involves the selection of naturally occurring variations in populations. Biological adaptations include changes in structures, behaviors, or physiology that enhance survival and reproductive success in a particular environment.

C.5.c   Extinction of a species occurs when the environment changes and the adaptive characteristics of a species are insufficient to allow its survival. Fossils indicate that many organisms that lived long ago are extinct. Extinction of species is common; most of the species that have lived on the earth no longer exist.

D. Earth and Space Science

As a result of their activities in grades 5–8, all students should develop an understanding of

Structure of the Earth System

D.1.a   The solid earth is layered with a lithosphere; hot, convecting mantle; and dense, metallic core.

D.1.b   Lithospheric plates on the scales of continents and oceans constantly move at rates of centimeters per year in response to movements in the mantle. Major geological events, such as earthquakes, volcanic eruptions, and mountain building, result from these plate motions.

D.1.c   Land forms are the result of a combination of constructive and destructive forces. Constructive forces include crustal deformation, volcanic eruption, and deposition of sediment, while destructive forces include weathering and erosion.

D.1.d   Some changes in the solid earth can be described as the "rock cycle." Old rocks at the earth's surface weather, forming sediments that are buried, then compacted, heated, and often recrystallized into new rock. Eventually, those new rocks may be brought to the surface by the forces that drive plate motions, and the rock cycle continues.

D.1.e   Soil consists of weathered rocks and decomposed organic material from dead plants, animals, and bacteria. Soils are often found in layers, with each having a different chemical composition and texture.

D.1.f   Water, which covers the majority of the earth's surface, circulates through the crust, oceans, and atmosphere in what is known as the "water cycle." Water evaporates from the earth's surface, rises and cools as it moves to higher elevations, condenses as rain or snow, and falls to the surface where it collects in lakes, oceans, soil, and in rocks underground.

D.1.g   Water is a solvent. As it passes through the water cycle it dissolves minerals and gases and carries them to the oceans.

D.1.h   The atmosphere is a mixture of nitrogen, oxygen, and trace gases that include water vapor. The atmosphere has different properties at different elevations.

D.1.i   Clouds, formed by the condensation of water vapor, affect weather and climate.

D.1.j   Global patterns of atmospheric movement influence local weather. Oceans have a major effect on climate, because water in the oceans holds a large amount of heat.

D.1.k   Living organisms have played many roles in the earth system, including affecting the composition of the atmosphere, producing some types of rocks, and contributing to the weathering of rocks.

Earth's History

D.2.a   The earth processes we see today, including erosion, movement of lithospheric plates, and changes in atmospheric composition, are similar to those that occurred in the past. Earth history is also influenced by occasional catastrophes, such as the impact of an asteroid or comet.

D.2.b   Fossils provide important evidence of how life and environmental conditions have changed.

Earth in the Solar System

D.3.a The earth is the third planet from the sun in a system that includes the moon, the sun, eight other planets and their moons, and smaller objects, such as asteroids and comets. The sun, an average star, is the central and largest body in the solar system.

D.3.b Most objects in the solar system are in regular and predictable motion. Those motions explain such phenomena as the day, the year, phases of the moon, and eclipses.

D.3.c Gravity is the force that keeps planets in orbit around the sun and governs the rest of the motion in the solar system. Gravity alone holds us to the earth's surface and explains the phenomena of the tides.

D.3.d The sun is the major source of energy for phenomena on the earth's surface, such as growth of plants, winds, ocean currents, and the water cycle. Seasons result from variations in the amount of the sun's energy hitting the surface, due to the tilt of the earth's rotation on its axis and the length of the day.

E. Science and Technology

As a result of activities in grades 5–8, all students should develop

Abilities of Technological Design

E.1 Identify appropriate problems for technological design. Students should develop their abilities by identifying a specified need, considering its various aspects, and talking to different potential users or beneficiaries. They should appreciate that for some needs, the cultural backgrounds and beliefs of different groups can affect the criteria for a suitable product.

E.2 Design a solution or product. Students should make and compare different proposals in the light of the criteria they have selected. They must consider constraints—such as cost, time, trade-offs, and materials needed—and communicate ideas with drawings and simple models.

E.3 Implement a proposed design. Students should organize materials and other resources, plan their work, make good use of group collaboration where appropriate, choose suitable tools and techniques, and work with appropriate measurement methods to ensure adequate accuracy.

E.4 Evaluate completed technological designs or products. Students should use criteria relevant to the original purpose or need, consider a variety of factors that might affect acceptability and suitability for intended users or beneficiaries, and develop measures of quality with respect to such criteria and factors; they should also suggest improvements and, for their own products, try proposed modifications.

E.5 Communicate the process of technological design. Students should review and describe any completed piece of work and identify the stages of problem identification, solution design, implementation, and evaluation.

Understandings about Science and Technology

E.6.a   Scientific inquiry and technological design have similarities and differences. Scientists propose explanations for questions about the natural world, and engineers propose solutions relating to human problems, needs, and aspirations. Technological solutions are temporary; technologies exist within nature and so they cannot contravene physical or biological principles; technological solutions have side effects; and technologies cost, carry risks, and provide benefits.

E.6.b   Many different people in different cultures have made and continue to make contributions to science and technology.

E.6.c   Science and technology are reciprocal. Science helps drive technology, as it addresses questions that demand more sophisticated instruments and provides principles for better instrumentation and technique. Technology is essential to science, because it provides instruments and techniques that enable observations of objects and phenomena that are otherwise unobservable due to factors such as quantity, distance, location, size, and speed. Technology also provides tools for investigations, inquiry, and analysis.

E.6.d   Perfectly designed solutions do not exist. All technological solutions have trade-offs, such as safety, cost, efficiency, and appearance. Engineers often build in back-up systems to provide safety. Risk is part of living in a highly technological world. Reducing risk often results in new technology.

E.6.e   Technological designs have constraints. Some constraints are unavoidable, for example, properties of materials, or effects of weather and friction; other constraints limit choices in the design, for example, environmental protection, human safety, and aesthetics.

E.6.f   Technological solutions have intended benefits and unintended consequences. Some consequences can be predicted, others cannot.

F. Science in Personal and Social Perspectives

As a result of activities in grades 5–8, all students should develop understanding of

Personal Health

F.1.a   Regular exercise is important to the maintenance and improvement of health. The benefits of physical fitness include maintaining healthy weight, having energy and strength for routine activities, good muscle tone, bone strength, strong heart/lung systems, and improved mental health. Personal exercise, especially developing cardiovascular endurance, is the foundation of physical fitness.

F.1.b   The potential for accidents and the existence of hazards imposes the need for injury prevention. Safe living involves the development and use of safety precautions and the recognition of risk in personal decisions. Injury prevention has personal and social dimensions.

F.1.c   The use of tobacco increases the risk of illness. Students should understand the influence of short-term social and psychological factors that lead to tobacco use, and the possible long-term detrimental effects of smoking and chewing tobacco.

F.1.d   Alcohol and other drugs are often abused substances. Such drugs change how the body functions and can lead to addiction.

F.1.e   Food provides energy and nutrients for growth and development. Nutrition requirements vary with body weight, age, sex, activity, and body functioning.

F.1.f   Sex drive is a natural human function that requires understanding. Sex is also a prominent means of transmitting diseases. The diseases can be prevented through a variety of precautions.

F.1.g   Natural environments may contain substances (for example, radon and lead) that are harmful to human beings. Maintaining environmental health involves establishing or monitoring quality standards related to use of soil, water, and air.

Populations, Resources, and Environments

F.2.a   When an area becomes overpopulated, the environment will become degraded due to the increased use of resources.

F.2.b   Causes of environmental degradation and resource depletion vary from region to region and from country to country.

Natural Hazards

F.3.a   Internal and external processes of the earth system cause natural hazards, events that change or destroy human and wildlife habitats, damage property, and harm or kill humans. Natural hazards include earthquakes, landslides, wildfires, volcanic eruptions, floods, storms, and even possible impacts of asteroids.

F.3.b   Human activities also can induce hazards through resource acquisition, urban growth, land-use decisions, and waste disposal. Such activities can accelerate many natural changes.

F.3.c   Natural hazards can present personal and societal challenges because misidentifying the change or incorrectly estimating the rate and scale of change may result in either too little attention and significant human costs or too much cost for unneeded preventive measures.

Risks and Benefits

F.4.a   Risk analysis considers the type of hazard and estimates the number of people that might be exposed and the number likely to suffer consequences. The results are used to determine the options for reducing or eliminating risks.

F.4.b   Students should understand the risks associated with natural hazards (fires, floods, tornadoes, hurricanes, earthquakes, and volcanic eruptions), with chemical hazards (pollutants in air, water, soil, and food), with biological hazards (pollen, viruses, bacterial, and parasites), social hazards (occupational safety and transportation), and with personal hazards (smoking, dieting, and drinking).

F.4.c   Individuals can use a systematic approach to thinking critically about risks and benefits. Examples include applying probability estimates to risks and comparing them to estimated personal and social benefits.

F.4.d   Important personal and social decisions are made based on perceptions of benefits and risks.

Science and Technology in Society

F.5.a   Science influences society through its knowledge and world view. Scientific knowledge and the procedures used by scientists influence the way many individuals in society think about themselves, others, and the environment. The effect of science on society is neither entirely beneficial nor entirely detrimental.

F.5.b   Societal challenges often inspire questions for scientific research, and social priorities often influence research priorities through the availability of funding for research.

F.5.c   Technology influences society through its products and processes. Technology influences the quality of life and the ways people act and interact. Technological changes are often accompanied by social, political, and economic changes that can be beneficial or detrimental to individuals and to society. Social needs, attitudes, and values influence the direction of technological development.

F.5.d   Science and technology have advanced through contributions of many different people, in different cultures, at different times in history. Science and technology have contributed enormously to economic growth and productivity among societies and groups within societies.

F.5.e   Scientists and engineers work in many different settings, including colleges and universities, businesses and industries, specific research institutes, and government agencies.

F.5.f   Scientists and engineers have ethical codes requiring that human subjects involved with research be fully informed about risks and benefits associated with the research before the individuals choose to participate. This ethic extends to potential risks to communities and property. In short, prior knowledge and consent are required for research involving human subjects or potential damage to property.

F.5.g   Science cannot answer all questions and technology cannot solve all human problems or meet all human needs. Students should understand the difference between scientific and other questions. They should appreciate what science and technology can reasonably contribute to society and what they cannot do. For example, new technologies often will decrease some risks and increase others.

G. History and Nature of Science

As a result of activities in grades 5–8, all students should develop understanding of

Science as a Human Endeavor

G.1.a   Women and men of various social and ethnic backgrounds—and with diverse interests, talents, qualities, and motivations—engage in the activities of science, engineering, and related fields such as the health professions. Some scientists work in teams, and some work alone, but all communicate extensively with others.

G.1.b   Science requires different abilities, depending on such factors as the field of study and type of inquiry. Science is very much a human endeavor, and the work of science relies on basic human qualities, such as reasoning, insight, energy, skill, and creativity—as well as on scientific habits of mind, such as intellectual honesty, tolerance of ambiguity, skepticism, and openness to new ideas.

Nature of Science

G.2.a   Scientists formulate and test their explanations of nature using observation, experiments, and theoretical and mathematical models. Although all scientific ideas are tentative and subject to change and improvement in principle, for most major ideas in science, there is much experimental and observational confirmation. Those ideas are not likely to change greatly in the future. Scientists do and have changed their ideas about nature when they encounter new experimental evidence that does not match their existing explanations.

G.2.b   In areas where active research is being pursued and in which there is not a great deal of experimental or observational evidence and understanding, it is normal for scientists to differ with one another about the interpretation of the evidence or theory being considered. Different scientists might publish conflicting experimental results or might draw different conclusions from the same data. Ideally, scientists acknowledge such conflict and work towards finding evidence that will resolve their disagreement.

G.2.c   It is part of scientific inquiry to evaluate the results of scientific investigations, experiments, observations, theoretical models, and the explanations proposed by other scientists. Evaluation includes reviewing the experimental procedures, examining the evidence, identifying faulty reasoning, pointing out statements that go beyond the evidence, and suggesting alternative explanations for the same observations. Although scientists may disagree about explanations of phenomena, about interpretations of data, or about the value of rival theories, they do agree that questioning, response to criticism, and open communication are integral to the process of science. As scientific knowledge evolves, major disagreements are eventually resolved through such interactions between scientists.

History of Science

G.3.a   Many individuals have contributed to the traditions of science. Studying some of these individuals provides further understanding of scientific inquiry, science as a human endeavor, the nature of science, and the relationships between science and society.

G.3.b   In historical perspective, science has been practiced by different individuals in different cultures. In looking at the history of many peoples, one finds that scientists and engineers of high achievement are considered to be among the most valued contributors to their culture.

G.3.c   Tracing the history of science can show how difficult it was for scientific innovators to break through the accepted ideas of their time to reach the conclusions that we currently take for granted.

1. The Nature of Science

By the end of the 8th grade, students should know that

1.A The Scientific World View

1.A.1 When similar investigations give different results, the scientific challenge is to judge whether the differences are trivial or significant, and it often takes further studies to decide. Even with similar results, scientists may wait until an investigation has been repeated many times before accepting the results as correct.

1.A.2 Scientific knowledge is subject to modification as new information challenges prevailing theories and as a new theory leads to looking at old observations in a new way.

1.A.3 Some scientific knowledge is very old and yet is still applicable today.

1.A.4 Some matters cannot be examined usefully in a scientific way. Among them are matters that by their nature cannot be tested objectively and those that are essentially matters of morality. Science can sometimes be used to inform ethical decisions by identifying the likely consequences of particular actions but cannot be used to establish that some action is either moral or immoral.

1.B Scientific Inquiry

1.B.1 Scientists differ greatly in what phenomena they study and how they go about their work. Although there is no fixed set of steps that all scientists follow, scientific investigations usually involve the collection of relevant evidence, the use of logical reasoning, and the application of imagination in devising hypotheses and explanations to make sense of the collected evidence.

1.B.2 If more than one variable changes at the same time in an experiment, the outcome of the experiment may not be clearly attributable to any one of the variables. It may not always be possible to prevent outside variables from influencing the outcome of an investigation (or even to identify all of the variables), but collaboration among investigators can often lead to research designs that are able to deal with such situations.

1.B.3 What people expect to observe often affects what they actually do observe. Strong beliefs about what should happen in particular circumstances can prevent them from detecting other results. Scientists know about this danger to objectivity and take steps to try and avoid it when designing investigations and examining data. One safeguard is to have different investigators conduct independent studies of the same questions.

1.C The Scientific Enterprise

1.C.1   Important contributions to the advancement of science, mathematics, and technology have been made by different kinds of people, in different cultures, at different times.

1.C.2   Until recently, women and racial minorities, because of restrictions on their education and employment opportunities, were essentially left out of much of the formal work of the science establishment; the remarkable few who overcame those obstacles were even then likely to have their work disregarded by the science establishment.

1.C.3   No matter who does science and mathematics or invents things, or when or where they do it, the knowledge and technology that result can eventually become available to everyone in the world.

1.C.4   Scientists are employed by colleges and universities, business and industry, hospitals, and many government agencies. Their places of work include offices, classrooms, laboratories, farms, factories, and natural field settings ranging from space to the ocean floor.

1.C.5   In research involving human subjects, the ethics of science require that potential subjects be fully informed about the risks and benefits associated with the research and of their right to refuse to participate. Science ethics also demand that scientists must not knowingly subject coworkers, students, the neighborhood, or the community to health or property risks without their prior knowledge and consent. Because animals cannot make informed choices, special care must be taken in using them in scientific research.

1.C.6   Computers have become invaluable in science because they speed up and extend people's ability to collect, store, compile, and analyze data, prepare research reports, and share data and ideas with investigators all over the world.

1.C.7   Accurate record-keeping, openness, and replication are essential for maintaining an investigator's credibility with other scientists and society.

3. The Nature of Technology

By the end of the 8th grade, students should know that

3.A Technology and Science

3.A.1   In earlier times, the accumulated information and techniques of each generation of workers were taught on the job directly to the next generation of workers. Today, the knowledge base for technology can be found as well in libraries of print and electronic resources and is often taught in the classroom.

3.A.2   Technology is essential to science for such purposes as access to outer space and other remote locations, sample collection and treatment, measurement, data collection and storage, computation, and communication of information.

3.A.3   Engineers, architects, and others who engage in design and technology use scientific knowledge to solve practical problems. But they usually have to take human values and limitations into account as well.

3.B Design and Systems

3.B.1 Design usually requires taking constraints into account. Some constraints, such as gravity or the properties of the materials to be used, are unavoidable. Other constraints, including economic, political, social, ethical, and aesthetic ones, limit choices.

3.B.2 All technologies have effects other than those intended by the design, some of which may have been predictable and some not. In either case, these side effects may turn out to be unacceptable to some of the population and therefore lead to conflict between groups.

3.B.3 Almost all control systems have inputs, outputs, and feedback. The essence of control is comparing information about what is happening to what people want to happen and then making appropriate adjustments. This procedure requires sensing information, processing it, and making changes. In almost all modern machines, microprocessors serve as centers of performance control.

3.B.4 Systems fail because they have faulty or poorly matched parts, are used in ways that exceed what was intended by the design, or were poorly designed to begin with. The most common ways to prevent failure are pretesting parts and procedures, overdesign, and redundancy.

3.C Issues in Technology

3.C.1 The human ability to shape the future comes from a capacity for generating knowledge and developing new technologies—and for communicating ideas to others.

3.C.2 Technology cannot always provide successful solutions for problems or fulfill every human need.

3.C.3 Throughout history, people have carried out impressive technological feats, some of which would be hard to duplicate today even with modern tools. The purposes served by these achievements have sometimes been practical, sometimes ceremonial.

3.C.4 Technology has strongly influenced the course of history and continues to do so. It is largely responsible for the great revolutions in agriculture, manufacturing, sanitation and medicine, warfare, transportation, information processing, and communications that have radically changed how people live.

3.C.5 New technologies increase some risks and decrease others. Some of the same technologies that have improved the length and quality of life for many people have also brought new risks.

3.C.6 Rarely are technology issues simple and one-sided. Relevant facts alone, even when known and available, usually do not settle matters entirely in favor of one side or another. That is because the contending groups may have different values and priorities. They may stand to gain or lose in different degrees, or may make very different predictions about what the future consequences of the proposed action will be.

3.C.7 Societies influence what aspects of technology are developed and how these are used. People control technology (as well as science) and are responsible for its effects.

4. The Physical Setting

By the end of the 8th grade, students should know that

4.A The Universe

4.A.1   The sun is a medium-sized star located near the edge of a disk-shaped galaxy of stars, part of which can be seen as a glowing band of light that spans the sky on a very clear night. The universe contains many billions of galaxies, and each galaxy contains many billions of stars. To the naked eye, even the closest of these galaxies is no more than a dim, fuzzy spot.

4.A.2   The sun is many thousands of times closer to the earth than any other star. Light from the sun takes a few minutes to reach the earth, but light from the next nearest star takes a few years to arrive. The trip to that star would take the fastest rocket thousands of years. Some distant galaxies are so far away that their light takes several billion years to reach the earth. People on earth, therefore, see them as they were that long ago in the past.

4.A.3   Nine planets of very different size, composition, and surface features move around the sun in nearly circular orbits. Some planets have a great variety of moons and even flat rings of rock and ice particles orbiting around them. Some of these planets and moons show evidence of geologic activity. The earth is orbited by one moon, many artificial satellites, and debris.

4.A.4   Large numbers of chunks of rock orbit the sun. Some of those that the earth meets in its yearly orbit around the sun glow and disintegrate from friction as they plunge through the atmosphere—and sometimes impact the ground. Other chunks of rocks mixed with ice have long, off-center orbits that carry them close to the sun, where the sun's radiation (of light and particles) boils off frozen material from their surfaces and pushes it into a long, illuminated tail.

4.B The Earth

4.B.1   We live on a relatively small planet, the third from the sun in the only system of planets definitely known to exist (although other, similar systems may be discovered in the universe).

4.B.2   The earth is mostly rock. Three-fourths of its surface is covered by a relatively thin layer of water (some of it frozen), and the entire planet is surrounded by a relatively thin blanket of air. It is the only body in the solar system that appears able to support life. The other planets have compositions and conditions very different from the earth's.

4.B.3   Everything on or anywhere near the earth is pulled toward the earth's center by gravitational force.

4.B.4   Because the earth turns daily on an axis that is tilted relative to the plane of the earth's yearly orbit around the sun, sunlight falls more intensely on different parts of the earth during the year. The difference in heating of the earth's surface produces the planet's seasons and weather patterns.

4.B.5   The moon's orbit around the earth once in about 28 days changes what part of the moon is lighted by the sun and how much of that part can be seen from the earth—the phases of the moon.

4.B.6   Climates have sometimes changed abruptly in the past as a result of changes in the earth's crust, such as volcanic eruptions or impacts of huge rocks from space. Even relatively small changes in atmospheric or ocean content can have widespread effects on climate if the change lasts long enough.

4.B.7   The cycling of water in and out of the atmosphere plays an important role in determining climatic patterns. Water evaporates from the surface of the earth, rises and cools, condenses into rain or snow, and falls again to the surface. The water falling on land collects in rivers and lakes, soil, and porous layers of rock, and much of it flows back into the ocean.

4.B.8   Fresh water, limited in supply, is essential for life and also for most industrial processes. Rivers, lakes, and groundwater can be depleted or polluted, becoming unavailable or unsuitable for life.

4.B.9   Heat energy carried by ocean currents has a strong influence on climate around the world.

4.B.10   Some minerals are very rare and some exist in great quantities, but—for practical purposes—the ability to recover them is just as important as their abundance. As minerals are depleted, obtaining them becomes more difficult. Recycling and the development of substitutes can reduce the rate of depletion but may also be costly.

4.B.11   The benefits of the earth's resources—such as fresh water, air, soil, and trees—can be reduced by using them wastefully or by deliberately or inadvertently destroying them. The atmosphere and the oceans have a limited capacity to absorb wastes and recycle materials naturally. Cleaning up polluted air, water, or soil or restoring depleted soil, forests, or fishing grounds can be very difficult and costly.

4.C Processes that Shape the Earth

4.C.1   The interior of the earth is hot. Heat flow and movement of material within the earth cause earthquakes and volcanic eruptions and create mountains and ocean basins. Gas and dust from large volcanoes can change the atmosphere.

4.C.2   Some changes in the earth's surface are abrupt (such as earthquakes and volcanic eruptions) while other changes happen very slowly (such as uplift and wearing down of mountains). The earth's surface is shaped in part by the motion of water and wind over very long times, which act to level mountain ranges.

4.C.3   Sediments of sand and smaller particles (sometimes containing the remains of organisms) are gradually buried and are cemented together by dissolved minerals to form solid rock again.

4.C.4   Sedimentary rock buried deep enough may be reformed by pressure and heat, perhaps melting and recrystallizing into different kinds of rock. These re-formed rock layers may be forced up again to become land surface and even mountains. Subsequently, this new rock too will erode. Rock bears evidence of the minerals, temperatures, and forces that created it.

4.C.5   Thousands of layers of sedimentary rock confirm the long history of the changing surface of the earth and the changing life forms whose remains are found in successive layers. The youngest layers are not always found on top, because of folding, breaking, and uplift of layers.

4.C.6   Although weathered rock is the basic component of soil, the composition and texture of soil and its fertility and resistance to erosion are greatly influenced by plant roots and debris, bacteria, fungi, worms, insects, rodents, and other organisms.

4.C.7   Human activities, such as reducing the amount of forest cover, increasing the amount and variety of chemicals released into the atmosphere, and intensive farming, have changed the earth's land, oceans, and atmosphere. Some of these changes have decreased the capacity of the environment to support some life forms.

4.D Structure of Matter

4.D.1 All matter is made up of atoms, which are far too small to see directly through a microscope. The atoms of any element are alike but are different from atoms of other elements. Atoms may stick together in well-defined molecules or may be packed together in large arrays. Different arrangements of atoms into groups compose all substances.

4.D.2 Equal volumes of different substances usually have different weights.

4.D.3 Atoms and molecules are perpetually in motion. Increased temperature means greater average energy, so most substances expand when heated. In solids, the atoms are closely locked in position and can only vibrate. In liquids, the atoms or molecules have higher energy, are more loosely connected, and can slide past one another; some molecules may get enough energy to escape into a gas. In gases, the atoms or molecules have still more energy and are free of one another except during occasional collisions.

4.D.4 The temperature and acidity of a solution influence reaction rates. Many substances dissolve in water, which may greatly facilitate reactions between them.

4.D.5 Scientific ideas about elements were borrowed from some Greek philosophers of 2,000 years earlier, who believed that everything was made from four basic substances: air, earth, fire, and water. It was the combinations of these "elements" in different proportions that gave other substances their observable properties. The Greeks were wrong about those four, but now over 100 different elements have been identified, some rare and some plentiful, out of which everything is made. Because most elements tend to combine with others, few elements are found in their pure form.

4.D.6 There are groups of elements that have similar properties, including highly reactive metals, less-reactive metals, highly reactive nonmetals (such as chlorine, fluorine, and oxygen), and some almost completely nonreactive gases (such as helium and neon). An especially important kind of reaction between substances involves combination of oxygen with something else—as in burning or rusting. Some elements don't fit into any of the categories; among them are carbon and hydrogen, essential elements of living matter.

4.D.7 No matter how substances within a closed system interact with one another, or how they combine or break apart, the total weight of the system remains the same. The idea of atoms explains the conservation of matter: If the number of atoms stays the same no matter how they are rearranged, then their total mass stays the same.

4.E Energy Transformations

4.E.1 Energy cannot be created or destroyed, but only changed from one form into another.

4.E.2 Most of what goes on in the universe—from exploding stars and biological growth to the operation of machines and the motion of people—involves some form of energy being transformed into another. Energy in the form of heat is almost always one of the products of an energy transformation.

4.E.3 Heat can be transferred through materials by the collisions of atoms or across space by radiation. If the material is fluid, currents will be set up in it that aid the transfer of heat.

4.E.4 Energy appears in different forms. Heat energy is in the disorderly motion of molecules; chemical energy is in the arrangement of atoms; mechanical energy is in moving bodies or in elastically distorted shapes; gravitational energy is in the separation of mutually attracting masses.

4.F Motion

4.F.1 Light from the sun is made up of a mixture of many different colors of light, even though to the eye the light looks almost white. Other things that give off or reflect light have a different mix of colors.

4.F.2 Something can be "seen" when light waves emitted or reflected by it enter the eye—just as something can be "heard" when sound waves from it enter the ear.

4.F.3 An unbalanced force acting on an object changes its speed or direction of motion, or both. If the force acts toward a single center, the object's path may curve into an orbit around the center.

4.F.4 Vibrations in materials set up wavelike disturbances that spread away from the source. Sound and earthquake waves are examples. These and other waves move at different speeds in different materials.

4.F.5 Human eyes respond to only a narrow range of wavelengths of electromagnetic radiation—visible light. Differences of wavelength within that range are perceived as differences in color.

4.G Forces of Nature

4.G.1 Every object exerts gravitational force on every other object. The force depends on how much mass the objects have and on how far apart they are. The force is hard to detect unless at least one of the objects has a lot of mass.

4.G.2 The sun's gravitational pull holds the earth and other planets in their orbits, just as the planets' gravitational pull keeps their moons in orbit around them.

4.G.3 Electric currents and magnets can exert a force on each other.

5. The Living Environment

By the end of the 8th grade, students should know that

5.A Diversity of Life

5.A.1 One of the most general distinctions among organisms is between plants, which use sunlight to make their own food, and animals, which consume energy-rich foods. Some kinds of organisms, many of them microscopic, cannot be neatly classified as either plants or animals.

5.A.2 Animals and plants have a great variety of body plans and internal structures that contribute to their being able to make or find food and reproduce.

5.A.3 Similarities among organisms are found in internal anatomical features, which can be used to infer the degree of relatedness among organisms. In classifying organisms, biologists consider details of internal and external structures to be more important than behavior or general appearance.

5.A.4 For sexually reproducing organisms, a species comprises all organisms that can mate with one another to produce fertile offspring.

5.A.5 All organisms, including the human species, are part of and depend on two main interconnected global food webs. One includes microscopic ocean plants, the animals that feed on them, and finally the animals that feed on those animals. The other web includes land plants, the animals that feed on them, and so forth. The cycles continue indefinitely because organisms decompose after death to return food material to the environment.

5.B Heredity

5.B.1 In some kinds of organisms, all the genes come from a single parent, whereas in organisms that have sexes, typically half of the genes come from each parent.

5.B.2 In sexual reproduction, a single specialized cell from a female merges with a specialized cell from a male. As the fertilized egg, carrying genetic information from each parent, multiplies to form the complete organism with about a trillion cells, the same genetic information is copied in each cell.

5.B.3 New varieties of cultivated plants and domestic animals have resulted from selective breeding for particular traits.

5.C Cells

5.C.1 All living things are composed of cells, from just one to many millions, whose details usually are visible only through a microscope. Different body tissues and organs are made up of different kinds of cells. The cells in similar tissues and organs in other animals are similar to those in human beings but differ somewhat from cells found in plants.

5.C.2 Cells repeatedly divide to make more cells for growth and repair. Various organs and tissues function to serve the needs of cells for food, air, and waste removal.

5.C.3 Within cells, many of the basic functions of organisms—such as extracting energy from food and getting rid of waste—are carried out. The way in which cells function is similar in all living organisms.

5.C.4 About two-thirds of the weight of cells is accounted for by water, which gives cells many of their properties.

5.D Interdependence of Life

5.D.1   In all environments—freshwater, marine, forest, desert, grassland, mountain, and others—organisms with similar needs may compete with one another for resources, including food, space, water, air, and shelter. In any particular environment, the growth and survival of organisms depend on the physical conditions.

5.D.2   Two types of organisms may interact with one another in several ways: They may be in a producer/consumer, predator/prey, or parasite/host relationship. Or one organism may scavenge or decompose another. Relationships may be competitive or mutually beneficial. Some species have become so adapted to each other that neither could survive without the other.

5.E Flow of Matter and Energy

5.E.1   Food provides molecules that serve as fuel and building material for all organisms. Plants use the energy in light to make sugars out of carbon dioxide and water. This food can be used immediately for fuel or materials or it may be stored for later use. Organisms that eat plants break down the plant structures to produce the materials and energy they need to survive. Then they are consumed by other organisms.

5.E.2   Over a long time, matter is transferred from one organism to another repeatedly and between organisms and their physical environment. As in all material systems, the total amount of matter remains constant, even though its form and location change.

5.E.3   Energy can change from one form to another in living things. Animals get energy from oxidizing their food, releasing some of its energy as heat. Almost all food energy comes originally from sunlight.

5.F Evolution of Life

5.F.1   Small differences between parents and offspring can accumulate (through selective breeding) in successive generations so that descendants are very different from their ancestors.

5.F.2   Individual organisms with certain traits are more likely than others to survive and have offspring. Changes in environmental conditions can affect the survival of individual organisms and entire species.

5.F.3   Many thousands of layers of sedimentary rock provide evidence for the long history of the earth and for the long history of changing life forms whose remains are found in the rocks. More recently deposited rock layers are more likely to contain fossils resembling existing species.

6. The Human Organism

By the end of the 8th grade, students should know that

6.A Human Identity

6.A.1 Like other animals, human beings have body systems for obtaining and providing energy, defense, reproduction, and the coordination of body functions.

6.A.2 Human beings have many similarities and differences. The similarities make it possible for human beings to reproduce and to donate blood and organs to one another throughout the world. Their differences enable them to create diverse social and cultural arrangements and to solve problems in a variety of ways.

6.A.3 Fossil evidence is consistent with the idea that human beings evolved from earlier species.

6.A.4 Specialized roles of individuals within other species are genetically programmed, whereas human beings are able to invent and modify a wider range of social behavior.

6.A.5 Human beings use technology to match or excel many of the abilities of other species. Technology has helped people with disabilities survive and live more conventional lives.

6.A.6 Technologies having to do with food production, sanitation, and disease prevention have dramatically changed how people live and work and have resulted in rapid increases in the human population.

6.B Human Development

6.B.1 Fertilization occurs when sperm cells from a male's testes are deposited near an egg cell from the female ovary, and one of the sperm cells enters the egg cell. Most of the time, by chance or design, a sperm never arrives or an egg isn't available.

6.B.2 Contraception measures may incapacitate sperm, block their way to the egg, prevent the release of eggs, or prevent the fertilized egg from implanting successfully.

6.B.3 Following fertilization, cell division produces a small cluster of cells that then differentiate by appearance and function to form the basic tissues of an embryo. During the first three months of pregnancy, organs begin to form. During the second three months, all organs and body features develop. During the last three months, the organs and features mature enough to function well after birth. Patterns of human development are similar to those of other vertebrates.

6.B.4 The developing embryo—and later the newborn infant—encounters many risks from faults in its genes, its mother's inadequate diet, her cigarette smoking or use of alcohol or other drugs, or from infection. Inadequate child care may lead to lower physical and mental ability.

6.B.5 Various body changes occur as adults age. Muscles and joints become less flexible, bones and muscles lose mass, energy levels diminish, and the senses become less acute. Women stop releasing eggs and hence can no longer reproduce. The length and quality of human life are influenced by many factors, including sanitation, diet, medical care, sex, genes, environmental conditions, and personal health behaviors.

6.C Basic Functions

6.C.1 Organs and organ systems are composed of cells and help to provide all cells with basic needs.

6.C.2 For the body to use food for energy and building materials, the food must first be digested into molecules that are absorbed and transported to cells.

6.C.3 To burn food for the release of energy stored in it, oxygen must be supplied to cells, and carbon dioxide removed. Lungs take in oxygen for the combustion of food and they eliminate the carbon dioxide produced. The urinary system disposes of dissolved waste molecules, the intestinal tract removes solid wastes, and the skin and lungs rid the body of heat energy. The circulatory system moves all these substances to or from cells where they are needed or produced, responding to changing demands.

6.C.4 Specialized cells and the molecules they produce identify and destroy microbes that get inside the body.

6.C.5 Hormones are chemicals from glands that affect other body parts. They are involved in helping the body respond to danger and in regulating human growth, development, and reproduction.

6.C.6 Interactions among the senses, nerves, and brain make possible the learning that enables human beings to cope with changes in their environment.

6.D Learning

6.D.1 Some animal species are limited to a repertoire of genetically determined behaviors; others have more complex brains and can learn a wide variety of behaviors. All behavior is affected by both inheritance and experience.

6.D.2 The level of skill a person can reach in any particular activity depends on innate abilities, the amount of practice, and the use of appropriate learning technologies.

6.D.3 Human beings can detect a tremendous range of visual and olfactory stimuli. The strongest stimulus they can tolerate may be more than a trillion times as intense as the weakest they can detect. Still, there are many kinds of signals in the world that people cannot detect directly.

6.D.4 Attending closely to any one input of information usually reduces the ability to attend to others at the same time.

6.D.5 Learning often results from two perceptions or actions occurring at about the same time. The more often the same combination occurs, the stronger the mental connection between them is likely to be. Occasionally a single vivid experience will connect two things permanently in people's minds.

6.D.6 Language and tools enable human beings to learn complicated and varied things from others.

6.E Physical Health

6.E.1   The amount of food energy (calories) a person requires varies with body weight, age, sex, activity level, and natural body efficiency. Regular exercise is important to maintain a healthy heart/lung system, good muscle tone, and bone strength.

6.E.2   Toxic substances, some dietary habits, and personal behavior may be bad for one's health. Some effects show up right away, others may not show up for many years. Avoiding toxic substances, such as tobacco, and changing dietary habits to reduce the intake of such things as animal fat increases the chances of living longer.

6.E.3   Viruses, bacteria, fungi, and parasites may infect the human body and interfere with normal body functions. A person can catch a cold many times because there are many varieties of cold viruses that cause similar symptoms.

6.E.4   White blood cells engulf invaders or produce antibodies that attack them or mark them for killing by other white cells. The antibodies produced will remain and can fight off subsequent invaders of the same kind.

6.E.5   The environment may contain dangerous levels of substances that are harmful to human beings. Therefore, the good health of individuals requires monitoring the soil, air, and water and taking steps to keep them safe.

6.F Mental Health

6.F.1   Individuals differ greatly in their ability to cope with stressful situations. Both external and internal conditions (chemistry, personal history, values) influence how people behave.

6.F.2   Often people react to mental distress by denying that they have any problem. Sometimes they don't know why they feel the way they do, but with help they can sometimes uncover the reasons.

8. The Designed World

By the end of the 8th grade, students should know that

8.A Agriculture

8.A.1   Early in human history, there was an agricultural revolution in which people changed from hunting and gathering to farming. This allowed changes in the division of labor between men and women and between children and adults, and the development of new patterns of government.

8.A.2   People control the characteristics of plants and animals they raise by selective breeding and by preserving varieties of seeds (old and new) to use if growing conditions change.

8.A.3   In agriculture, as in all technologies, there are always trade-offs to be made. Getting food from many different places makes people less dependent on weather in any one place, yet more dependent on transportation and communication among far-flung markets. Specializing in one crop may risk disaster if changes in weather or increases in pest populations wipe out that crop. Also, the soil may be exhausted of some nutrients, which can be replenished by rotating the right crops.

8.A.4   Many people work to bring food, fiber, and fuel to U.S. markets. With improved technology, only a small fraction of workers in the United States actually plant and harvest the products that people use. Most workers are engaged in processing, packaging, transporting, and selling what is produced.

8.B Materials and Manufacturing

8.B.1   The choice of materials for a job depends on their properties and on how they interact with other materials. Similarly, the usefulness of some manufactured parts of an object depends on how well they fit together with the other parts.

8.B.2   Manufacturing usually involves a series of steps, such as designing a product, obtaining and preparing raw materials, processing the materials mechanically or chemically, and assembling, testing, inspecting, and packaging. The sequence of these steps is also often important.

8.B.3   Modern technology reduces manufacturing costs, produces more uniform products, and creates new synthetic materials that can help reduce the depletion of some natural resources.

8.B.4   Automation, including the use of robots, has changed the nature of work in most fields, including manufacturing. As a result, high-skill, high-knowledge jobs in engineering, computer programming, quality control, supervision, and maintenance are replacing many routine, manual-labor jobs. Workers therefore need better learning skills and flexibility to take on new and rapidly changing jobs.

8.C Energy Sources and Use

8.C.1   Energy can change from one form to another, although in the process some energy is always converted to heat. Some systems transform energy with less loss of heat than others.

8.C.2   Different ways of obtaining, transforming, and distributing energy have different environmental consequences.

8.C.3   In many instances, manufacturing and other technological activities are performed at a site close to an energy source. Some forms of energy are transported easily, others are not.

8.C.4   Electrical energy can be produced from a variety of energy sources and can be transformed into almost any other form of energy. Moreover, electricity is used to distribute energy quickly and conveniently to distant locations.

8.C.5   Energy from the sun (and the wind and water energy derived from it) is available indefinitely. Because the flow of energy is weak and variable, very large collection systems are needed. Other sources don't renew or renew only slowly.

8.C.6   Different parts of the world have different amounts and kinds of energy resources to use and use them for different purposes.

8.D Communication

8.D.1   Errors can occur in coding, transmitting, or decoding information, and some means of checking for accuracy is needed. Repeating the message is a frequently used method.

8.D.2   Information can be carried by many media, including sound, light, and objects. In this century, the ability to code information as electric currents in wires, electromagnetic waves in space, and light in glass fibers has made communication millions of times faster than is possible by mail or sound.

8.E Information Processing

8.E.1 Most computers use digital codes containing only two symbols, 0 and 1, to perform all operations. Continuous signals (analog) must be transformed into digital codes before they can be processed by a computer.

8.E.2 What use can be made of a large collection of information depends upon how it is organized. One of the values of computers is that they are able, on command, to reorganize information in a variety of ways, thereby enabling people to make more and better uses of the collection.

8.E.3 Computer control of mechanical systems can be much quicker than human control. In situations where events happen faster than people can react, there is little choice but to rely on computers. Most complex systems still require human oversight, however, to make certain kinds of judgments about the readiness of the parts of the system (including the computers) and the system as a whole to operate properly, to react to unexpected failures, and to evaluate how well the system is serving its intended purposes.

8.E.4 An increasing number of people work at jobs that involve processing or distributing information. Because computers can do these tasks faster and more reliably, they have become standard tools both in the workplace and at home.

8.F Health Technology

8.F.1 Sanitation measures such as the use of sewers, landfills, quarantines, and safe food handling are important in controlling the spread of organisms that cause disease. Improving sanitation to prevent disease has contributed more to saving human life than any advance in medical treatment.

8.F.2 The ability to measure the level of substances in body fluids has made it possible for physicians to make comparisons with normal levels, make very sophisticated diagnoses, and monitor the effects of the treatments they prescribe.

8.F.3 It is becoming increasingly possible to manufacture chemical substances such as insulin and hormones that are normally found in the body. They can be used by individuals whose own bodies cannot produce the amounts required for good health.

9. The Mathematical World

By the end of the 8th grade, students should know that

9.A Numbers

9.A.1 There have been systems for writing numbers other than the Arabic system of place values based on tens. The very old Roman numerals are now used only for dates, clock faces, or ordering chapters in a book. Numbers based on 60 are still used for describing time and angles.

9.A.2 A number line can be extended on the other side of zero to represent negative numbers. Negative numbers allow subtraction of a bigger number from a smaller number to make sense, and are often used when something can be measured on either side of some reference point (time, ground level, temperature, budget).

9.A.3 Numbers can be written in different forms, depending on how they are being used. How fractions or decimals based on measured quantities should be written depends on how precise the measurements are and how precise an answer is needed.

9.A.4 The operations + and − are inverses of each other—one undoes what the other does; likewise x and ÷ .

9.A.5 The expression a/b can mean different things: a parts of size $1/b$ each, a divided by b, or a compared to b.

9.A.6 Numbers can be represented by using sequences of only two symbols (such as 1 and 0, on and off); computers work this way.

9.A.7 Computations (as on calculators) can give more digits than make sense or are useful.

9.B Symbolic Relationships

9.B.1 An equation containing a variable may be true for just one value of the variable.

9.B.2 Mathematical statements can be used to describe how one quantity changes when another changes. Rates of change can be computed from differences in magnitudes and vice versa.

9.B.3 Graphs can show a variety of possible relationships between two variables. As one variable increases uniformly, the other may do one of the following: increase or decrease steadily, increase or decrease faster and faster, get closer and closer to some limiting value, reach some intermediate maximum or minimum, alternately increase and decrease indefinitely, increase or decrease in steps, or do something different from any of these.

9.C Shapes

9.C.1 Some shapes have special properties: triangular shapes tend to make structures rigid, and round shapes give the least possible boundary for a given amount of interior area. Shapes can match exactly or have the same shape in different sizes.

9.C.2 Lines can be parallel, perpendicular, or oblique.

9.C.3 Shapes on a sphere like the earth cannot be depicted on a flat surface without some distortion.

9.C.4 The graphic display of numbers may help to show patterns such as trends, varying rates of change, gaps, or clusters. Such patterns sometimes can be used to make predictions about the phenomena being graphed.

9.C.5 It takes two numbers to locate a point on a map or any other flat surface. The numbers may be two perpendicular distances from a point, or an angle and a distance from a point.

9.C.6 The scale chosen for a graph or drawing makes a big difference in how useful it is.

9.D Uncertainty

9.D.1 How probability is estimated depends on what is known about the situation. Estimates can be based on data from similar conditions in the past or on the assumption that all the possibilities are known.

9.D.2 Probabilities are ratios and can be expressed as fractions, percentages, or odds.

9.D.3 The mean, median, and mode tell different things about the middle of a data set.

9.D.4 Comparison of data from two groups should involve comparing both their middles and the spreads around them.

9.D.5 The larger a well-chosen sample is, the more accurately it is likely to represent the whole. But there are many ways of choosing a sample that can make it unrepresentative of the whole.

9.D.6 Events can be described in terms of being more or less likely, impossible, or certain.

9.E Reasoning

9.E.1 Some aspects of reasoning have fairly rigid rules for what makes sense; other aspects don't. If people have rules that always hold, and good information about a particular situation, then logic can help them to figure out what is true about it. This kind of reasoning requires care in the use of key words such as if, and, not, or, all, and some. Reasoning by similarities can suggest ideas but can't prove them one way or the other.

9.E.2 Practical reasoning, such as diagnosing or troubleshooting almost anything, may require many-step, branching logic. Because computers can keep track of complicated logic, as well as a lot of information, they are useful in a lot of problem-solving situations.

9.E.3 Sometimes people invent a general rule to explain how something works by summarizing observations. But people tend to overgeneralize, imagining general rules on the basis of only a few observations.

9.E.4 People are using incorrect logic when they make a statement such as "If *A* is true, then *B* is true; but *A* isn't true, therefore *B* isn't true either."

9.E.5 A single example can never prove that something is always true, but sometimes a single example can prove that something is not always true.

9.E.6 An analogy has some likenesses to but also some differences from the real thing.

10. Historical Perspectives

By the end of the 8th grade, students should know that

10.A Displacing the Earth from the Center of the Universe

10.A.1 The motion of an object is always judged with respect to some other object or point and so the idea of absolute motion or rest is misleading.

10.A.2 Telescopes reveal that there are many more stars in the night sky than are evident to the unaided eye, the surface of the moon has many craters and mountains, the sun has dark spots, and Jupiter and some other planets have their own moons.

10.F Understanding Fire

10.F.1 From the earliest times until now, people have believed that even though millions of different kinds of material seem to exist in the world, most things must be made up of combinations of just a few basic kinds of things. There has not always been agreement, however, on what those basic kinds of things are. One theory long ago was that the basic substances were earth, water, air, and fire. Scientists now know that these are not the basic substances. But the old theory seemed to explain many observations about the world.

10.F.2 Today, scientists are still working out the details of what the basic kinds of matter are and of how they combine, or can be made to combine, to make other substances.

10.F.3 Experimental and theoretical work done by French scientist Antoine Lavoisier in the decade between the American and French revolutions led to the modern science of chemistry.

10.F.4 Lavoisier's work was based on the idea that when materials react with each other many changes can take place but that in every case the total amount of matter afterward is the same as before. He successfully tested the concept of conservation of matter by conducting a series of experiments in which he carefully measured all the substances involved in burning, including the gases used and those given off.

10.F.5 Alchemy was chiefly an effort to change base metals like lead into gold and to produce an elixir that would enable people to live forever. It failed to do that or to create much knowledge of how substances react with each other. The more scientific study of chemistry that began in Lavoisier's time has gone far beyond alchemy in understanding reactions and producing new materials.

10.G Splitting the Atom

10.G.1 The accidental discovery that minerals containing uranium darken photographic film, as light does, led to the idea of radioactivity.

10.G.2 In their laboratory in France, Marie Curie and her husband, Pierre Curie, isolated two new elements that caused most of the radioactivity of the uranium mineral. They named one radium because it gave off powerful, invisible rays, and the other polonium in honor of Madame Curie's country of birth. Marie Curie was the first scientist ever to win the Nobel prize in two different fields—in physics, shared with her husband, and later in chemistry.

10.I Discovering Germs

10.I.1 Throughout history, people have created explanations for disease. Some have held that disease has spiritual causes, but the most persistent biological theory over the centuries was that illness resulted from an imbalance in the body fluids. The introduction of germ theory by Louis Pasteur and others in the 19th century led to the modern belief that many diseases are caused by microorganisms—bacteria, viruses, yeasts, and parasites.

10.I.2 Pasteur wanted to find out what causes milk and wine to spoil. He demonstrated that spoilage and fermentation occur when microorganisms enter from the air, multiply rapidly, and produce waste products. After showing that spoilage could be avoided by keeping germs out or by destroying them with heat, he investigated animal diseases and showed that microorganisms were involved. Other investigators later showed that specific kinds of germs caused specific diseases.

10.I.3 Pasteur found that infection by disease organisms—germs—caused the body to build up an immunity against subsequent infection by the same organisms. He then demonstrated that it was possible to produce vaccines that would induce the body to build immunity to a disease without actually causing the disease itself.

10.I.4 Changes in health practices have resulted from the acceptance of the germ theory of disease. Before germ theory, illness was treated by appeals to supernatural powers or by trying to adjust body fluids through induced vomiting, bleeding, or purging. The modern approach emphasizes sanitation, the safe handling of food and water, the pasteurization of milk, quarantine, and aseptic surgical techniques to keep germs out of the body; vaccinations to strengthen the body's immune system against subsequent infection by the same kind of microorganisms; and antibiotics and other chemicals and processes to destroy microorganisms.

10.I.5 In medicine, as in other fields of science, discoveries are sometimes made unexpectedly, even by accident. But knowledge and creative insight are usually required to recognize the meaning of the unexpected.

10.J Harnessing Power

10.J.1 Until the 1800s, most manufacturing was done in homes, using small, handmade machines that were powered by muscle, wind, or running water. New machinery and steam engines to drive them made it possible to replace craftsmanship with factories, using fuels as a source of energy. In the factory system, workers, materials, and energy could be brought together efficiently.

10.J.2 The invention of the steam engine was at the center of the Industrial Revolution. It converted the chemical energy stored in wood and coal, which were plentiful, into mechanical work. The steam engine was invented to solve the urgent problem of pumping water out of coal mines. As improved by James Watt, it was soon used to move coal, drive manufacturing machinery, and power locomotives, ships, and even the first automobiles.

11. Common Themes

By the end of the 8th grade, students should know that

11.A Systems

11.A.1 A system can include processes as well as things.

11.A.2 Thinking about things as systems means looking for how every part relates to others. The output from one part of a system (which can include material, energy, or information) can become the input to other parts. Such feedback can serve to control what goes on in the system as a whole.

11.A.3 Any system is usually connected to other systems, both internally and externally. Thus a system may be thought of as containing subsystems and as being a subsystem of a larger system.

11.B Models

11.B.1 Models are often used to think about processes that happen too slowly, too quickly, or on too small a scale to observe directly, or that are too vast to be changed deliberately, or that are potentially dangerous.

11.B.2 Mathematical models can be displayed on a computer and then modified to see what happens.

11.B.3 Different models can be used to represent the same thing. What kind of a model to use and how complex it should be depends on its purpose. The usefulness of a model may be limited if it is too simple or if it is needlessly complicated. Choosing a useful model is one of the instances in which intuition and creativity come into play in science, mathematics, and engineering.

11.C Constancy and Change

11.C.1 Physical and biological systems tend to change until they become stable and then remain that way unless their surroundings change.

11.C.2 A system may stay the same because nothing is happening or because things are happening but exactly counterbalance one another.

11.C.3 Many systems contain feedback mechanisms that serve to keep changes within specified limits.

11.C.4 Symbolic equations can be used to summarize how the quantity of something changes over time or in response to other changes.

11.C.5 Symmetry (or the lack of it) may determine properties of many objects, from molecules and crystals to organisms and designed structures.

11.C.6 Cycles, such as the seasons or body temperature, can be described by their cycle length or frequency, what their highest and lowest values are, and when these values occur. Different cycles range from many thousands of years down to less than a billionth of a second.

11.D Scale

11.D.1 Properties of systems that depend on volume, such as capacity and weight, change out of proportion to properties that depend on area, such as strength or surface processes.

11.D.2 As the complexity of any system increases, gaining an understanding of it depends increasingly on summaries, such as averages and ranges, and on descriptions of typical examples of that system.

12. Habits of Mind

By the end of the 8th grade, students should know that

12.A Values and Attitudes

12.A.1 Know why it is important in science to keep honest, clear, and accurate records.

12.A.2 Know that hypotheses are valuable, even if they turn out not to be true, if they lead to fruitful investigations.

12.A.3 Know that often different explanations can be given for the same evidence, and it is not always possible to tell which one is correct.

12.B Computation and Estimation

12.B.1 Find what percentage one number is of another and figure any percentage of any number.

12.B.2 Use, interpret, and compare numbers in several equivalent forms such as integers, fractions, decimals, and percents.

12.B.3 Calculate the circumferences and areas of rectangles, triangles, and circles, and the volumes of rectangular solids.

12.B.4 Find the mean and median of a set of data.

12.B.5 Estimate distances and travel times from maps and the actual size of objects from scale drawings.

12.B.6 Insert instructions into computer spreadsheet cells to program arithmetic calculations.

12.B.7 Determine what unit (such as seconds, square inches, or dollars per tankful) an answer should be expressed in from the units of the inputs to the calculation, and be able to convert compound units (such as yen per dollar into dollar per yen, or miles per hour into feet per second).

12.B.8 Decide what degree of precision is adequate and round off the result of calculator operations to enough significant figures to reasonably reflect those of the inputs.

12.B.9 Express numbers like 100, 1,000, and 1,000,000 as powers of 10.

12.B.10 Estimate probabilities of outcomes in familiar situations, on the basis of history or the number of possible outcomes.

12.C Manipulation and Observation

12.C.1 Use calculators to compare amounts proportionally.

12.C.2 Use computers to store and retrieve information in topical, alphabetical, numerical, and key-word files, and create simple files of their own devising.

12.C.3 Read analog and digital meters on instruments used to make direct measurements of length, volume, weight, elapsed time, rates, and temperature, and choose appropriate units for reporting various magnitudes.

12.C.4 Use cameras and tape recorders for capturing information.

12.C.5 Inspect, disassemble, and reassemble simple mechanical devices and describe what the various parts are for; estimate what the effect that making a change in one part of a system is likely to have on the system as a whole.

12.D Communication Skills

12.D.1 Organize information in simple tables and graphs and identify relationships they reveal.

12.D.2 Read simple tables and graphs produced by others and describe in words what they show.

12.D.3 Locate information in reference books, back issues of newspapers and magazines, compact disks, and computer databases.

12.D.4 Understand writing that incorporates circle charts, bar and line graphs, two-way data tables, diagrams, and symbols.

12.D.5 Find and describe locations on maps with rectangular and polar coordinates.

12.E Critical-Response Skills

12.E.1 Question claims based on vague attributions (such as "Leading doctors say...") or on statements made by celebrities or others outside the area of their particular expertise.

12.E.2 Compare consumer products and consider reasonable personal trade-offs among them on the basis of features, performance, durability, and cost.

12.E.3 Be skeptical of arguments based on very small samples of data, biased samples, or samples for which there was no control sample.

12.E.4 Be aware that there may be more than one good way to interpret a given set of findings.

12.E.5 Notice and criticize the reasoning in arguments in which (1) fact and opinion are intermingled or the conclusions do not follow logically from the evidence given, (2) an analogy is not apt, (3) no mention is made of whether the control groups are very much like the experimental group, or (4) all members of a group (such as teenagers or chemists) are implied to have nearly identical characteristics that differ from those of other groups.